REVERSE SUPPLY CHAINS

CHAINS

ISSUES AND ANALYSIS

REVERSE SUPPLY CHAINS

ISSUES AND ANALYSIS

Edited by

Surendra M. Gupta

CRC Press
Taylor & Francis Group
Boca Raton   London   New York

CRC Press is an imprint of the
Taylor & Francis Group, an **informa** business

MATLAB® is a trademark of The MathWorks, Inc. and is used with permission. The MathWorks does not warrant the accuracy of the text or exercises in this book. This book's use or discussion of MATLAB® software or related products does not constitute endorsement or sponsorship by The MathWorks of a particular pedagogical approach or particular use of the MATLAB® software.

CRC Press
Taylor & Francis Group
6000 Broken Sound Parkway NW, Suite 300
Boca Raton, FL 33487-2742

© 2013 by Taylor & Francis Group, LLC
CRC Press is an imprint of Taylor & Francis Group, an Informa business

No claim to original U.S. Government works

Printed in the United States of America on acid-free paper
Version Date: 20121105

International Standard Book Number: 978-1-4398-9902-1 (Hardback)

Library of Congress Cataloging-in-Publication Data

Reverse supply chains : issues and analysis / editor, Surendra M. Gupta.
    p. cm.
  Includes bibliographical references and index.
  ISBN 978-1-4398-9902-1 (hardback)
    1. Recycling (Waste, etc.) 2. Remanufacturing. 3. Business logistics. I. Gupta, Surendra M.

  TD794.5.R47 2013
  658.7--dc23   2012024126

Visit the Taylor & Francis Web site at
http://www.taylorandfrancis.com

and the CRC Press Web site at
http://www.crcpress.com

To my family:

Sharda Gupta, Monica Gupta, and Neil M. Gupta

Contents

Preface

Reverse supply chains consist of a series of activities required to collect used products from consumers and reprocess them to either recover their leftover market values or dispose of them. It has become common for companies involved in a traditional (forward) supply chain (series of activities required to produce new products from virgin materials and distribute them to consumers) to also carry out collection and reprocessing of used products (reverse supply chain). Strict environmental regulations and diminishing raw material resources have intensified the importance of reverse supply chains at an increasing rate. In addition to being environment friendly, effective management of reverse supply chain operations leads to higher profitability by reducing transportation, inventory, and warehousing costs. Moreover, reverse supply chain operations have a strong impact on the operations of a forward supply chain such as occupancy of storage spaces and transportation capacity. The introduction of reverse supply chains has created many challenges in the areas of network design, transportation, selection of used products, selection and evaluation of suppliers, performance measurement, marketing-related issues, end-of-life (EOL) alternative selection, remanufacturing, disassembly, and product acquisition management to name a few.

This book provides comprehensive coverage of a variety of topics within reverse supply chains. Students, academicians, scholars, consultants, and practitioners worldwide would benefit from this book. It is my hope that it will inspire further research in reverse supply chains and motivate new researchers to get interested in this all-too-important field of study.

The book is organized into 15 chapters. Chapter 1 by Ilgin and Gupta presents an introduction to the basic concepts of reverse logistics, which is an element of reverse supply chain, and systematically analyzes the literature by classifying over 400 published references into five major types of product returns. Finally, some avenues for future research are also discussed. Chapter 2 by Srivastava identifies the basic activities and scope of reverse logistics together with its drivers and barriers as well as the major issues and challenges. The author also describes a few initiatives and suggests frameworks and models for better reverse logistics design and practices. Chapter 3 by McGovern and Gupta presents metrics for quantitatively comparing competing new-product designs for end-of-life disassembly on a reverse-production line. A case study consisting of three design alternatives (each equally desirable and efficient in terms of assembly) of a notional consumer product is analyzed to illustrate application of the metrics. The new-product design metrics are shown to lead to better decisions than decisions made without the metrics. Chapter 4 by Yüksel makes use of the theory of constraints' thinking processes to determine the core problems in the reverse logistics of an electronics firm. Chapter 5 by Nakashima and Gupta develops an integrated multicriteria decision-making methodology using Taguchi loss functions, AHP (analytic hierarchy process), and fuzzy programming that evaluates suppliers and determines the order quantities under different degrees of information vagueness in the decision parameters in a reverse supply chain network.

Chapters 6 through 8 address various issues associated with remanufacturing, which is an important element of reverse supply chain. Chapter 6 by Ghoreishi et al. deals with a general modeling framework for cost/benefit analysis of remanufacturing. The model consists of three phases, viz., take back, disassembly and reassembly, and resale. The first phase considers the process of buying back the used product from customers; the second phase focuses on modeling the disassembly of the taken back product into its cores and reassembly of the recovered cores into the remanufactured product; the last phase models marketing of the remanufactured product. These three phases are modeled separately using the transfer pricing mechanism. Chapter 7 by Liu et al. develops mathematical models for determining optimal decisions involving inventory replenishment, retail pricing, and reimbursement to customers for returns. These decisions are made in an integrated manner for a single manufacturer and a single retailer dealing with a single recoverable item under deterministic conditions. A numerical example is presented to illustrate the methodology. Chapter 8 by Ondemir and Gupta presents a fuzzy multiobjective ARTODTO (advanced remanufacturing-to-order and disassembly-to-order) model. The model deals with products that are embedded with sensors and RFID (radio-frequency identification) tags. The goal of the proposed model is to determine how to process each and every EOLP (end-of-life product) on hand to meet used product and component demands as well as recycled material demand given all are uncertain. The model considers remanufacturing, disassembly, recycling, disposal, and storage options for each EOLP in order to attain uncertain aspiration levels on a number of physical and financial objectives, viz., total cost, total disposal weight, total recycled weight, and customer satisfaction level. Outside component procurement option is also assumed to be available.

Chapter 9 by Azevedo et al. explores the importance of green and resilient supply chain management practices in the competitiveness of the automotive supply chain. To attain this objective, a worldwide panel of academics and professionals from Portugal, Belgium, China, Germany, Switzerland, and the United States, involved in the automotive industry, was used to evaluate these paradigms in varying countries using descriptive and multivariate statistics. The results, contrary to expectation, indicate that the resilient paradigm is considered more important than the green paradigm. Moreover, the importance given to green and resilient paradigms does not vary between academics and professionals or among countries. Chapter 10 by Agrawal et al. discusses a balanced principal solution to address a green supply chain model subjected to governmental regulations. Their results indicate that the production quantities and the negotiation prices at equilibrium decrease with a rise in government's financial instruments. Chapter 11 by Kaliyan et al. highlights the results of a survey that was carried out among the industries to evaluate the extent of green in industries. Ten barriers were considered for the survey, and experts from various departments of the industries were asked to score each barrier through questionnaires. The results of the analysis of this survey are reported in this chapter.

Chapter 12 by Kalayci and Gupta considers a sequence-dependent disassembly line balancing problem (SDDLBP) that requires the assignment of disassembly tasks to a set of ordered disassembly workstations while satisfying the disassembly precedence constraints and optimizing the effectiveness of several measures, considering sequence-dependent time increments between tasks. It presents a river formation dynamics approach for obtaining (near) optimal solutions. Different scenarios are considered and a comparison with ant colony optimization approach is provided to show the effectiveness of the methodology. Chapter 13 by Giudice proposes system modeling based on graph theory and network flows application to analyze material resource flows in the life cycle of a product. Chapter 14 by Chen and Wang reports three strategies and four derived schemes for delivery and pickup problems. The pros and cons of these schemes are also discussed with the help of examples. Chapter 15 by Lambert et al. presents results of an ongoing quantitative study on the historical evolution of all materials flows in the Dutch economy.

This book would not have been possible without the devotion and commitment of the contributing authors. They have been very thorough in preparing their manuscripts. We would also like to express our appreciation to Taylor & Francis Group and its staff for providing seamless support in making it possible to complete this timely and important manuscript.

MATLAB® is a registered trademark of The MathWorks, Inc. For product information, please contact:

The MathWorks, Inc.
3 Apple Hill Drive
Natick, MA, 01760-2098 USA
Tel: 508-647-7000
Fax: 508-647-7001
E-mail: info@mathworks.com
Web: www.mathworks.com

Editor

Surendra M. Gupta, PhD, PE, is a professor of mechanical and industrial engineering and the director of the Laboratory for Responsible Manufacturing, Northeastern University, Boston, Massachusetts. He received his BE in electronics engineering from Birla Institute of Technology and Science, an MBA from Bryant University, and an MSIE and PhD in industrial engineering from Purdue University.

Dr. Gupta is a registered professional engineer in the State of Massachusetts. His research interests are in the areas of production/manufacturing systems and operations research. He is most interested in environmentally conscious manufacturing, reverse and closed-loop supply chains, disassembly modeling, and remanufacturing.

Dr. Gupta has authored or coauthored well over 425 technical papers published in books, journals, and international conference proceedings. His publications have been cited by thousands of researchers all over the world in journals, proceedings, books, and dissertations. He has traveled to all seven continents, viz., Africa, Antarctica, Asia, Australia, Europe, North America, and South America, and presented his work at international conferences on six continents. Dr. Gupta has taught over 100 courses in such areas as operations research, inventory theory, queuing theory, engineering economy, supply chain management, and production planning and control. Among the many recognitions received, he is the recipient of the Outstanding Research Award and the Outstanding Industrial Engineering Professor Award (in recognition of teaching excellence) from Northeastern University as well as a national outstanding doctoral dissertation advisor award.

Contributors

Lovelesh Agarwal
Department of Humanities and Social
  Sciences
Indian Institute of Technology
Kharagpur, India

Neelesh Agrawal
Department of Civil Engineering
Indian Institute of Technology
Kharagpur, India

Susana G. Azevedo
Department of Business and Economics
University of Beira Interior
Covilhã, Portugal

Avijit Banerjee
Department of Decision Sciences
Drexel University
Philadelphia, Pennsylvania

F.T.S. Chan
Department of Industrial and Systems
  Engineering
Hong Kong Polytechnic University
Hung Hom, Kowloon, Hong Kong

Ying-Yen Chen
Department of Industrial
  Engineering and Engineering
  Management
National Tsing Hua University
Hsinchu, Taiwan, Republic of China

V. Cruz-Machado
Department of Mechanical and
  Industrial Engineering
Universidade Nova de Lisboa
Caparica, Portugal

Elizabeth A. Cudney
Department of Engineering
  Management and Systems
  Engineering
Missouri University of Science and
  Technology
Rolla, Missouri

Ruo Du
School of Statistics
Southwestern University of Finance and
  Economics
Chengdu, Sichuan, People's Republic
  of China

Niloufar Ghoreishi
Department of Mechanical
  Engineering and Materials
  Science
Washington University in St. Louis
St. Louis, Missouri

Fabio Giudice
Department of Industrial
  Engineering
University of Catania
Catania, Italy

Kannan Govindan
Department of Business and Economics
Syddansk Universitet
Odense, Denmark

Surendra M. Gupta
Laboratory of Responsible Manufacturing
Department of Mechanical and
  Industrial Engineering
Northeastern University
Boston, Massachusetts

Noorul Haq
Department of Production Engineering
National Institute of Technology
Tiruchirappalli, India

Joerg S. Hofstetter
Chair of Logistics Management
University of St. Gallen
St. Gallen, Switzerland

W.H.P.M. van Hooff
Department of Innovation Sciences
Eindhoven University of Technology
Eindhoven, the Netherlands

Mehmet Ali Ilgin
Department of Industrial
    Engineering
Celal Bayar University
Manisa, Turkey

Mark J. Jakiela
Department of Mechanical
    Engineering and Materials
    Science
Washington University in St. Louis
St. Louis, Missouri

Can B. Kalayci
Department of Industrial
    Engineering
Pamukkale University
Denizli, Turkey

Mathiyazhagan Kaliyan
Department of Production Engineering
National Institute of Technology
Tiruchirappalli, India

Seung-Lae Kim
Department of Decision Sciences
Drexel University
Philadelphia, Pennsylvania

A.J.D. Lambert
Department of Innovation Sciences
Eindhoven University of Technology
Eindhoven, the Netherlands

H.W. Lintsen
Department of Innovation Sciences
Eindhoven University of Technology
Eindhoven, the Netherlands

Xiangrong Liu
Department of Management
Bridgewater State University
Bridgewater, Massachusetts

Seamus M. McGovern
Laboratory of Responsible Manufacturing
Department of Mechanical and
    Industrial Engineering
Northeastern University
Boston, Massachusetts, USA

Kenichi Nakashima
Department of Industrial
    Engineering and
    Management
Kanagawa University
Yokohama, Japan

Ali Nekouzadeh
Department of Biomedical
    Engineering
Washington University in St. Louis
St. Louis, Missouri

Onder Ondemir
Department of Industrial
    Engineering
Yildiz Technical University
Istanbul, Turkey

J.L. Schippers
Department of Innovation Sciences
Eindhoven University of Technology
Eindhoven, the Netherlands

Samir K. Srivastava
Operations Management Group
Indian Institute of Management
Lucknow, India

M.K. Tiwari
Department of Industrial
  Engineering and
  Management
Indian Institute of Technology
Kharagpur, India

F.C.A. Veraart
Department of Innovation Sciences
Eindhoven University of Technology
Eindhoven, the Netherlands

Hsiao-Fan Wang
Department of Industrial
  Engineering and Engineering
  Management
National Tsing Hua University
Hsinchu, Taiwan, Republic of China

Tian Yihui
School of Business Management
Dalian University of Technology
Dalian, Liaoning, People's Republic
  of China

Hilmi Yüksel
Faculty of Economics and
  Administrative Sciences
Dokuz Eylul University
Izmir, Turkey

1 Reverse Logistics

Mehmet Ali Ilgin and Surendra M. Gupta

CONTENTS

1.1 INTRODUCTION

Reverse logistics (RL) involves all the activities required for the retrieval of products returned by customers for any reason (end of life [EOL], repair, end of lease, and warranty) (Rogers and Tibben-Lembke 1999). In recent years, RL is receiving increasing attention from both academia and industry. There are environmental as well as economic reasons behind this trend.

We can cite saturated landfill areas, global warming, and rapid depletion of raw materials as the main environmental concerns. In order to deal with these problems, governments impose new and stricter environmental regulations which require manufacturers to take back their EOL products through a RL network. Besides complying with legal regulations, firms can utilize the remaining economical value contained in EOL products through different product recovery options, viz., reuse, recycling, and remanufacturing.

Another popular economical concern associated with RL is the increasing amount of customer returns mainly due to more liberal return policies. Rise in the volume of Internet marketing is another reason for this phenomenon. A well-designed and well-operated RL network is a must for profitable handling of customer returns, which in turn results in higher profit levels and increased customer retention rates.

Although there are studies in the literature using the terms "reverse logistics" and "reverse supply chains" interchangeably, there is a slight difference between them.

1

RL mainly deals with transportation, production planning, and inventory management while reverse supply chain has a broader focus involving additional elements such as coordination and collaboration among channel partners (Prahinski and Kocabasoglu 2006). In other words, RL is one of the elements of a reverse supply chain.

The previous reviews only analyze the EOL product returns-related reverse logistic issues (Pokharel and Mutha 2009, Jayant et al. 2011). In this chapter, we try to present a holistic view of RL by simultaneously considering EOL product returns and other types of product returns.

In the following section, differences between reverse and forward logistics are analyzed. Section 1.3 discusses the components and working mechanism of a typical RL system. Various issues in RL are explained by providing studies from literature in Section 1.4. Section 1.5 presents the conclusions.

1.2   DIFFERENCES BETWEEN REVERSE AND FORWARD LOGISTICS

RL differs from forward logistics in many aspects (Tibben-Lembke and Rogers 2002, Pochampally et al. 2009c). In this section, we investigate these differences. Table 1.1 gives a summary of the differences.

Traditional forecasting techniques can be directly applied to forecast the demand for a product type in forward supply chains. However, these techniques may need to be modified in RL case considering the higher level of uncertainty associated with product returns.

In forward logistics, new products produced in a facility are transported to many distributors. In RL, the returned products collected from many collection centers are transported to the producer or to a product recovery (remanufacturing, recycling, or disposal) facility. In other words, the transportation flows in forward supply chain are one-to-many while they are many-to-one in RL.

New products have complete packaging which protects them during transportation and provides ease of handling and identification. However, returned products rarely have complete packaging. This creates problems in the transportation, handling, and identification of returned products.

If a firm is not able to deliver a new product to a customer on time, the customer can switch to one of the competitors of this firm. That is why a forward supply chain must be fast enough to prevent stock-out instances. In RL, the returned products are received by the firm itself. Hence, slow delivery of returned products to the firm does not create any stock-outs and loss of customer goodwill.

New products have a fixed structure determined based on a bill of materials document. They are also subject to strict quality inspections to ensure conformance to certain quality standards. However, returned products, especially EOL returns, have many missing, modified, or damaged parts. As a result, more time has to be spent on inspection and sorting. Thus, the prediction of reusable part yield is very difficult. In addition, processing steps and times vary widely depending on the condition of the returned product.

There are many inventory models developed considering the characteristics of forward supply chains. Some of the assumptions of these models such as unlimited supply and deterministic demand cannot be applied to RL systems in which

TABLE 1.1
Differences between Reverse and Forward Supply Chains

Forward	Reverse
Based on profit and cost optimization	Based on environmentally conscious principles and laws as well as on profit and cost optimization
Relatively easier and straightforward forecasting for product demand	More difficult forecasting for product returns
Less variation in product quality	Highly stochastic product quality
Traditional marketing techniques can be applied	There are factors complicating marketing
Processing times and steps are well defined	Processing times and steps depend on the condition of the returned product
Goods are transported from one location to many other locations	Returned products collected from many locations arrive in one processing facility
Speed is a competitive advantage	Speed is not a critical factor
Standard product packaging	Highly variable packaging/lack of packaging
Standard product structure	Modified product structure
Cost estimation is easier due to accounting systems	Determination and visualization of cost factors is complicated
Disposition alternatives are clear	Disposition options for a returned product depend on its condition
Consistent inventory management	Inconsistent inventory management
Financial implications are clear	Financial implications are not clear
Highly visible processes due to real-time product tracking	Less visible processes due to lack of information system capabilities for product tracking
Relatively easier management of product life cycle changes	Adjusting to the product life cycle changes is more difficult
Relatively more deterministic	Relatively more stochastic
Primary importance to manufacturers	Primary importance to EOL processors (i.e., remanufacturers, recyclers)

product returns have a high level of uncertainty. Hence, lack of proven and effective inventory management systems makes the inventory management very inconsistent and chaotic.

Firms install information system infrastructure to track the flow of products through the forward supply chain. Such information system capabilities are usually not available in their RL networks since RL is given secondary importance. Due to the unavailability of critical information on product returns such as the number of in-transit returns or the number of in-store returns, operational planning becomes very difficult in RL networks.

For manufacturing companies, the primary importance is forward supply chain because an important portion of their revenues comes from the sale of new products that are distributed using a forward supply chain. For remanufacturing or recycling companies, the primary importance is reverse supply chain since they recover parts or materials from EOL products that are obtained using a reverse supply chain.

Forward supply chains are mainly designed based on cost minimization and profit maximization, whereas in reverse supply chains, environmental laws and directives are as important as cost minimization and profit maximization.

Final disposition decision of a product in a forward supply chain is the sale of the product to a customer. In reverse supply chains, this decision depends on the type (viz., EOL, customer, repair/service, reusable container, leased) and condition of the returned product. For instance, an EOL product can be reused, remanufactured, recycled, or disposed depending on its condition.

1.3 REVERSE LOGISTICS PROCESS

The stages in a RL process are mainly determined by the type of returns (viz., customer returns, leased product returns, repair/service returns, reusable container returns, EOL product returns). Collection, sorting, and inspection stages are common to all return types. For the cases of customer, leased product, repair/service, and reusable container returns, if a returned product is found to be in a very bad condition (non-refurbishable, non-repairable, nonreusable) at the end of inspection operation, then it is regarded as an EOL product return. If it is found to be in a good condition, then a series of refurbishing or repair operations are carried. These operations are presented in Figure 1.1 for each returned product type.

1.4 ISSUES IN REVERSE LOGISTICS

As can be seen in Figure 1.2, we distinguish five different types of product returns in RL: customer returns, repair/service returns, EOL returns, reusable container returns, and leased product returns. In this study, we investigate each of these return types by providing related literature.

1.4.1 CUSTOMER RETURNS

Due to liberal return policies, customers may return a purchased product within a certain time frame. Dissatisfaction with the product and finding better deals in other stores are just some of the reasons presented by the customers.

Upon receiving the returned product, the firm has to decide on the appropriate disposition strategy. This decision is largely dependent on the condition of the product. If the customer returned the product without even opening the package, it can be resold as new. If the product package is opened, following testing and refurbishing operations, it can be resold as a refurbished product. In this case, another option is selling the product in an outlet dedicated to the resale of returned products. If the returned product is severely damaged due to customer misuse and/or improper handling during transportation, then it has to be sent to an EOL product processing facility for component or material recovery. In this case, customer return can be regarded as EOL product return. For the studies on RL issues related with EOL product returns, we refer the reader to Section 1.4.3. In this section, we focus on the studies investigating RL issues associated with non-EOL customer returns.

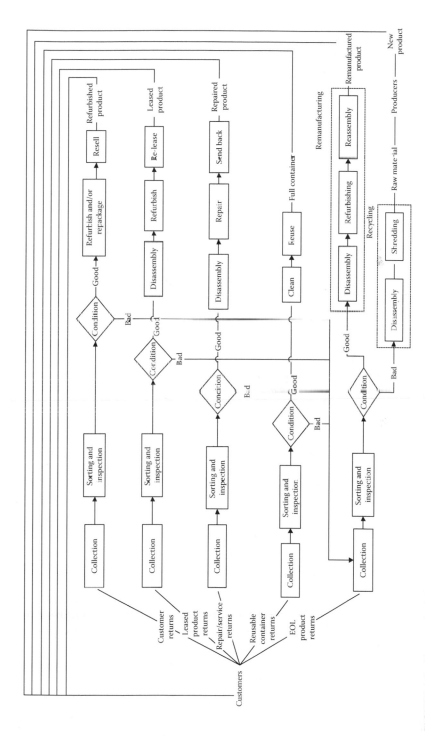

FIGURE 1.1 Reverse logistics process.

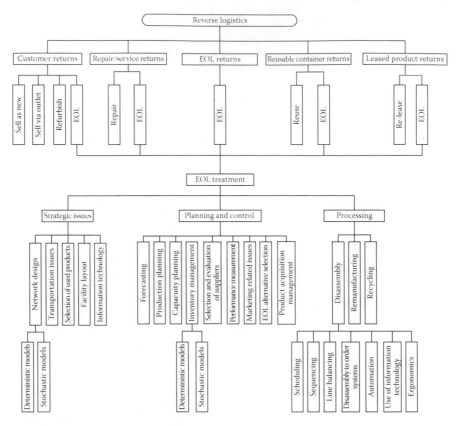

FIGURE 1.2   Issues in reverse logistics.

Autry et al. (2001) performed a survey analysis for catalog retailers to determine how RL performance and satisfaction is influenced by various factors including industry type, firm size/sales volume, and internal or external assignment of responsibility for disposition. They concluded that industry type significantly affects satisfaction while it has no significant impact on performance. Location of responsibility for disposition does not have any significant impact on performance or satisfaction. Sales volume has significant effect on performance while it does not have any significant effect on satisfaction.

Stuart et al. (2005) presented the results of a performance improvement study for the return processing operations of a fashion catalog distributor. The proposed algorithm determines disposal decision considering inventory level, demand pattern, cost, and lead-time factors in addition to the typical factors considered in catalog return processing such as the condition of the returned item, fashion obsolescence, and back-order status.

1.4.2 REPAIR/SERVICE RETURNS

A product can be returned to a firm for repair when it fails to perform its function. If the repair activities are successful, the product is returned back to the customer. If the product cannot be repaired, EOL processing operations must be carried on it. In this section, we only investigate studies on the former case. The discussion on the studies associated with the latter case can be found in Section 1.4.3.

Du and Evans (2008) consider a RL problem involving a manufacturer outsourcing its post-sale services to a third-party logistics (3PLs) provider which collects defective products returned by customers, transports the returned products to repair facilities, and delivers repaired products back to collection sites. In addition to product flow, replaced defective parts are sent to the plants of the manufacturer for remanufacturing or for other purposes, and the new spare parts are transported to the repair facilities. For this system, the authors develop an optimization model considering two objectives: minimization of the total cost and minimization of the total tardiness of cycle time. This optimization problem is solved using a methodology which integrates scatter search, the dual simplex method, and the constraint method.

Tan et al. (2003) consider a U.S.-based computer manufacturer which provides post-sale services to its Asia Pacific customers. Any defective parts from products that are under warranty are returned to U.S. headquarters for refurbishment or repair operations. They are returned back to Asia Pacific upon completion of the required operation. After analyzing the current RL system, the authors proposed several modifications in order to improve the performance of the RL operations. Piplani and Saraswat (2012) develop a mixed integer linear programming (MILP) model to design the service network of a company providing repair and refurbishment services for its products (laptops and desktops) in the Asia Pacific region.

1.4.3 EOL Returns

Due to rapid development in technology and customers' desire for newer product models, many products reach their EOL prematurely. In other words, although they are functional, consumers dispose of them whenever they can buy a similar product having more advanced technology and more features. In some parts of the world such as Europe and Japan, firms have to collect their EOL products and treat them in an environmentally responsible manner. In other areas, such as the United States, EOL products are collected mainly due to their material content and/or their functional components. In both cases, firms have to have a RL network in order to collect the EOL products from customers.

It must be noted that some of the products in other return types can also be regarded as EOL products depending on the condition of the products. For instance, if a leased product returned to a leasing company is out of date or is not functional at all, then this leased product return is considered as an EOL product return. Likewise, if a reusable container is damaged or broken, it must be treated as an EOL product when it is returned.

In this section, we present a review of the studies on EOL product-related RL issues. First, the papers associated with strategic issues such as network design, transportation, and selection of used products are considered. Then the studies on planning and control issues (viz., forecasting, inventory management, supplier selection, performance evaluation, marketing, EOL alternative selection, and product acquisition management) are reviewed. In the last section, the articles on the processing issues such as remanufacturing, recycling, and disassembly are discussed.

We analyze EOL returns-related RL issues under three main categories: strategic, planning, and processing issues. Strategic issues are about the structure of a RL network. That is why any decision on these issues will affect the operation and profitability of an RL network in the long term. Network design, selection of used products, and facility layout are considered under this category. Planning issues involve medium- and short-term decisions on the operation of an RL network. The issues considered under this category include forecasting, inventory management, production planning and control, supplier selection and evaluation, performance measurement, marketing-related issues, and product acquisition management. The processing issues are related with the physical processing of EOL products. Cleaning, disassembly, and reassembly of EOL products can be evaluated under this category.

1.4.3.1   Strategic Issues

1.4.3.1.1   Network Design

We can classify network design models into two categories: deterministic and stochastic. In deterministic models, the uncertainty associated with RL and closed-loop networks is not explicitly considered in model building. However, in stochastic models, the uncertain characteristics of RL and closed-loop networks are integrated into modeling process.

1.4.3.1.1.1   Deterministic Models   The most commonly used modeling technique in deterministic modeling of network design problem is MILP. Barros et al. (1998) propose a heuristic procedure based on a linear relaxation strengthened by valid inequalities to generate a lower bound for a two-level location model for recycling sand from construction waste. The nonlinear programming model associated with a carpet waste management network is solved using a linear approximation solution procedure in Louwers et al. (1999). The MILP model proposed by Krikke et al. (1999a) is applied to a copier manufacturer while the MILP models developed by Realff et al. (1999, 2000b) consider carpet recycling. In Jayaraman et al. (1999), the optimum quantities of transshipment, production, and stocking for cores and remanufactured products are determined together with the location of remanufacturing/distribution facilities using a binary mixed integer programming (MIP) model. By considering case studies from different industries, Fleischmann et al. (2000) investigate the general characteristics of product recovery network design problem. Shih (2001) develops an MIP model for a computer and appliance recycling network. The discrete-time linear analytical model developed by Hu et al. (2002) minimizes the operating costs of a multi-time-step, multi-type hazardous-waste RL system. Schultmann et al. (2003) design a spent-battery RL network by combining facility-location planning and flow-sheeting-based process simulation. Jayaraman et al. (2003) determine the number and location of the collection centers and refurbishing sites and the corresponding flow of hazardous products. Their heuristic concentration procedures have the ability of solving relatively large problems with up to 40 collection sites and 30 refurbishment sites. However, they did not consider the multiple period problems, freight rate discounts, and inventory cost savings resulting from consolidation of returned products. Wang and Yang (2007) compare their heuristics with the heuristics proposed by Jayaraman et al. (2003) by considering the location-allocation problem of recycling e-waste. The MIP model presented in this study is a modified version of the model proposed by Jayaraman et al. (2003). The real-world parameters used in Shih (2001) are also exploited. Amini et al. (2005) develop a binary IP considering the repair operations of a major international medical diagnostics manufacturer. In Du and Evans (2008), an MIP model is constructed for the design of the RL network for a 3PLs company. They develop a solution methodology by integrating scatter search, the dual simplex method, and the constraint method. Pati et al. (2008) determine the facility location, route, and flow of different varieties of recyclable wastepaper by developing a multi-item, multi-echelon, and multi-facility decision-making framework based on a mixed integer GP model. A multi-period two-level hierarchical optimization model is proposed by Srivastava (2008a,b). The opening decision for collection centers is determined by the first MILP optimization model based on the minimization of investment (fixed and running costs of facilities and transportation costs). The second MILP model determines disposition, location, and capacity addition decisions for rework sites at different time periods together with the flows to them from collection centers based on maximization of profit. Min and Ko (2008) employ a GA to solve an MIP model associated with the location-allocation problem of a 3PLs service provider. A novel GA is proposed by Lee et al. (2009) for an RL network while Dehghanian and Mansour (2009) develop a multi-objective GA to design a product recovery network. Simulated annealing (SA) is used in Pishvaee et al. (2010b) to solve the MILP model associated with an RL network design problem. Sasikumar et al. (2010) develop a mixed integer nonlinear programming (MINLP) model to design the RL network of a truck tire remanufacturer. Ren and Ye (2011) propose an improved particle swarm optimization procedure for RL network design problem. Alumur et al. (2012) present an MILP formulation for multi-period RL network design.

Tuzkaya et al. (2011) propose a multi-objective decision-making methodology involving two stages. In the first stage, ANP and fuzzy-TOPSIS (Technique for Order Preference by Similarity to Ideal Solution) are integrated for the evaluation of centralized return centers (CRCs). In the second stage, a RL network design problem is constructed using the CRC weights obtained in the first stage. A GA is developed to solve this model.

In all of the aforementioned studies, only reverse flows are considered while modeling network design problem. However, in some cases, simultaneous consideration of forward and reverse flows may be required. Considering this need, in recent years, several deterministic network design models have been developed for closed-loop supply chains that involve both reverse and forward logistics components.

One of the earliest models on the integration of forward and RL was presented by Fleischmann et al. (2001). In this study, they compare the traditional logistic design with the simultaneous design of forward and reverse network by employing an MILP-based generic recovery network model (RNM). According to the results obtained from the two cases studied, it is concluded that product recovery can be integrated with existing logistic structure in an efficient way for many cases. However, the integrated design of reverse and forward logistic networks may be required in some other cases.

Salema et al. (2007) improve Fleischmann et al.'s (2001) study by considering capacity limits on production/storage, multiproduct production, and uncertainty in demand/return flows. They extend the RNM and develop an MILP formulation for a capacitated multiproduct RL model by considering forward flows. Considering the same problem, Salema et al. (2005) present a two-level approach. The multi-period MILP model proposed by Beamon and Fernandes (2004) for a closed-loop supply chain determines the location and sorting capabilities of warehouses together with the amount of material to be transported between each pair of sites. A linear programming (LP)-based GA is proposed by Sim et al. (2004) for the design of a closed-loop supply chain while a linear optimization model is developed by Sheu et al. (2005). Zhou et al. (2005) develop a GA to solve the MINLP model associated with a distribution problem with forward and reverse flows. The location problem of a remanufacturing RL network with forward flows is solved in Lu and Bostel (2007) by employing a Lagrangian heuristic approach. The forward/reverse network design problem of a 3PLs provider is solved by Ko and Evans (2007) using a heuristic procedure developed based on GAs. In Uster et al. (2007), an exact solution is obtained for the large-scale MILP model of a closed-loop supply chain network using Benders' decomposition with alternative multiple cuts. Easwaran and Uster (2010) extend Uster et al. (2007) by considering the following details: (1) Hybrid (manufacturing and remanufacturing) production plants and hybrid centers for inter-mediate storage and handling (as opposed to separate distribution centers and col-lection centers). (2) Determination of the locations of facilities in both forward and reverse channel networks. (3) Finite processing/storage capacity restrictions. In addi-tion, their Benders' framework involves simultaneous use of strengthened Benders' cuts obtained from alternative formulations. The bi-level programming model pro-posed by Wang et al. (2007) determines the location and inventory policies for a closed-loop supply chain. Lee et al. (2007a) consider two objectives: maximization of the returned products shipped from customers back to the collection facilities and minimization of the total costs associated with the forward and RL operations. They integrate fuzzy GP and GAs to solve the problem. Lee et al. (2007b) develop a GA to solve the MILP model associated with the closed-loop supply chain design of a 3PL. Lee and Dong (2008) integrate Tabu Search (TS) and network simplex algorithm to solve the location-allocation problem of an end-of-lease computer recovery network. Optimal manufacturing, remanufacturing, transportation quantities together with the optimal locations of disassembly, collection, and distribution facilities are deter-mined in Demirel and Gökçen (2008) using an MIP model. Mutha and Pokharel (2009) propose a mathematical model for the design of a multi-echelon closed-loop supply chain. GA and particle swarm optimization are integrated in Kannan et al. (2009b) to design a multi-echelon closed-loop supply chain in a build-to-order envi-ronment. A spanning-tree-based GA is employed in Wang and Hsu (2009) to solve an IP model associated with the design of a closed-loop supply chain. Salema et al. (2010) propose a generic model for the simultaneous design and planning of supply chains with reverse flows by developing a graph approach based on the conventional concepts of nodes and arcs. In Kannan et al. (2010), a GA-based heuristic is pro-posed to solve an MILP model associated with a closed-loop supply chain network design problem. Yang et al. (2009) formulate and optimize the equilibrium state of a closed-loop supply chain network by using theory of variational inequalities. Pishvaee et al. (2010a) propose a memetic algorithm-based methodology for the integrated design of forward and RL networks.

In some studies, only location issues associated with collection centers are considered. The collection point location problem is formulated as a set covering and MAX-SAT problem in Bautista and Pereira (2006). Then, GAs and GRASP (Greedy Randomized Adaptive Search Procedure) methodologies are employed to solve the set covering and MAX-SAT formulation, respectively. Min et al. (2006b) develop a nonlinear integer program to solve a multi-echelon RL problem. However, they ignore temporal consolidation issues in a multiple planning horizon. The mixed integer nonlinear model proposed by Min et al. (2006a) determines the number and location of initial collection points and CRCs. This model allows for the determination of the exact length of holding time for consolidation at the initial collection points and total RL costs associated with product returns in a multiple planning horizon. They solve the model using a GA-based solution procedure. The analytical model developed by Wojanowski et al. (2007) for the collection facility network design and pricing policy considers the impact of the deposit-refund on the sales rate and return rate. In Aras and Aksen (2008), an MINLP model is developed for collection center location problem with distance and incentive-dependent returns under a dropoff policy. In a follow-up study, Aras et al. (2008) consider a pickup policy with capacitated vehicles. The analytical model proposed by de Figueiredo and Mayerle (2008) allows for the design of minimum-cost recycling collection networks with required throughput. Cruz-Rivera and Ertel (2009) construct an uncapacitated facility-location model in order to design a collection network for EOL vehicles in Mexico.

The evaluation of potential collection center locations is another active research area. Bian and Yu (2006) develop an AHP-based approach for an international electrical manufacturer. An integrated ANP-fuzzy technique is proposed by Tuzkaya and Gülsün (2008). In Pochampally and Gupta (2008), AHP and fuzzy set theory are integrated to determine potential facilities from a set of candidate recovery facilities. AHP and fuzzy AHP are applied to determine collection center location in an RL network in Kannan et al. (2008). The approach proposed by Pochampally and Gupta (2009) evaluates the efficiencies of collection and recovery facilities in four phases, viz., (1) determination of the criteria, (2) use of fuzzy ratings of existing facilities to construct a neural network that gives the importance value for each criterion, (3) calculation of the overall ratings of the facilities of interest using a fuzzy-TOPSIS approach, and (4) calculation of the maximized consensus ratings of the facilities of interest by employing Borda's choice rule. Pochampally and Gupta (2012) use linear physical programming (LPP) to select efficient collection centers.

Another important issue in RL network design is the determination of the appropriate reverse channel structure for the collection of used products from customers. This problem is studied by Savaskan et al. (2004) for a single manufacturer single retailer case. Savaskan and Van Wassenhove (2006) consider two retailers in a competitive retailing environment. Bhattacharya et al. (2006) consider a retailer ordering product from a manufacturer which makes new products and orders remanufactured products from a remanufacturer. They analyze the impact of centralized and decentralized channel structures in the optimal order quantity. Centralized as well as remanufacturer- and collector-driven decentralized channels are studied in Karakayali et al. (2007). Hong et al. (2008) compare centralized (i.e., a decision maker gives decisions for the entire system) and decentralized (i.e., several independent entities are individually operated by self-interested parties) decision making. The negotiation-based coordination mechanism proposed by Walther et al. (2008) assigns recycling tasks to the companies of a recycling network in a decentralized way. Lee et al. (2011) consider a decentralized RL system with retailer collection. They determine a profitable apportionment of effort between the manufacturer and retailer for different product recovery processes. The reverse logistics channel (RLC) design framework proposed by El Korchi and Millet (2011) involves two stages. In the first stage, current RLC structure is evaluated by comparing the current structure with the alternatives. Several criteria (viz., feasibility assessment, economic assessment, environmental assessment, and social assessment) are considered in the second stage in order to select potential generic RLC structure from among the 18 generic structures.

1.4.3.1.1.2 Stochastic Models   There is a high degree of uncertainty associated with quality and quantity of returns. In order to deal with this uncertainty, various stochastic RL network design models have been developed. A commonly used technique is robust optimization. The robust MILP model proposed by Realff et al. (2000a, 2004) can search for solutions close to the mathematically optimal solutions for a set of alternative scenarios identified by a decision maker. In Hong et al. (2006), the proposed robust MILP model maximizes system net profit for specified deterministic parameter values in each scenario. A robust solution for all of the scenarios is then found using a min–max robust optimization methodology. Pishvaee et al. (2011) develop a robust MIP model for designing a closed-loop supply chain network by using the recent extensions in robust optimization theory. Hasani et al. (2012) propose a robust closed-loop supply chain network design model for perishable goods in agile manufacturing.

Another popular technique is stochastic programming. Considering the sand recycling case study presented in Barros et al. (1998), Listes and Dekker (2005) propose a stochastic programming-based RL network design methodology. Listes (2007) develop a generic stochastic model for closed-loop supply chains. In Lee et al. (2007c), a stochastic programming-based methodology is developed for the design of a product recovery network. The stochastic programming model developed by Chouinard et al. (2008) considers randomness associated with recovery, processing, and demand volumes in a closed-loop supply chain. A sample average approximation-based heuristic is employed to solve the problem. Location and allocation decisions of an RL network under uncertainty are modeled by Lee and Dong (2009) based on stochastic programming. This model is then solved by combining SA and sample average approximation. Pishvaee et al. (2009) propose a scenario-based stochastic optimization model for the design of a closed-loop integrated forward/reverse logistics network including production/recovery, hybrid distribution-collection, customer, and disposal centers. Kara and Onut (2010) propose a two-stage stochastic and robust programming approach to a paper recycling network design problem involving optimal recycling center locations and optimal flow amounts between the nodes.

El-Sayed et al. (2010) propose a stochastic mixed integer linear programming (SMILP) model for the design of a closed-loop supply chain by considering multi-period stochastic demand with three echelons (suppliers, facilities, and distributors) in the forward direction and two echelons (disassemblies and redistributors) in the reverse direction.

Lee et al. (2010) integrate the sample average approximation scheme with an importance sampling strategy to solve the stochastic sampling formulation developed for a large-scale RL network in the Asia Pacific region.

In order to deal with the dynamic and stochastic aspects of RL networks, Lieckens and Vandaele (2007) develop an MINLP model by combining a conventional RL MILP model with a queueing model. The model is solved using a GA-based technique, Differential Evolution. Lieckens and Vandaele (2012) extend Lieckens and Vandaele (2007) by considering multiple levels, quality-dependent routings, and stochastic transportation delays.

In some studies, fuzzy logic is used to model uncertain factors. Qin and Ji (2010) integrate fuzzy simulation and GA for the design of an RL network. Pishvaee and Torabi (2010) develop a fuzzy solution approach by combining a number of efficient solution approaches for the design of a closed-loop supply chain network. Zarandi et al. (2011) use interactive fuzzy goal programming in order to solve the network design problem of a closed-loop supply chain while Pishvaee and Razmi (2012) employ multi-objective fuzzy mathematical programming.

Pishvaee et al. (2012) propose a bi-objective (viz., minimization of environmental impacts and total cost) credibility-based fuzzy mathematical programming model.

Swarnkar and Harding (2009) develop a GA-based simulation optimization methodology for the design of a product recovery network.

A comprehensive review of the studies on RL network design can be found in Akcali et al. (2009) and Wang and Bai (2010).

1.4.3.1.2 Transportation Issues

Cost-effective management of an RL network highly depends on the efficient and effective planning of transportation activities. Among transportation issues, determination of vehicle routes using different versions of well-known vehicle routing problem (VRP) is a very active research topic. In some studies, only return flows are considered. Mixed capacitated arc routing problem of a refuse collection network is solved in Mourao and Almeida (2000) and Mourao and Amado (2005) by developing heuristic methods. Reliable estimates of transportation costs in a recycling network redesign problem are obtained in Blanc et al. (2004) with the help of a VRP. In Blanc et al. (2006), a multi-depot pickup and delivery model with capacitated vehicles and alternative delivery locations is proposed for the collection of containers from EOL vehicle dismantlers in the Netherlands. The model is then solved using an integrated heuristic based on set partitioning and route generation. Considering EOL auto RL network, Schultmann et al. (2006) propose a symmetric capacitated VRP to generate a tour schedule with minimal cost. Optimum schedule is determined using a TS-based methodology. VRP associated with the low-frequency collection of materials disassembled from EOL vehicles is solved in Krikke et al. (2008) by using a methodology integrating route generation and set partitioning. In Kim et al. (2009d), an RL network in South Korea is modeled as a VRP. The problem is then solved using a TS algorithm. Sasikumar et al. (2009) develop a TS-based heuristic procedure to solve the VRP associated with a third-party RL provider.

Some researchers develop vehicle routing plans by simultaneously considering return and delivery flows. In the RL system considered by Dethloff (2001), customers have both pickup and delivery demands. First, this system is modeled as a vehicle routing problem with simultaneous delivery and pickup (VRPSDP). Then the problem is solved using a heuristic construction procedure. The heuristic procedure proposed in Dethloff (2001) is used by Dethloff (2002) to solve the VRP with backhauls. After modeling VRPSDP as an MILP model, general solutions are developed using conventional construction and improvement heuristics and TS in Gribkovskaia et al. (2007). Blood distribution network of the American Red Cross is analyzed by Alshamrani et al. (2007). In this network, products are delivered from a central processing point to customers (stops) in one period and are available for return to the central point in the following period. Route design and pickup strategies are determined simultaneously through the development of a heuristic procedure. Çatay (2010) develops a saving-based ant colony algorithm for VRPSDP.

Shaik and Abdul-Kader (2011) propose a methodology for comprehensive performance measurement of transportation system in RL. This methodology complements and integrates the two frameworks (viz., BSC and performance prism [PP]) and employs AHP to understand the importance and priority of various performance criteria.

1.4.3.1.3 Selection of Used Products

There are many third-party firms collecting used products to make profit. While selecting used products, these firms compare the revenues from recycle or resale of products' components and collection and reprocessing costs of the used products (Pochampally et al. 2009c). Construction of a cost-benefit function is the most commonly used technique in the selection of used products for reprocessing. The value of cost-benefit function proposed by Veerakamolmal and Gupta (1999) is calculated by subtracting the sum of revenue terms from the sum of cost terms. This cost-benefit function is improved by Pochampally and Gupta (2005) and Pochampally et al. (2009b) by considering two important details associated with a used product of interest: the probability of breakage and the probability of missing components. Then they develop an integer LP model with the aim of maximizing the modified cost-benefit function. The uncertainty associated with revenues and costs is considered by Pochampally and Gupta (2008) through the development of a fuzzy cost-benefit function. The application of cost-benefit function technique requires the evaluation criteria to be presented in terms of classical numerical constraints. An LPP formulation is presented by Pochampally et al. (2009c) for the case of presentation of evaluation criteria in terms of range of different degrees of desirability.

1.4.3.1.4 Facility Layout

Selection of facility layout is another strategic issue. Lim and Noble (2006) present a simulation and design of experiments-based analysis for the performance evaluation of four different remanufacturing layouts, namely, cellular, fractal, holonic, and job shop. They conclude that the best layout type depends on the operating conditions of the remanufacturing system. For instance, if there is a need for a low mean flow-time, low WIP level, and moderate production volume, the cellular layout is found to be the best choice. Opalic et al. (2004) propose a disassembly line layout for appliance recycling. The movement of EOL appliances through the line is provided by a closed-loop conveyor that allows the operator to pick a unit which is similar to the previous unit the operator disassembled. By this way, the tools in the station can be used in an organized and efficient manner. They also introduce some other practical concepts to improve disassembly speed while reducing lifting, contamination risk, and overloading of sorting operator. Topcu et al. (2008) use simulation and stochastic programming to study the facility and storage space design issues that come up due to higher level of uncertainty associated with remanufacturing systems. They specifically consider uncertainty and variability due to (1) the number of returned products, (2) the type and number of parts reclaimed from each returned product, (3) the type of processes required to remanufacture a part, (4) the flow of parts and materials, and (5) the demand for the remanufactured part or the final product.

1.4.3.1.5 Information Technology

An effective information technology (IT) infrastructure is a must in an RL system considering the need for accurate projection of time and amount of returned products. Moreover, the coordination between the various parties involved in an RL system is provided by the IT infrastructure. Researchers studied the impact of IT on RL operations. Dhanda and Hill (2005) present a case study to investigate the role of IT in RL. Daugherty et al. (2005) analyze a survey of businesses in the automobile aftermarket industry to emphasize the importance of resource commitment to IT in RL. Olorunniwo and Li (2010) analyze the IT types used in RL by focusing on the impact of these technologies on RL performance. It is concluded that the operational attributes derived from the use of IT (e.g., efficient tracking and effective planning) have a positive impact on RL performance. Chouinard et al. (2005) develop an information system architecture for the integration of RL activities within a supply chain information system. Kumar and Chan (2011) integrate the impact of RFID (radio-frequency identification) technology in the easy counting of returned products into a mathematical model, which simultaneously determines the amount of products/parts processed at the RL facilities and quantity of virgin parts purchased from external suppliers. Condea et al. (2010) develop an analytical model to analyze the monetary benefits of RFID-enabled returns management processes.

1.4.3.2 Planning and Control

1.4.3.2.1 Forecasting

The high level of uncertainty in the timing and quantity of returns makes the use of traditional forecasting methods impossible (Marx-Gomez et al. 2002) for RL systems. That is why various novel forecasting methods have been developed to predict the product returns. Kelle and Silver (1989) predict the container returns throughout the lead time by proposing four forecasting methods with different information requirements. Toktay et al. (2000) estimate the total number of circuit boards for Kodak's single-use camera return network by using the methods developed by Kelle and Silver (1989). Considering the impact of imperfect information on inventory-related costs, the performance of the forecasting methods proposed in Kelle and Silver (1989) is analyzed by de Brito and van der Laan (2009) and Toktay et al. (2004).

The waste stream resulting from disposal of the CRTs in the United States for the period between the years 2000 and 2050 is estimated by Linton and Yeomans (2003) and Linton et al. (2002, 2005). First, a waste disposal model is developed to capture the uncertainty associated with the television life cycle, the CRT weight in the televisions, the time between television failure and actual entrance time to the waste stream, and the proportion of televisions that are reclaimed. Then, the forecasting for future television sales is carried out under three technological change scenarios: no technological change, moderate change, and aggressive change. Monte Carlo simulation is employed to investigate each scenario.

Marx-Gomez et al. (2002) integrate FL, simulation, and neural networks to estimate scrapped product returns. At the first phase, a simulation model is developed for the generation of data on return amounts, sales, and failures. The second phase involves the design of a fuzzy inference system to estimate the return amounts for a specific planning period. At the last phase, multi-period forecasting of product returns is achieved using a neuro-fuzzy system.

1.4.3.2.2 Production Planning

Production planning involves decisions on how much and when to disassemble, how much and when to remanufacture, how much to produce and/or order for new materials, and how to coordinate disassembly and reassembly (Guide et al. 1999). The material requirements planning (MRP)-based methodology developed by Ferrer and Whybark (2001) assists managers in answering the following questions: How many and which cores to buy, what mix of cores to disassemble, and which components should be assembled to meet demand. Gupta and Veerakamolmal (2001) determine the number of products to disassemble in order to fulfill the demand for various components for remanufacturing in different time periods using an IP-based algorithm. Souza and Ketzenberg (2002) and Souza et al. (2002) use a two-stage GI/G/1 queueing network model to find the optimal, long-run production mix that maximizes profit for a firm that meets demand for an order with remanufactured products, new products, or a mix of both. They also use a DES (discrete event simulation) model involving some real-life issues such as stochastic product returns and stochastic production yield to test the robustness of the model. The number of units of core type with a nominal quality level that is disassembled, disposed, remanufactured, and acquired in a given time period is determined in Jayaraman (2006) by developing a mathematical programming model. This model can also determine the inventory of modules and cores that remain at the end of a given time period. The MIP model developed by Kim et al. (2006c) maximizes the total remanufacturing cost saving while determining the quantity of products/parts processed in the remanufacturing facilities/subcontractors and the amount of parts purchased from the external suppliers. What products to accept, process, and reprocess is determined in Lu et al. (2006) using a short-term bulk recycling planning model. The expected number of remanufactured units to be completed in each future period and components needed to be purchased to avoid any projected shortages are estimated by DePuy et al. (2007).

Li et al. (2009) optimize the production planning and control policies for dedicated remanufacturing by integrating a hybrid cell evaluated GA with a DES model. The amount of EOL products and components to be collected, nondestructively or destructively disassembled, recycled, remanufactured, stored, backordered, and disposed in each period is determined by Xanthopoulos and Iakovou (2009) based on an MILP-based aggregate production planning model. Denizel et al. (2010) develop a multi-period remanufacturing planning model considering the uncertain quality of product returns. A generic mixed IP model incorporating setup costs and times is proposed by Doh and Lee (2010). Shi et al. (2011b) propose a mathematical model for the simultaneous optimization of production quantities of brand-new products, the remanufactured quantities, and the acquisition prices of the used products based on the maximization of profit in a multiproduct closed-loop system.

1.4.3.2.3 Capacity Planning

Unique characteristics of reverse and closed-loop supply chains forced researchers to develop new capacity planning methodologies. Guide and Spencer (1997) consider probabilistic material replacement and probabilistic routing files while developing a rough cut capacity planning (RCCP) method for remanufacturing firms. Guide et al. (1997) conclude that traditional techniques tend to perform poorly in a recoverable environment after comparing the modified RCCP techniques with traditional RCCP techniques.

In some studies, capacity planning models were developed using LP and/or simulation. The mathematical model presented by Kim et al. (2005) develops a capacity plan considering the maximization of the saving from the investment on remanufacturing facilities. An integrated capacity planning methodology based on LP and DES is developed in Franke et al. (2006). System dynamics simulation (SDS)-based closed-loop supply chain capacity planning models are developed in Georgiadis et al. (2006) and Vlachos et al. (2007). Georgiadis and Athanasiou (2010) extend Georgiadis et al. (2006) in two ways. First, two product types with two sequential product life cycles are considered. Second, two scenarios created based on customer preferences over the product types are analyzed.

1.4.3.2.4 Inventory Management

RL causes the following two complexities in traditional inventory management approaches developed for forward logistics systems (Inderfurth and van der Laan 2001):

- The level of uncertainty is higher due to uncertain product returns.
- Remanufacturing and regular mode of procurement must be carried out in a coordinated manner.

Various inventory models have been developed to deal with these complexities. In this section, these models are analyzed. We can classify inventory models into two main groups: deterministic and stochastic. Deterministic models can further be classified as stationary and dynamic demand models while the two main models that can be analyzed under stochastic models category are periodic review and continuous review models.

1.4.3.2.4.1   Deterministic Models   These models search for an optimal balance between fixed setup costs and variable inventory holding costs by assuming that demand and return quantities are known for entire planning horizon.

1.4.3.2.4.1.1   Stationary Demand   The logic of economic order quantity (EOQ) is exploited by deterministic models in case of stationary demand (Fleischmann et al. 1997). Schrady (1967) proposed the first EOQ model with item returns by assuming infinite production rates for manufacturing and remanufacturing. As an extension to Schrady (1967), finite remanufacturing rates are considered by Nahmias and Rivera (1979). Mabini et al. (1992) also extend Schrady (1967) by considering stock-out service level constraints and a multi-item system. EOQ waste disposal and repair models with variable remanufacturing and return rates are presented by Richter (1996a,b, 1997). The optimal number of remanufacturing and production batches is determined for the different values of the return rate. Integer nonlinear models for the analysis of EOQ repair and waste disposal problem with integer setup numbers are developed by Richter and Dobos (1999) and Dobos and Richter (2000). They state that the pure strategy (total repair or total waste disposal) is optimal. Assuming that there is only one recycling lot and one production lot, a production-recycling system is analyzed by Dobos and Richter (2003). The results of Dobos and Richter (2003) are generalized by Dobos and Richter (2004) by considering multiple production and recycling lots. As an extension to the model of Dobos and Richter (2004), Dobos and Richter (2006) relax the assumption of perfect quality of the returned items. Extending Richter (1996a,b), El Saadany and Jaber (2008) consider the costs associated with switching between production and recovery runs. It is also pointed out that ignoring the first time interval causes an overestimation of holding costs due to an unnecessary residual inventory. Jaber and Rosen (2008) consider the EOQ repair and waste disposal model of Richter (1996a,b) and apply the first and second laws of thermodynamics to reduce system entropy. Jaber and El Saadany (2009) extend the inventory models presented in Richter (1996a,b) by incorporating the possibility of lost sales. The models of Dobos and Richter (2003, 2004) are extended by El Saadany and Jaber (2010) by considering a price- and quality-dependent return rate. In Jaber and El Saadany (2011), the learning effects in production and remanufacturing (repair) are incorporated into the model of Dobos and Richter (2003, 2004).

EOQ formulae are developed by Teunter (2001) by using different holding cost rates for manufactured and recovered items. In the joint EOQ and EPQ model proposed by Koh et al. (2002), remanufacturing or procurement can be used to satisfy the stationary demand. As an extension to previous studies, a capacitated repair facility is considered. Extending Koh et al. (2002), Wee et al. (2006) develop a search procedure for the optimal ordering and recovery policy by allowing shortage backorders. In Teunter (2004), simple expressions are developed for the determination of optimal lot sizes for the production/procurement of new items and for the recovery of returned items. These expressions are more general than those in the literature since they can be used for finite and infinite production rates as well as finite and infinite recovery rates. In a follow-up study, Konstantaras and Papachristos (2006) consider backordering and develop a mathematically rigorous approach which leads to overall optimal policy within a specific set of policies. The performance of cycle order policy and dual sourcing ordering policy is compared in Tang and Grubbström (2005) by considering stochastic lead times for manufacturing and remanufacturing. Optimal policy parameters for a recycling system in which returned items are used as raw material in the production of new products are developed by Oh and Hwang (2006). Tang and Teunter (2006) formulate an MIP problem to find the cycle time and the production start times for (re)manufacturing lots based on the minimization of the total cost per time unit considering a hybrid remanufacturing/manufacturing system in which manufacturing and remanufacturing operations for multiple product types are performed on the same production line. In a follow-up paper, Teunter et al. (2008) consider the case of dedicated lines for manufacturing and remanufacturing. Chung et al. (2008) simultaneously consider the concerns of the supplier, the manufacturer, the retailer, and the third-party recycler while developing an optimal production and replenishment policy for a multi-echelon inventory system with remanufacturing. A model cycle involves one manufacturing cycle and one remanufacturing cycle. Yuan and Gao (2010) extend the model of Chung et al. (2008) to the more general $(1, R)$ (i.e., one manufacturing cycle and R remanufacturing cycle) and $(P, 1)$ (i.e., P manufacturing cycle and one remanufacturing cycle) policies. In the hybrid remanufacturing-production system considered by Roy et al. (2009), defective units are continuously transferred to the remanufacturing and the constant demand is met by the perfect items from production and remanufactured units. Rate of defectiveness of the production system is modeled as a fuzzy parameter, whereas the remanufactured units are treated as perfect items. The total number of cycles in the time horizon, the duration for which the defective items are collected, and the cycle length after the first cycle are determined using a GA based on the maximization of total profit. Rubio and Corominas (2008) determine the optimal values for manufacturing and remanufacturing capacities, return rates, and use rates for EOL products by considering a lean production-remanufacturing environment in which capacities of manufacturing and remanufacturing can be adjusted according to constant demand in order to prevent the excessive inventory levels.

1.4.3.2.4.1.2 Dynamic Demand   Earlier studies deal with the dynamic demand by modifying classical Wagner-Whitin algorithms while recent studies determine the optimal parameter values by developing DP-based algorithms. The extended Wagner-Whitin algorithm developed by Richter and Sombrutzki (2000) for a deterministic recovery system assumes a linear cost model with no backordering and negligible lead times. The applicability of the algorithm is limited to large quantities of used products. In other words, it is assumed that the quantity of used products matches the demand of remanufactured goods. This model is extended by Richter and Weber (2001) for variable manufacturing and remanufacturing costs. An application of Richter and Sombrutzki's (2000) model in a just-in-time framework was presented by Richter and Gobsch (2003). In Minner and Kleber (2001), control theory is used to find an optimal policy for a remanufacturing system with dynamic demand by assuming no backorders and lead times. As an extension to Minner and Kleber (2001), Kiesmüller (2003b) determine an optimal policy by considering positive and different lead times for production and remanufacturing. The lot-sizing problem with remanufacturing is modeled as a network flow problem in

Golany et al. (2001). A polynomial time algorithm is presented for the case of linear costs. For a similar problem, a polynomial time algorithm for the case of concave costs is presented in Yang et al. (2005). Kleber et al. (2002) consider multiple remanufacturing options and determine the optimal policy using Pontryagin's Maximum Principle with the assumption of no backorders and zero lead times. Considering the dynamic lot-sizing problem with directly saleable returns, Beltran and Krass (2002) determine the manufacturing and disposal decisions by developing a DP algorithm with a $O(N^3)$ complexity for the case of concave costs. In Teunter et al. (2006), two setup cost settings are considered: a joint setup cost for manufacturing and remanufacturing (single production line) or separate setup costs (dedicated production lines). After modeling both problems as MIP programs, a DP algorithm is proposed for the joint setup cost setting. Modified versions of Silver Meal (SM), Least Unit Cost (LUC), and Part Period Balancing (PPB) heuristics are also provided for both settings. Schulz (2011) provides a generalization of the SM-based heuristic proposed by Teunter et al. (2006). He applies methods known from the corresponding static problem by considering separate setup cost setting (without disposal option and restricted capacities). He also develops a simple improvement heuristic to enhance the heuristic's performance. An optimal policy specifying the period of switching from remanufacturing to manufacturing, the periods where remanufacturing and manufacturing activities take place, and the corresponding lot sizes is presented by Konstantaras and Papachristos (2007). Bera et al. (2008) assume stochastic product defectiveness and fuzzy upper bounds for production, remanufacturing, and disposal while investigating a production-remanufacturing control problem.

1.4.3.2.4.2 Stochastic Models   Stochastic models employ stochastic processes while modeling demand and returns. We can distinguish two common stochastic modeling approaches, viz., continuous and periodic review policies.

1.4.3.2.4.2.1   Continuous Review Models   Using a continuous time axis, these models determine the optimal static control policies by minimizing the long-run average costs per unit of time (Fleischmann et al. 1997). A continuous review strategy for a single-item inventory system with remanufacturing and disposal is proposed by Heyman (1977). An optimum disposal level is determined with the assumption of no fixed ordering costs and instantaneous outside procurement. Muckstadt and Isaac (1981) extend Heyman (1977) by developing a model involving nonzero lead times for repair and procurement and nonzero fixed costs. However, they develop an approximate numerical procedure to determine the optimal parameter values by ignoring disposal of the products. As an extension to Muckstadt and Isaac (1981), van der Laan et al. (1996a,b) consider a disposal option and compare a number of policies numerically. Van der Laan and Salomon (1997) and van der Laan et al. (1999b) consider nonzero lead times for serviceable and recoverable stock while presenting a detailed analysis of different policies in the aforementioned setting. Two policies are mainly considered: a push- and a pull-driven recovery policy. Extending van der Laan et al. (1999b), van der Laan et al. (1999a) consider stochastic lead times for manufacturing and remanufacturing. Parameters of a (s, Q) policy for a basic inventory model involving Poisson demand and returns are optimized in Fleischmann et al. (2002). Fleischmann and Kuik (2003) develop an average cost optimal (s, S) policy using general results on Markov decision processes for an inventory system involving independent stochastic demand and item returns. Using certain extensions of (s, Q) policy and assuming the equality of manufacturing and remanufacturing lead times, van der Laan and Teunter (2006) propose closed-form expressions for each policy to calculate near-optimal policy parameters. In Zanoni et al. (2006), some inventory control policies extended from the traditional inventory control models such as (s, Q) and (s, S) are analyzed for a hybrid manufacturing/remanufacturing system with stochastic demand, return rate, and lead times. Different inventory control policies are compared based on the total cost using DES. Planning stability of production and remanufacturing setups in a product recovery system are discussed in Heisig and Fleischmann (2001). In the models mentioned earlier, an average cost comparison is done while giving a priority decision between manufacturing and remanufacturing. Questioning the reliability of this technique, Aras et al. (2006) develop two alternative strategies in which demand is satisfied using either manufacturing or remanufacturing.

The behavior of a multi-echelon inventory system with returns is analyzed in Korugan and Gupta (1998) using a queueing network model. Toktay et al. (2000) investigate the procurement of new components for recyclable products by developing a closed queueing network model. A queueing network model involving manufacturing/remanufacturing operations, supplier's operations for the new parts and useful lifetime of the product is presented in Bayindir et al. (2003). The conditions on different system parameters (lifetime of the product, supplier lead time, lead time and value added of manufacturing and remanufacturing operations, capacity of the production facilities) that make remanufacturing alternative attractive are investigated using this model based on the total cost. A closed-form solution for the system steady-state probability distribution for an inventory model with returns and lateral transshipments between inventory systems is developed in Ching et al. (2003). Nakashima et al. (2002, 2004) analyze the behavior of stochastic remanufacturing systems by developing Markov chain models. Takahashi et al. (2007) develop a Markov chain model to evaluate the policies proposed for a decomposition process in which recovered products are decomposed into parts, materials and waste. Mitra (2009) develops a deterministic model as well as a stochastic model under continuous review for a two-echelon system with returns. Teng et al. (2011) extend Mitra (2009) by optimizing the partial backordering inventory model with product returns and excess stocks.

1.4.3.2.4.2.2 Periodic Review Models In these models, optimal policies are determined by minimizing the expected costs over a finite planning horizon (Fleischmann et al. 1997). The optimal policy proposed by Simpson (1978) involves three parameters (S, M, U), where S is produce-up-to level, M is remanufacture-up-to level, and U is the dispose-down-to level. They assume zero lead times and a disposal option. Inderfurth (1997) shows the optimality of the (S, M, U) policy for the case of equal lead times. An exact computation method and two approximations are provided by Kiesmüller and Scherer (2003) to determine the optimal parameter values for the periodic review policy studied by Simpson (1978) and Inderfurth (1997). Assuming that all available recoverables can be remanufactured, Mahadevan et al. (2003) develop heuristics to determine only produce-up-to level for a pull policy. Simple expressions for computing the produce-up-to level and the remanufacture-up-to level for the cases of identical and nonidentical lead times are presented by Kiesmüller (2003a) and Kiesmuller and Minner (2003). Ahiska and King (2010) extend Kiesmüller (2003a), Kiesmuller and Minner (2003), and Kiesmüller and Scherer (2003) by considering setup costs and different lead time cases for manufacturing and remanufacturing. An approximation algorithm for the determination of optimal policy parameters of a stochastic remanufacturing system with multiple reuse options is developed by Inderfurth et al. (2001).

In some studies, multi-echelon systems are considered. The model studied by Simpson (1978) and Inderfurth (1997) is extended to a series system with no disposal in DeCroix (2006). Considering an infinite-horizon series system where returns go directly to stock, optimality of an echelon base-stock policy is showed by DeCroix et al. (2005).

A special case of periodic review models with only one period is Newsboy problem (Dong et al. 2005). In Vlachos and Dekker (2003) and Mostard and Teunter (2006), the classical newsboy problem is extended to incorporate returns with the aim of determining the initial order quantity. In Vlachos and Dekker (2003), it is assumed that a constant portion of the sold products is returned and a returned product can be resold at most once. In the newsboy problem presented in Vlachos and Dekker (2003), a sold product is returned with a certain probability and it can be sold as long as there is no damage on it.

1.4.3.2.5 Selection and Evaluation of Suppliers

Reverse and forward flows differ from each other in many aspects including the cost and complexity of transportation, storage and/or handling operations. Instead of dealing with the issues created by these differences, many firms prefer outsourcing their RL operations to a 3PL provider (Meade and Sarkis 2002, Efendigil et al. 2008). Many multiple criteria decision-making (MCDM) methodologies have been developed for the selection of the 3PLs. An ANP-based 3PL evaluation and selection methodology is presented by Meade and Sarkis (2002). Presley et al. (2007) combine four techniques, viz., activity-based costing (ABC), balanced scorecard, AHP, and QFD, for the selection of the best 3PL. The holistic approach presented by Efendigil et al. (2008) allows for selecting a 3PL in the presence of vagueness by integrating NN and FL. Tsai and Hung (2009) solve the treatment supplier selection problem of an electronic equipment manufacturer by developing a preemptive GP with environmental goals, ABC goals, and supply chain goals. AHP is used to calculate the performance weights of suppliers. Kannan et al. (2009a) integrate AHP and LP to consider both tangible and intangible factors while selecting the best 3PRLPs and placing the optimum quantities among them. Candidate companies collecting and selling used products are rated in Pochampally et al. (2009b) with the use of a TOPSIS-based methodology. The concerns of consumers, local government officials, and manufacturers are considered in this methodology. For the 3PL selection problem, Kannan et al. (2009c) present a methodology by combining fuzzy-TOPSIS and interpretive structural modeling. AHP and fuzzy AHP are used by Kannan (2009) for the same problem, while DEA-based methodologies are proposed by Saen (2009, 2010, 2011). Kannan and Murugesan (2011) use fuzzy extent analysis. Azadi and Saen (2011) propose a chance-constrained data envelopment analysis approach considering both dual-role factors and stochastic data. A conceptual framework based on a review of the literature of the factors that influence 3PL is developed by Sharif et al. (2012). They also evaluate and discuss the requirements for performant 3PL components using a fuzzy logic-based model.

1.4.3.2.6   Performance Measurement

Analysis of the impact of different factors and/or policies on the performance of a reverse or closed-loop supply chain is a developing research area. Due to its suitability for realistic modeling of reverse/closed-loop supply chain systems, the most commonly used technique is simulation. An SDS model is developed in Georgiadis and Vlachos (2004) to investigate the long-term behavior of a closed-loop supply chain with respect to alternative environmental protection policies concerning take-back obligation, proper collection campaigns, and green image effect. Biehl et al. (2007) analyze the impact of various system design factors together with the environmental factors on the operational performance of a carpet RL system. After developing a DES model of the system, an experimental design study is carried out. Kara et al. (2007) use DES modeling to investigate the issues associated with the RL network of EOL white goods in Sydney Metropolitan Area. The most important factors in the performance of the RL system are determined by conducting a what-if analysis. The impact of legislation, green image factor, and design for environment on the long-term behavior of a closed-loop supply chain with recycling activities is investigated in Georgiadis and Besiou (2008) through the use of an SDS model.

Pochampally et al. (2009a) develop a mathematical model based on QFD and LPP to measure a reverse/closed-loop supply chain's performance. Paksoy et al. (2011) investigate the effects of various exogenous parameters (viz., demand, product types, return rates, unit profits of the products, transportation capacities, and emission rates) on the performance measures of a closed-loop supply chain.

After reviewing some studies on green supply chain performance measurement, environmental management, traditional supply chain performance measurement, and automobile supply chain management, Olugu et al. (2011) propose various performance measures to be used in forward and reverse supply chains of automotive industry. Olugu and Wong (2011b) apply fuzzy logic to evaluate the performance of the RL process in the automotive industry. In a follow-up study, Olugu and Wong (2011a) develop an expert fuzzy rule-based system for closed-loop supply chain performance assessment in the automotive industry.

1.4.3.2.7   Marketing-Related Issues

Pricing, competition, and determination of an optimal return policy are the fundamental marketing issues in RL systems. The heuristic proposed in Guide et al. (2003) determines the optimal acquisition price of the used phones and selling price of the remanufactured phones based on the maximization of the profit which is given as the difference between the total revenue and acquisition and remanufacturing costs.

In the remanufacturing system considered in this study, used phones with different quality levels are remanufactured to a single quality level and are sold at a certain price. Since it is assumed that demand and return flows are perfectly matched, the selling price of remanufactured products could be completely determined by the acquisition prices of returns. It is also assumed that demand is a function of the price. Guide et al.'s (2003) study is extended by Mitra (2007) in four ways. First, acquisition prices are avoided since he considers a manufacturer which is responsible to recover the returns. Second, he considers more than one quality level for remanufactured products. Third, he considers a probabilistic situation where not all items may be sold, and the unsold items may have to be disposed of. Finally, demand is assumed to be a function of not only price but also availability or supply of recovered goods. Pricing policies of reusable and recyclable components for third-party firms involved in discarded product processing are investigated by Vadde et al. (2007) for the case of strict environmental regulations. Vadde et al. (2010) determine the prices of reusable and recyclable components and acquisition price of discarded products in a multi-criteria environment which involves the maximization of sales revenue and minimization of product recovery costs (viz., disposal cost, disassembly cost, preparation cost, holding cost, acquisition cost, and sorting cost). Qiaolun et al. (2008) determine collection, wholesale, and retail prices in a closed-loop supply chain using game theory. Considering an automotive shredder, two hulk pricing strategies are compared in Qu and Williams (2008) under constant, increasing and decreasing trends for ferrous metal and hulk prices. Liang et al. (2009) use the options framework and the geometric Brownian motion followed by the sales price of cores in order to determine the acquisition price of cores in an open market. Karakayali et al. (2007) determine the optimal acquisition price of EOL products and the selling price of remanufactured parts under centralized as well as remanufacturer- and collector-driven decentralized channels. Price and return policies in terms of certain market reaction parameters are determined in the model developed by Mukhopadhyay and Setoputro (2004) for e-business. The joint pricing and production technology selection problem faced by an original equipment manufacturer (OEM) operating in a market where customers differentiate between the new and the remanufactured products is investigated by Debo et al. (2005). Vorasayan and Ryan (2006) determine the optimal price and proportion of refurbished products by modeling the sale, return, refurbishment, and resale processes as an open queuing network. Shi et al. (2011a) consider a closed-loop system in which a manufacturer satisfies the demand using two channels: manufacturing brand-new products and remanufacturing returns into as-new products. They simultaneously determine the selling price, the production quantities for brand-new products and remanufactured products, and the acquisition price of used products based on the maximization of profit.

OEMs may have to compete with an independent operator who may intercept EOL products produced by OEM to sell the remanufactured products in future periods. In the two-period model proposed by Majumder and Groenevelt (2001), OEM may or may not remanufacture in the second period. Extending Majumder and Groenevelt (2001), Ferrer and Swaminathan (2006) consider a multi-period setting where there is competition between the independent operator and the OEM in the second and subsequent periods. After finding closed-form solutions for prices and quantities, an optimum solution region which was numerically explored by Majumder and Groenevelt (2001) is characterized. New product pricing decisions and recovery strategy of an OEM in a two-period model are investigated by Ferguson and Toktay (2006) by making two assumptions. First, it is assumed that OEM has an easier access to the used product. Second, the average variable cost of remanufacturing is assumed to be increasing with the remanufacturing quantity. Extending Majumder and Groenevelt (2001) and Ferguson and Toktay (2006), two periods with differentiated remanufactured products are considered in Ferrer and Swaminathan (2010). Jung and Hwang (2011) develop mathematical models to determine the optimal pricing policies under two cases, cooperation or competition between an OEM and a remanufacturer. The impact of take-back laws and government subsidies on competitive remanufacturing strategy is analyzed by Webster and Mitra (2007) and Mitra and Webster (2008), respectively. A three-stage game involving sequential decisions of two OEMs whether to take back used products during the first two stages is investigated in Heese et al. (2005). Both firms simultaneously determine the discount offered for returned products together with the price for their new products in the third stage. Wei and Zhao (2011) investigate the pricing decisions in a closed-loop supply chain with retail competition by considering the fuzziness associated with consumer demand, remanufacturing cost, and collecting cost. Closed-form expressions are developed using fuzzy theory and game theory in order to understand how the manufacturer and two competitive retailers make their own decisions about wholesale price, collecting rate, and retail prices.

An effective return policy can be used as a marketing tool to increase sales. There are studies in the literature analyzing return policies within the RL context. Mukhopadhyay and Setoputro (2005) develop a profit maximization model to determine the optimal return policy for build-to-order products. Yao et al. (2005) investigate the role of return policy in the coordination of supply chain by using a game theory-based methodology. Yalabik et al. (2005) develop an integrated product returns model with logistics and marketing coordination for a retailer servicing two distinct market segments. Mukhopadhyay and Setaputra (2006) investigate the role of a fourth-party logistics (4PL) as a return service provider and propose optimal decision policies for both the seller and the 4PL.

The use of remanufactured products as a tool to satisfy the demand arising from secondary markets is studied by Robotis et al. (2005). It is stated that the reseller reduces the number of units procured from the advanced market by using remanufactured products to satisfy the demand form secondary markets.

1.4.3.2.8   EOL Alternative Selection

EOL products collected through a RL network are processed according to one of the following five options: direct reuse, repair, remanufacturing, recycling, and disposal. We can define direct reuse as the reuse of the whole product as is for its original task. Repair option involves the replacement of damaged parts in order to have a fully functional product. Remanufacturing is the process of bringing used products up to a quality level similar to a new product by replacing, rebuilding, and upgrading its components. In recycling, the material content of returned products is recovered. Landfill or incineration of the used products is considered in disposal option. Various qualitative and quantitative factors including environmental impact, quality, legislative factors, and cost must be considered while developing a decision model for EOL option selection. Researchers have developed many mathematical programming-based EOL option selection methodologies. In the stochastic dynamic programming (DP) model presented in Krikke et al. (1998), a product recovery and disposal strategy for one product type is determined by maximizing the net profit considering relevant technical, ecological, and commercial feasibility criteria at the product level. The methodology proposed in Krikke et al. (1998) is applied to real-life cases on the recycling of copiers and monitors by Krikke et al. (1999a,b), respectively. An extension of Krikke et al.'s (1998) model is presented in Teunter (2006) by considering partial disassembly and multiple disassembly processes. Lee et al. (2001) define their objective function as the weighted sum of economic value and environmental impact to determine the EOL option of each part. The mixed integer program developed by Das and Yedlarajiah (2002) determines the optimal part disposal strategy by maximizing the net profit. Optimal allocation of disassembled parts to five disposal options (refurbish, resell, reuse, recycle, landfill) is carried out in Jorjani et al. (2004) through the development of a piecewise linear concave program which maximizes the overall return. The mathematical model proposed by Ritchey et al. (2005) evaluates the economic viability of remanufacturing option under a government mandated take-back program. The impact of reductions in the expected disassembly time and cost on the optimal EOL strategy is analyzed by Willems et al. (2006) with the use of an LP model. An LP model is developed to evaluate three EOL options for each part, namely, repair, repackage, or scrap, in Tan and Kumar (2008).

Various MCDM methodologies have been developed for the simultaneous consideration of several factors in EOL option selection process. In order to consider the trade-offs between environmental and economic variables in the selection of EOL alternatives, Hula et al. (2003) present a multi-objective GA. Bufardi et al. (2004) use ELECTRE III MCDM methodology to obtain a partial ranking of EOL options. Extending Chan (2008), Bufardi et al. (2004) consider complete ranking of EOL options under uncertainty environment in their GRA-based MCDM methodology. Multi-objective evolutionary algorithm proposed by Jun et al. (2007) maximizes the recovery value of an EOL product including recovery cost and quality in the selection of the best EOL options of parts. Fernandez et al. (2008) consider product value, recovery value, useful life and level of sophistication as criteria in their fuzzy approach evaluating five recovery options and one disposal option. Knowledge of experts (evaluators or sortation specialists) is used in the FL-based MCDM methodology proposed by Wadhwa et al. (2009) for the selection of most appropriate alternative(s) for product reprocessing. Iakovou et al. (2009) consider residual value, environmental burden, weight, quantity, and ease of disassembly of each component in the evaluation of EOL alternatives for a product in their MCDM methodology, called "Multicriteria Matrix." The most attractive subassemblies and components to be disassembled for recovery from a set of different types of EOL products are determined using GP in Xanthopoulos and Iakovou (2009).

Rahimifard et al. (2004) and Bakar and Rahimifard (2007) develop computer-aided decision support systems to support the EOL option selection process. Staikos and

Rahimifard (2007a,b) integrate AHP, LCA, and cost-benefit analysis to determine the most appropriate reuse, recovery, and recycling option for postconsumer shoes. Gonzalez and Adenso-Diaz (2005) simultaneously determine the depth of disassembly and EOL option for the disassembled parts based on the maximization of the profit by developing a bill of material-based method. Considering three different new product development projects (design for single use, design for reuse, and design for reuse with stock-keeping) that require different EOL recovery strategies, dynamic policies are developed in Kleber (2006). In Shih et al. (2006), EOL strategy is determined using a case-based reasoning (CBR)-based methodology.

There is a strong correlation between a product's design and the best EOL option for it. That is why, in some studies, EOL option selection problem and product design have been considered at the same time. The tools proposed by Rose and Ishii (1999) and Gehin et al. (2008) allow for the identification of appropriate EOL strategies in the early design phase. In Mangun and Thurston (2002), planning for component reuse, remanufacture, and recycle concepts is incorporated into product portfolio design with the development of a mathematical model. The value flow model proposed by Kumar et al. (2007) helps decision makers select the best EOL option for a product considering different product life cycle stages. Innovations in product design and recovery technologies are taken into consideration in Zuidwijk and Krikke (2008) in order to improve product recovery strategy.

1.4.3.2.9 Product Acquisition Management

There is a high level of uncertainty associated with the quantity, quality, and timing of EOL product returns. In order to deal with this uncertainty, firms must develop effective product acquisition policies which can prevent excessive inventory levels or low customer satisfaction (i.e., stockouts due to insufficient used products). Product acquisition management acts as an interface between RL activities and production planning and control activities for firms (Guide and Jayaraman 2000). Waste stream system and market-driven system are the two most commonly used product acquisition systems (Guide and Van Wassenhove 2001, Guide and Pentico 2003). In a waste stream system, all product returns are passively accepted by firms encouraged by the legislation while financial incentives are employed in a market-driven system to encourage users to return their products.

Deposit systems, cash paid for a specified level of quality, and credit toward a new unit are the most commonly used financial incentives in market-driven systems (Guide and Van Wassenhove 2001). Research in product acquisition management mainly deals with the implementation of different forms of financial incentives and their impact on the performance of the RL activities. An implementation of buy-back programs in power-tools industry was presented by Klausner and Hendrickson (2000). The implementation of a quality-dependent incentive policy in which predetermined prices are offered for products with a specific nominal quality level is illustrated in Guide and Van Wassenhove (2001) through the use of a real-life case study. Optimal incentive values are determined under a quality-dependent incentive policy in Guide et al. (2003), Aras and Aksen (2008), and Aras et al. (2008). As an extension to Aras et al. (2008), a government subsidized collection system is analyzed in Aksen et al. (2009). In the deposit-refund system presented by Wojanowski et al. (2007), customers paying a certain deposit at the time of purchase are refunded upon the return of the used product. The optimal incentive value is determined in Kaya (2010) by considering partial substitution between original and remanufactured products together with stochastic demand.

Various closed-loop relationship forms including ownership based, service contract, direct order, deposit based, credit based, buyback, and voluntary based are investigated in Ostlin et al. (2008).

1.4.3.3   Processing

1.4.3.3.1   Disassembly

In reverse supply chains, selective separation of desired parts and materials from returned products is achieved by means of disassembly which is the systematic separation of an assembly into its components, subassemblies or other groupings (Moore et al. 2001, Pan and Zeid 2001). More information on the general area of disassembly can be obtained from a recent book by Lambert and Gupta (2005). In this section, we first investigate the two important phases of disassembly process, scheduling and sequencing. Then other important issues such as disassembly line balancing, disassembly-to-order systems, use of IT in disassembly, ergonomics, and automation of disassembly systems are discussed.

1.4.3.3.1.1   Scheduling   We can define disassembly scheduling as the scheduling of the ordering and disassembly of EOL products with the aim of fulfilling part or component demand over a planning horizon (Veerakamolmal and Gupta 1998, Lee and Xirouchakis 2004). Two general types of disassembly scheduling problems can be distinguished: uncapacitated and capacitated. For the first case, an MRP-like algorithm is proposed by Gupta and Taleb (1994) considering the disassembly scheduling of a discrete, well-defined product structure. Quantity and timing of disassembly of a single product to fulfill the demand for its various parts can be determined using this algorithm. Extending Gupta and Taleb (1994), Taleb et al. (1997) and Taleb and Gupta (1997) consider components/materials commonality and the disassembly of multiple product types. The two-phase heuristic algorithm presented in Lee and Xirouchakis (2004) minimizes various costs associated with the disassembly process. In the first phase, an initial solution is determined using Gupta and Taleb's (1994) algorithm. The second phase uses backward move to improve the initial solution. A lot-sizing methodology for reverse MRP algorithm of Gupta and Taleb (1994) is developed by Barba-Gutierrez et al. (2008). Considering multiple product types with parts commonality, Kim et al. (2003) propose an LP-relaxation-based heuristic algorithm with the aim of minimizing the sum of setup, disassembly operation, and inventory holding costs. As an extension to Kim et al.'s (2003) study, a two-phase heuristic is developed by Kim et al. (2006b). In the first phase, an initial solution is developed using the LP-relaxation heuristic suggested by Kim et al. (2003). The initial solution is improved in the second phase using DP. Considering three cases of the uncapacitated disassembly scheduling problem, i.e., a single product type without parts commonality and single and multiple product types with parts commonality, three IP models are presented in Lee et al. (2004).

A branch-and-bound algorithm is developed for the case of single product type without parts commonality in Kim et al. (2009c).

Several heuristic algorithms have been developed for the capacitated case. In Meacham et al. (1999), an optimization algorithm is presented by considering common components among products, and limited inventory of products available for disassembly. Lee et al. (2002) minimize the sum of disassembly operation and inventory holding costs by developing an IP model requiring excessive computation times to obtain optimal solutions for practical-sized problems. The Lagrangian heuristic proposed by Kim et al. (2006a) finds an optimal solution for practical problems in a reasonable amount of time. In this study, the objective function also involves disassembly setup costs. In Kim et al. (2006c), an optimal algorithm is developed considering single product type without parts commonality by minimizing the number of disassembled products. In this algorithm, the feasibility of the initial solution obtained using Gupta and Taleb's (1994) algorithm is checked. In order to satisfy the capacity constraints, any infeasible solution is modified.

Some rules for the scheduling of disassembly and bulk recycling are defined in Stuart and Christina (2003) considering the product turnover in incoming staging space. In a follow-up study, Rios and Stuart (2004) consider product turnover together with the outgoing plastics demand. Both studies employ DES models to evaluate scheduling rules. Sequence-dependent setups are considered in the cyclic lot scheduling heuristic developed by Brander and Forsberg (2005).

1.4.3.3.1.2 Sequencing Determination of the best order of operations in the separation of a product into its constituent parts or other groupings is the main concern of disassembly sequencing (Moore et al. 1998, Dong and Arndt 2003). Graphical approaches have been extensively used to solve the disassembly sequencing problem. An AND/OR graph-based methodology is presented by Lambert (1997). Kaebernick et al. (2000) sort the components of a product into different levels based on their accessibility for disassembly to develop a cluster graph. Torres et al. (2003) establish a partial nondestructive disassembly sequence of a product by developing an algorithm based on the product representation. Disassembly sequences can be automatically generated from a hierarchical attributed liaison graph using the method developed by Dong et al. (2006). Possible disassembly sequences for maintenance are generated using a disassembly constraint graph (DCG) in Li et al. (2006). Ma et al. (2011) develop an extended AND/OR algorithm and suggest a two-phase algorithm which considers various practical constraints (i.e., reuse probability and environmental impacts of parts or subassemblies, sequence-dependent setup costs, regulation on recovery rate and incineration capacity).

There is an increasing trend in the use of Petri net (PN) modeling to solve disassembly sequencing problems. The PN-based approach developed by Moore et al. (1998, 2001) allows for the automatic generation of disassembly process plans for products with complex AND/OR precedence relationships. The disassembly Petri nets (DPNs) for the design and implementation of adaptive disassembly systems have been proposed in Zussman and Zhou (1999, 2000). Zha and Lim (2000) develop an expert PN model by integrating expert systems and ordinary PNs. Tang et al. (2001) employ PN models for workstation status, product disassembly sequences, and scheduling while developing an integrated disassembly planning and demanufacturing scheduling approach. In order to determine an effective disassembly sequencing strategy, PNs are integrated with cost-based indices in Tiwari et al. (2001). A PN-based heuristic approach is developed in Rai et al. (2002). Kumar et al. (2003) and Singh et al. (2003) propose an expert enhanced colored stochastic PN involving a knowledge base, graphic characteristics, and artificial intelligence to deal with the unmanageable complexity of normal PNs. Considering the uncertainty associated with the disassembly process, Gao et al. (2004) propose a fuzzy reasoning PN. Tang et al. (2006) propose a fuzzy attributed PN to address the human factor-related uncertainty in disassembly planning. Grochowski and Tang (2009) determine the optimal disassembly action without human assistance by developing an expert system based on a DPN and a hybrid Bayesian network.

Another popular approach is mathematical programming. Lambert (1999) determines optimal disassembly sequences by developing an algorithm based on straightforward LP. Considering sequence-dependent costs and disassembly precedence graph representation, a binary integer linear programming (BILP)-based methodology is presented by (Lambert 2006). The same methodology is applied for the problems with AND/OR representation in Lambert (2007).

Combinatorial nature of the disassembly sequencing problem has encouraged many researchers to develop metaheuristics-based solution methodologies. Seo et al. (2001) consider both economic and environmental aspects while developing a GA-based heuristic algorithm to determine the optimal disassembly sequence. Li et al. (2005) develop an object-oriented intelligent disassembly sequence planner by integrating DCG and a GA. GA-based approaches for disassembly sequencing of EOL products are presented in Kongar and Gupta (2006a), Giudice and Fargione (2007), Duta et al. (2008b), Hui et al. (2008), and Gupta and Imtanavanich (2010). Gonzalez and Adenso-Diaz (2006) develop a scatter search-based methodology for complex products with sequence-dependent disassembly costs. It is assumed that only one component can be released at each time. Chung and Peng (2006) consider batch disassembly and tool accessibility while developing a GA to generate a feasible selective disassembly plan. Shimizu et al. (2007) derive an optimal disassembly sequence by using genetic programming as a resolution method. (Near-) optimal disassembly sequences are developed using a reinforcement-learning-based approach in Reveliotis (2007). Considering the uncertainty associated with the quality of the returned products, a fuzzy disassembly sequencing problem formulation is presented in Tripathi et al. (2009). Optimal disassembly sequence as well as the optimal depth of disassembly is determined using an ant colony optimization (ACO)-based metaheuristic. A multi-objective TS algorithm is developed in Kongar and Gupta (2009a) for near-optimal/optimal disassembly sequence generation. Tseng et al. (2010) propose a GA-based approach for integrated assembly and disassembly sequencing. ElSayed et al. (2012a) utilize a GA to generate feasible sequences for selective disassembly.

CBR applications have been presented by some researchers. In Zeid et al. (1997), CBR is applied to develop a disassembly plan for a single product. Veerakamolmal and Gupta (2002) extend Zeid et al. (1997) by automatically generating disassembly process plans for multiple products using a CBR approach. Disassembly sequences can be indexed and retrieved by the knowledge base developed by Pan and Zeid (2001).

Some researchers have developed heuristic procedures. Near-optimal disassembly sequences are determined in Gungor and Gupta (1997) by using a heuristic procedure. They also develop a methodology for the evaluation of different disassembly strategies. Gungor and Gupta (1998) propose a methodology for disassembly sequence planning for products with defective parts in product recovery by addressing the uncertainty-related difficulties in disassembly sequence planning. A disassembly sequence and cost analysis study for the electromechanical products during the design stage is presented by Kuo (2000). Disassembly planning is divided into four stages: geometric assembly representation, cut-vertex search analysis, disassembly precedence matrix analysis, and disassembly sequences and plan generation. Three types of disassembly cost are considered, viz., target disassembly cost, full disassembly cost, and optimal disassembly cost. A branch-and-bound algorithm is developed for disassembly sequence plan generation in Gungor and Gupta (2001a). After developing a heuristic to discover the subassemblies within the product structure, Erdos et al. (2001) use a shortest hyper-path calculation to determine the optimal disassembly sequence. Considering interval profit values in the objective function, Kang et al. (2003) develop a mini-max regret criterion-based algorithm. Mascle and Balasoiu (2003) use wave propagation to develop an algorithm which can determine the disassembly sequence of a specific component of a product. The heuristic algorithm proposed by Lambert and Gupta (2008) has the ability of detecting "good enough" solutions for the case of sequence-dependent costs. Both the heuristic algorithm and the iterative BILP method (Lambert 2006) are applied to the disassembly precedence graph of a cell phone. A three-phase iterative solution procedure is proposed for a precedence-constrained asymmetric traveling salesman problem formulation in Sarin et al. (2006). A bi-criteria disassembly planning problem is solved in Adenso-Diaz et al. (2008) by integrating GRASP and path relinking.

Disassembly sequence generation problem is solved by developing a neural network in Hsin-Hao et al. (2000).

1.4.3.3.1.3 Line Balancing Due to its high productivity and suitability for automation, disassembly line is the most suitable layout for disassembly operations. A disassembly line must be balanced to optimize the use of resources (i.e., labor, money, and time). We can define disassembly line balancing problem (DLBP) as the assignment of disassembly tasks to a set of ordered disassembly stations while satisfying the disassembly precedence constraints and minimizing the number of stations needed and the variation in idle times between all stations (McGovern and Gupta 2007b, Altekin et al. 2008). The first examples of disassembly line balancing algorithms were presented by Gungor and Gupta (2001b, 2002). Disassembly line balancing problem in the presence of task failures (DLBP-F) is investigated by Gungor and Gupta (2001b). They present a disassembly line balancing algorithm which assign tasks to workstations by minimizing the effect of the defective parts on the disassembly line. Disassembly line-related complications and their effects are discussed in Gungor and Gupta (2002). They also modify the existing concepts of assembly line balancing to balance a paced disassembly line. Considering line balance and different process flows and meeting different order due dates, Tang and Zhou (2006) develop a two-phase PNs and DES-based methodology which maximizes system throughput and system revenue by dynamically configuring the disassembly system into many disassembly lines.

Several disassembly line balancing algorithms have been developed using metaheuristics. Optimal or near-optimal solution is obtained by developing an ACO algorithm in McGovern and Gupta (2006). Agrawal and Tiwari (2006) develop a collaborative ant colony algorithm for the balancing of a stochastic mixed-model U-shaped disassembly line. McGovern and Gupta (2007b) obtain near-optimal solutions by employing several combinatorial optimization techniques (exhaustive search, GA and ACO metaheuristics, a greedy algorithm, and greedy/hill-climbing and greedy/2-optimal hybrid heuristics). They illustrate the implementation of the methodologies, measure performance, and enable comparisons by developing a known, optimal, varying size dataset. After presenting a new formula for quantifying the level of balancing, McGovern and Gupta (2007a) present a first-ever set of a priori instances to be used in the evaluation of any disassembly line balancing solution technique. They also develop a GA which can be used to obtain optimal or near-optimal solutions. Ding et al. (2010) propose a novel multi-objective ACO algorithm for DLBP.

In some studies, the DLBP is solved using mathematical programming techniques. Considering profit maximization in partial DLBP, Altekin et al. (2008) develop an MIP formulation which simultaneously determines the parts and tasks, the number of stations, and the cycle time. Altekin and Akkan (2012) propose a MIP-based two-step procedure for predictive-reactive disassembly line balancing. First, a predictive balance is created. Then, given a task failure, the tasks of the disassembled product with that task failure are re-selected and reassigned to the stations. Duta et al. (2008a) integrate integer quadratic programming and branch-and-cut algorithm to solve the problem of disassembly line balancing in real time (DLBP-R). Koc et al. (2009) develop IP and DP formulations which check the feasibility of the precedence relations among the tasks using an AND/OR graph.

1.4.3.3.1.4 Disassembly-to-Order Systems In disassembly-to-order systems, components and materials from different types of EOL products are disassembled to satisfy all the demands for the components and materials considering various physical, financial, and environmental constraints. The studies in the first group develop heuristic procedures by assuming that disassembly yield is deterministic. Kongar and Gupta (2002) determine the best combination of multiple products to selectively disassemble to meet the demand for items and materials under a variety of physical, financial, and environmental constraints and goals by developing a single period integer GP model. Extending Kongar and Gupta (2002), Kongar and Gupta (2006b) model the fuzzy aspiration levels of various goals by using fuzzy GP. Lambert and Gupta (2002) consider a multiproduct demand-driven disassembly system with commonality and multiplicity and develop a method called tree network model by modifying the disassembly graph. Holding costs and external procurement of items are considered in the multi-period heuristic developed by Langella (2007).

In the LPP-based solution methodology developed by Kongar and Gupta (2009b), tangible or intangible financial, environmental, and performance-related measures of DTO systems are satisfied. Multiple objective functions, viz., maximizing the total profit, maximizing the resale/recycling percentage, and minimizing the disposal percentage, are considered in Kongar and Gupta (2009a) through the development of a multi-objective TS algorithm. An NN-based approach is developed in Gupta et al. (2010).

Stochastic nature of disassembly yields is considered in the second group of studies. The effect of stochastic yields on the DTO system is investigated by developing two heuristic procedures (i.e., one-to-one, one-to-many) in Inderfurth and Langella (2006). These heuristic procedures are used by Imtanavanich and Gupta (2006) to deal with the stochastic elements of the DTO system. Then, the number of returned products that satisfy various goals is determined by using a GP procedure.

1.4.3.3.1.5 Automation Disassembly tasks are usually carried out using manual labor. However, firms are adopting automated disassembly systems at an increasing pace in order to deal with the increase in the amount of electronic scrap as well as achieving higher productivity levels and reducing labor cost (Santochi et al. 2002, Kopacek and Kopacek 2006). That is why different aspects of automated disassembly systems have been studied by researchers in recent years. Considering an integrated disassembly cell controlled by product accompanying information systems, Seliger et al. (2002) develop modular disassembly processes and tools. There are several subsystems in the semi-automated personal computer disassembly cell designed by Torres et al. (2004). Recognition and localization of the product and of each of its components are provided by a computer vision subsystem while a modeling subsystem is used for disassembly sequence and planning of the disassembly movements. After discussing the required equipment and suitable strategies for the automated disassembly, Weigl-Seitz et al. (2006) propose a disassembly line layout with the suitable software structures for the disassembly video camera recorders and PCs. Kopacek and Kopacek (2006) investigate robotized, semi-automated, flexible disassembly cells for minidisks, PCBs, and mobile phones in industrial use. They also integrate disassembly families, mobile robots, and multi-agent systems (MAS) to develop a modular approach for disassembly cells. The advantages of emulation in control logic development and validation of new conceptual disassembly systems are discussed in Kim et al. (2009b). Kim et al. (2007) generate automatic control sequences for a partly automated system by developing an adaptive and modular control system which considers the availability of disassembly tools and the technological feasibility of disassembly processes and tools. Several solutions for the design of robotic disassembly cells are investigated in Duta and Filip (2008). In the dynamic process planning procedure developed by Kim et al. (2009a), available alternatives are generated by a database-supported procedure and the best suitable among them is selected if a device or tool is not available. ElSayed et al. (2012b) propose an intelligent automated disassembly cell for online (real time) selective disassembly. An industrial robotic manipulator, a camera, range sensing and component segmentation visual algorithms are the components of this cell. Santochi et al. (2002) discuss the software tools developed to optimize the disassembly process of discarded goods. An overview of layouts and modules of automated disassembly systems developed at various companies and research institutes is presented in Wiendahl et al. (2001).

1.4.3.3.1.6 Use of Information Technology in Disassembly There is a high level of uncertainty associated with disassembly yield due to missing and/or nonfunctional components in returned products. Recent developments in IT such as embedded sensors and RFID tags can reduce this uncertainty by providing information on the condition, type, and remaining lives of components in a returned product prior to disassembly. However, the use of these technologies must be economically justified. In other words, the economical benefits obtained from the use of these technologies must be higher than the cost of installing them into products. In order to address this need, researchers have presented cost-benefit analyses for the use of IT considering different disassembly scenarios.

Effectiveness of embedding sensors in computers is investigated by Vadde et al. (2008). They compare two scenarios (viz., with embedded sensors and without embedded sensors) considering several performance measures (viz., average life cycle cost, average maintenance cost, average disassembly cost, and average downtime of a computer). However, quantitative assessment of the impact of sensor-embedded products (SEPs) on these performance measures is not provided. Furthermore, since they only consider one component of a computer (hard disk), the disassembly setting does not represent the complexity of a disassembly line. Extending Vadde et al. (2008), Ilgin and Gupta (2011c) investigate the quantitative impact of SEPs on different performance measures of a disassembly line used to disassemble three components from EOL computers. First, they perform separate design of experiments studies based on orthogonal arrays for conventional products and SEPs. Then, the results of pairwise t-tests comparing the two cases based on different performance measures are presented. They also determine the range of monetary resources that can be invested in SEPs by utilizing the improvements achieved by the SEPs on profit for different experiments. In follow-up studies, Ilgin and Gupta (2010, 2011a,b) present a similar analysis methodology by considering multiple precedence relationships, precedence relationships, and component discriminating demands, respectively. The mathematical model developed by Ondemir et al. (2012) uses sensor information to determine how to process each and every EOL product on hand to meet used product and component demands as well as recycled material demand. Zhou and Piramuthu (2010) develop an adaptive knowledge-based system to utilize RFID item-level information for product remanufacturing including inspection, disassembly, and reassembly. The implementability of the proposed system is shown by means of a cost-benefit analysis. Gonnuru (2010) determines the true condition of the product components by analyzing the data from RFID tags by using a Bayesian approach. He also develops a fuzzy logic model synthesizing three input variables (i.e., product usage, component usage, and biographical data) into a solution that maximizes the recovery value while minimizing disassembly costs with an optimal disassembly sequence. Genetic algorithms are used to solve the model. Cao et al. (2011) integrate

MAS design approach and RFID technology to develop a shop-floor control system, which provides life cycle information for returned products. Ferrer et al. (2011) evaluate the use of RFID technology for improving remanufacturing efficiency based on the results of a DES study.

1.4.3.3.1.7 Ergonomics

Incorporation of ergonomic factors in the design of disassembly lines is an important issue due to the hands-on nature of disassembly tasks. However, the literature on disassembly ergonomics is very limited. Kazmierczak et al. (2004) use several explorative methods such as site visits and interviews to analyze the current situation and future perspectives for the ergonomics of car disassembly in Sweden. In Kazmierczak et al. (2005), disassembly work is analyzed considering time and physical work load requirements of constituent tasks. Kazmierczak et al. (2007) predict the performance of alternative system configurations in terms of productivity and ergonomics for a serial-flow car disassembly line by combining human and flow simulations. Takata et al. (2001) and Bley et al. (2004) investigate the human involvement in disassembly. Tang et al. (2006) and Tang and Zhou (2008) define the effect of several human factors (e.g., disassembly time, quality of disassembled components, and labor cost) as membership functions in their fuzzy attributed PN models to consider the uncertainty in manual disassembly operations.

Difficulty scores of standard disassembly tasks are determined using Maynard Operation Sequence Technique (MOST) in Kroll (1996). Methods time measurement (MTM) is employed to calculate the ease of disassembly scores for disassembly tasks in Desai and Mital (2005).

1.4.3.3.2 Remanufacturing

Remanufacturing involves the transformation of used products into products having same warranty conditions with the brand-new products. A typical remanufacturing process starts with the arrival of used products to a remanufacturing facility where they are disassembled into parts. After cleaning and inspection, disassembled parts are repaired and/or refurbished depending on their condition. Finally, remanufactured products are obtained by reassembling all parts. Besides repair and/or refurbishing, upgrading of some parts and/or modules can also be carried out in a remanufacturing process.

Remanufacturing is the most environment-friendly and the most profitable product recovery option. In remanufacturing, labor, energy, and material used in the manufacturing process can be recovered since the returned products preserve their current form. However, in recycling, returned products are simply shredded. In other words, only the material content of a returned product can be recovered. In refurbishment/repair, a returned product is kept functional by changing and/or repairing some components. The resultant product cannot be given the same warranty conditions with a brand-new product.

However, remanufactured products' warranty conditions can be the same with brand-new products since they are completely reassembled using disassembled, brand-new, and upgraded components.

1.4.3.3.3   Recycling

Recycling involves the collection, sorting, and processing of returned products in order to recover materials that are used as raw materials in the production process of new products. Recycling provides important saving in energy usage since processing new materials requires more energy than recycling materials from returned products. It saves the space by minimizing the quantity of returned products sent to landfills. Being an important source of various raw materials (e.g., metals, glass, paper), it reduces imports and material costs. In addition to the said economical benefits, it has many environmental benefits including the conservation of natural resources and minimization of carbon emissions to the atmosphere.

1.4.4   REUSABLE CONTAINER RETURNS

Reusable containers are used by various companies. Bottles/cans in beverages industry and cylindrical tubes in liquid gas industry are some examples of reusable containers.

Although a reusable container can be used many times, it has to be disposed after some time depending on the usage. In this section, we focus on the issues observed during the life cycle of a reusable container. For the issues associated with EOL treatment of reusable containers, we refer the reader to Section 1.4.3.

Kelle and Silver (1989) developed four methods for forecasting the returns of reusable containers. Each method requires different levels of information. Anbuudayasankar et al. (2010) consider a RL problem in which bottles/cans delivered from a processing depot to customers in one period are available for return to the depot in the following period. They modeled this problem as simultaneous delivery and pickup problem with constrained capacity (SDPC). Three unified heuristics based on extended branch-and-bound heuristic, genetic algorithm, and SA were developed to solve SDPC. Atamer et al. (2012) investigate pricing and production decisions in utilizing reusable containers with stochastic customer demand.

One of the most important problems in the management of returnable containers is the loss of containers due to theft, undocumented damage, or the failure of customers to return empty containers (Thoroe et al. 2009). The development of RFID-based container tracking systems is a popular solution approach for this problem. Johansson and Hellström (2007) use simulation analysis to investigate the potential effect of an RFID-based container management system. In Thoroe et al. (2009), a deterministic inventory model is employed in order to analyze the impact of RFID on container management.

1.4.5   LEASED PRODUCT RETURNS

In today's business environment, there is a widespread usage of leased equipment. In particular, most of the electronic office equipment is leased. The main reason for this trend is the short life cycle of electronic devices as a result of rapid technological changes. When electronic equipment becomes outdated, a firm has two options: using the equipment in another department or disposing it. If the decision is disposal, the firm has to pay high disposal fees due to hazardous materials involved in electronic equipment. This option also involves substantial storage and logistics costs. By leasing electronic equipment, the firm can minimize the costs associated with the short life cycle of electronic products. Because proper disposal of an equipment at the end-of-lease term is the responsibility of the leasing company, leasing companies also have to manage the disposal of end-of-lease equipment returns in a cost-effective way. This requires the joint consideration of RL and leasing decisions.

Sharma et al. (2007) develop a mathematical model for the simultaneous evaluation of the RL and equipment replacement-related decisions of a leasing company. They develop an MILP formulation to help the company in determining length of leases, utilization of logistics facilities, and EOL disposal options. Thurston and De La Torre (2007) present a mathematical model to explore the impact of leasing on the effectiveness of product take-back programs. This model assists decision makers in determination of the leasing period and which computer components are remanufactured or recycled for a portfolio of three market segments.

1.5 CONCLUSIONS

In this study, we presented an overview of current issues in RL. After analyzing the unique characteristics and working mechanism of a typical RL system, the papers from the RL literature were reviewed by considering five types of product returns (viz., customer returns, repair/service returns, EOL returns, reusable container returns, and leased product returns). Based on this review, we can present the following general points on the current and future research directions in RL:

- RL issues related with EOL returns were heavily addressed by researchers. However, the number of studies on RL issues associated with the other return types is very limited.
- Majority of the proposed heuristics and models were developed considering one particular return type. There is a need for the development of models and/or heuristics that consider more than one return type simultaneously.
- Network design and inventory models received considerable attention from researchers. More research is necessary on other areas such as facility layout, IT, marketing, and transportation issues.
- Disassembly is the most active research area considering processing issues in RL. Scheduling and sequencing are the two most commonly studied areas within disassembly. Ergonomics-related issues in disassembly need more attention from researchers. In addition, a recently emerging research issue, use of IT in disassembly, should be studied more extensively in order to deal with the uncertainty associated with the arrival time and condition of product returns.

REFERENCES

Adenso-Diaz, B., Garcia-Carbajal, S., and Gupta, S.M. 2008. A path-relinking approach for a bi-criteria disassembly sequencing problem. *Computers & Operations Research* 35 (12):3989–3997.

Agrawal, S. and Tiwari, M.K. 2006. A collaborative ant colony algorithm to stochastic mixed-model U-shaped disassembly line balancing and sequencing problem. *International Journal of Production Research* 46 (6):1405–1429.

Ahiska, S.S. and King, R.E. 2010. Inventory optimization in a one product recoverable manufacturing system. *International Journal of Production Economics* 124 (1):11–19.

Akcali, E., Cetinkaya, S., and Uster, H. 2009. Network design for reverse and closed-loop supply chains: An annotated bibliography of models and solution approaches. *Networks* 53 (3):231–248.

Aksen, D., Aras, N., and Karaarslan, A.G. 2009. Design and analysis of government subsidized collection systems for incentive-dependent returns. *International Journal of Production Economics* 119 (2):308–327.

Alshamrani, A., Mathur, K., and Ballou, R.H. 2007. Reverse logistics: Simultaneous design of delivery routes and returns strategies. *Computers & Operations Research* 34 (2):595–619.

Altekin, F.T. and Akkan, C. 2012. Task-failure-driven rebalancing of disassembly lines. *International Journal of Production Research* 50 (18):4955–4976.

Altekin, F.T., Kandiller, L., and Ozdemirel, N.E. 2008. Profit-oriented disassembly-line balancing. *International Journal of Production Research* 46 (10):2675–2693.

Alumur, S.A., Nickel, S., Saldanha-da-Gama, F., and Verter, V. 2012. Multi-period reverse logistics network design. *European Journal of Operational Research* 220 (1):67–78.

Amini, M.M., Retzlaff-Roberts, D., and Bienstock, C.C. 2005. Designing a reverse logistics operation for short cycle time repair services. *International Journal of Production Economics* 96 (3):367–380.

Anbuudayasankar, S.P., Ganesh, K., Lenny Koh, S.C., and Mohandas, K. 2010. Unified heuristics to solve routing problem of reverse logistics in sustainable supply chain. *International Journal of Systems Science* 41 (3):337–351.

Aras, N. and Aksen, D. 2008. Locating collection centers for distance- and incentive-dependent returns. *International Journal of Production Economics* 111 (2):316–333.

Aras, N., Aksen, D., and Tanugur, A.G. 2008. Locating collection centers for incentive-dependent returns under a pick-up policy with capacitated vehicles. *European Journal of Operational Research* 191 (3):1223–1240.

Aras, N., Verter, V., and Boyaci, T. 2006. Coordination and priority decisions in hybrid manufacturing/remanufacturing systems. *Production & Operations Management* 15 (4):528–543.

Atamer, B., Bakal, I.S., and Bayindir, Z.P. 2012. Optimal pricing and production decisions in utilizing reusable containers. *International Journal of Production Economics* (in press).

Autry, C.W., Daugherty, P.J., and Richey, R.G. 2001. The challenge of reverse logistics in catalog retailing. *International Journal of Physical Distribution & Logistics Management* 31 (1):26–37.

Azadi, M. and Saen, R.F. 2011. A new chance-constrained data envelopment analysis for selecting third-party reverse logistics providers in the existence of dual-role factors. *Expert Systems with Applications* 38 (10):12231–12236.

Bakar, M.S.A. and Rahimifard, S. 2007. Computer-aided recycling process planning for end-of-life electrical and electronic equipment. *Proceedings of the Institution of Mechanical Engineers, Part B: Engineering Manufacture* 221 (8):1369–1374.

Barba-Gutierrez, Y., Adenso-Diaz, B., and Gupta, S.M. 2008. Lot sizing in reverse MRP for scheduling disassembly. *International Journal of Production Economics* 111 (2):741–751.

Barros, A.I., Dekker, R., and Scholten, V. 1998. A two-level network for recycling sand: A case study. *European Journal of Operational Research* 110 (2):199–214.

Bautista, J. and Pereira, J. 2006. Modeling the problem of locating collection areas for urban waste management. An application to the metropolitan area of Barcelona. *Omega* 34 (6):617–629.

Bayindir, Z.P., Erkip, N., and Gullu, R. 2003. A model to evaluate inventory costs in a remanufacturing environment. *International Journal of Production Economics* 81:597–607.

Beamon, B.M. and Fernandes, C. 2004. Supply-chain network configuration for product recovery. *Production Planning & Control* 15 (3):270–281.

Beltran, J.L. and Krass, D. 2002. Dynamic lot sizing with returning items and disposals. *IIE Transactions* 34 (5):437–448.

Bera, U.K., Maity, K., and Maiti, M. 2008. Production-remanufacturing control problem for defective item under possibility constrains. *International Journal of Operational Research* 3 (5):515–532.

Bhattacharya, S., Guide, V.D.R., and Van Wassenhove, L.N. 2006. Optimal order quantities with remanufacturing across new product generations. *Production & Operations Management* 15 (3):421–431.

Bian, W. and Yu, M. 2006. Location analysis of reverse logistics operations for an international electrical manufacturer in Asia Pacific region using the analytic hierarchy process. *International Journal of Services Operations and Informatics* 1 (1):187–201.

Biehl, M., Prater, E., and Realff, M.J. 2007. Assessing performance and uncertainty in developing carpet reverse logistics systems. *Computers & Operations Research* 34 (2):443–463.

Blanc, H.M., Fleuren, H.A., and Krikke, H.R. 2004. Redesign of a recycling system for LPG-tanks. *OR Spectrum* 26 (2):283–304.

Blanc, L.L., Krieken, M.V., Krikke, H., and Fleuren, H. 2006. Vehicle routing concepts in the closed-loop container network of ARN—A case study. *OR Spectrum* 28 (1):52–70.

Bley, H., Reinhart, G., Seliger, G., Bernardi, M, and Korne, T. 2004. Appropriate human involvement in assembly and disassembly. *CIRP Annals—Manufacturing Technology* 53 (2):487–509.

Brander, P. and Forsberg, R. 2005. Cyclic lot scheduling with sequence-dependent set-ups: A heuristic for disassembly processes. *International Journal of Production Research* 43 (2):295–310.

de Brito, M.P. and van der Laan, E.A. 2009. Inventory control with product returns: The impact of imperfect information. *European Journal of Operational Research* 194 (1):85–101.

Bufardi, A., Gheorghe, R., Kiritsis, D., and Xirouchakis, P. 2004. Multicriteria decision-aid approach for product end-of-life alternative selection. *International Journal of Production Research* 42 (16):3139–3157.

Cao, H., Folan, P., Potter, D., and Browne, J. 2011. Knowledge-enriched shop floor control in end-of-life business. *Production Planning & Control: The Management of Operations* 22 (2):174–193.

Çatay, B. 2010. A new saving-based ant algorithm for the vehicle routing problem with simultaneous pickup and delivery. *Expert Systems with Applications* 37 (10):6809–6817.

Chan, J.W.K. 2008. Product end-of-life options selection: Grey relational analysis approach. *International Journal of Production Research* 46 (11):2889–2912.

Ching, W.K., Yuen, W.O., and Loh, A.W. 2003. An inventory model with returns and lateral transshipments. *Journal of the Operational Research Society* 54 (6):636–641.

Chouinard, M., D'Amours, S., and Ait-Kadi, D. 2005. Integration of reverse logistics activities within a supply chain information system. *Computers in Industry* 56 (1):105–124.

Chouinard, M., D'Amours, S., and Aït-Kadi, D. 2008. A stochastic programming approach for designing supply loops. *International Journal of Production Economics* 113 (2):657–677.

Chung, C. and Peng, Q. 2006. Evolutionary sequence planning for selective disassembly in de-manufacturing. *International Journal of Computer Integrated Manufacturing* 19 (3):278–286.

Chung, S.-L., Wee, H.-M., and Yang, P.-C. 2008. Optimal policy for a closed-loop supply chain inventory system with remanufacturing. *Mathematical and Computer Modelling* 48 (5–6):867–881.

Condea, C., Thiesse, F., and Fleisch, E. 2010. Assessing the impact of RFID and sensor technologies on the returns management of time-sensitive products. *Business Process Management Journal* 16 (6):954–971.

Cruz-Rivera, R. and Ertel, J. 2009. Reverse logistics network design for the collection of end-of-life vehicles in Mexico. *European Journal of Operational Research* 196 (3):930–939.

Das, S. and Yedlarajiah, D. 2002. An integer programming model for prescribing material recovery strategies. Paper read at *Proceedings of the IEEE International Symposium on Electronics and the Environment*, San Francisco, CA, pp. 118–122.

Daugherty, P.J., Richey, R.G., Genchev, S.E., and Chen, H. 2005. Reverse logistics: Superior performance through focused resource commitments to information technology. *Transportation Research Part E* 41 (2):77–92.

Debo, L.G., Totkay, L.B., and Van Wassenhove, L.N. 2005. Market segmentation and product technology selection for remanufacturable products. *Management Science* 51 (8):1193–1205.

DeCroix, G.A. 2006. Optimal policy for a multiechelon inventory system with remanufacturing. *Operations Research* 54 (3):532–543.

DeCroix, G., Jing-Sheng, S., and Zipkin, P. 2005. A series system with returns: Stationary analysis. *Operations Research* 53 (2):350–362.

Dehghanian, F. and Mansour, S. 2009. Designing sustainable recovery network of end-of-life products using genetic algorithm. *Resources, Conservation and Recycling* 53 (10):559–570.

Demirel, N. and Gökçen, H. 2008. A mixed integer programming model for remanufacturing in reverse logistics environment. *International Journal of Advanced Manufacturing Technology* 39 (11–12):1197–1206.

Denizel, M., Ferguson, M., and Souza, G.G.C. 2010. Multiperiod remanufacturing planning with uncertain quality of inputs. *IEEE Transactions on Engineering Management* 57 (3):394–404.

DePuy, G.W., Usher, J.S., Walker, R.L., and Taylor, G.D. 2007. Production planning for remanufactured products. *Production Planning & Control* 18 (7):573–583.

Desai, A. and Mital, A. 2005. Incorporating work factors in design for disassembly in product design. *Journal of Manufacturing Technology Management* 16 (7):712–732.

Dethloff, J. 2001. Vehicle routing and reverse logistics: The vehicle routing problem with simultaneous delivery and pick-up. *OR Spectrum* 23 (1):79–96.

Dethloff, J. 2002. Relation between vehicle routing problems: An Insertion heuristic for the vehicle routing problem with simultaneous delivery and pick-up applied to the vehicle routing problem with backhauls. *Journal of the Operational Research Society* 53 (1):115–118.

Dhanda, K.K. and Hill, R.P. 2005. The role of information technology and systems in reverse logistics: A case study. *International Journal of Technology Management* 31 (1):140–151.

Ding, L.-P., Feng, Y.-X., Tan, J.-R., and Gao, Y.-C. 2010. A new multi-objective ant colony algorithm for solving the disassembly line balancing problem. *International Journal of Advanced Manufacturing Technology* 48 (5):761–771.

Dobos, I. and Richter, K. 2000. The integer EOQ repair and waste disposal model—Further analysis. *Central European Journal of Operations Research* 8 (2):173–194.

Dobos, I. and Richter, K. 2003. A production/recycling model with stationary demand and return rates. *Central European Journal of Operations Research* 11 (1):35–46.

Dobos, I. and Richter, K. 2004. An extended production/recycling model with stationary demand and return rates. *International Journal of Production Economics* 90 (3):311–323.

Dobos, I. and Richter, K. 2006. A production/recycling model with quality consideration. *International Journal of Production Economics* 104 (2):571–579.

Doh, H.-H. and Lee, D.-H. 2010. Generic production planning model for remanufacturing systems. *Proceedings of the Institution of Mechanical Engineers, Part B: Journal of Engineering Manufacture* 224 (1):159–168.

Dong, J. and Arndt, G. 2003. A review of current research on disassembly sequence generation and computer aided design for disassembly. *Proceedings of the Institution of Mechanical Engineers, Part B: Engineering Manufacture* 217 (3):299–312.

Dong, Y., Kaku, I., and Tang, J. 2005. Inventory management in reverse logistics: A survey. In *Proceedings of the International Conference on Services Systems and Services Management*, June 13–15, Chongqing, China, pp. 352–356.

Dong, T., Zhang, L., Tong, R., and Dong, J. 2006. A hierarchical approach to disassembly sequence planning for mechanical product. *International Journal of Advanced Manufacturing Technology* 30 (5):507–520.

Du, F. and Evans, G.W. 2008. A bi-objective reverse logistics network analysis for post-sale service. *Computers & Operations Research* 35 (8):2617–2634.

Duta, L. and Filip, F.G. 2008. Control and decision-making process in disassembling used electronic products. *Studies in Informatics and Control* 17 (1):17–26.

Duta, L., Filip, F.G., and Caciula, I. 2008a. Real time balancing of complex disassembly lines. In *Proceedings of the 17th IFAC World Congress on Automatic Control (IFAC 2008)*, Seoul, Korea, pp. 913–918.

Duta, L., Filip, F.G., and Popescu, C. 2008b. Evolutionary programming in disassembly decision making. *International Journal of Computers, Communications & Control* 3 (3):282–286.

Easwaran, G. and Uster, H. 2010. A closed-loop supply chain network design problem with integrated forward and reverse channel decisions. *IIE Transactions* 42 (11):779–792.

Efendigil, T., Önüt, S., and Kongar, E. 2008. A holistic approach for selecting a third-party reverse logistics provider in the presence of vagueness. *Computers & Industrial Engineering* 54 (2):269–287.

El Korchi, A. and Millet, D. 2011. Designing a sustainable reverse logistics channel: The 18 generic structures framework. *Journal of Cleaner Production* 19 (6–7):588–597.

El Saadany, A.M.A. and Jaber, M.Y. 2008. The EOQ repair and waste disposal model with switching costs. *Computers & Industrial Engineering* 55 (1):219–233.

El Saadany, A.M.A. and Jaber, M.Y. 2010. A production/remanufacturing inventory model with price and quality dependant return rate. *Computers & Industrial Engineering* 58 (3):352–362.

El-Sayed, M., Afia, N., and El-Kharbotly, A. 2010. A stochastic model for forward-reverse logistics network design under risk. *Computers & Industrial Engineering* 58 (3):423–431.

ElSayed, A., Kongar, E., and Gupta, S.M. 2012a. An evolutionary algorithm for selective disassembly of end-of-life products. *International Journal of Swarm Intelligence and Evolutionary Computation* 1:1–7.

ElSayed, A., Kongar, E., Gupta, S.M., and Sobh, T. 2012b. A robotic-driven disassembly sequence generator for end-of-life electronic products. *Journal of Intelligent & Robotic Systems* 68 (1):43–52.

Erdos, G., Kis, T., and Xirouchakis, P. 2001. Modelling and evaluating product end-of-life options. *International Journal of Production Research* 39 (6):1203–1220.

Ferguson, M.E. and Toktay, L.B. 2006. The effect of competition on recovery strategies. *Production & Operations Management* 15 (3):351–368.

Fernandez, I., Puente, J., Garcia, N., and Gomez, A. 2008. A decision-making support system on a products recovery management framework: A fuzzy approach. *Concurrent Engineering* 16 (2):129–138.

Ferrer, G., Heath, S.K., and Dew, N. 2011. An RFID application in large job shop remanufacturing operations. *International Journal of Production Economics* 133 (2):612–621.

Ferrer, G. and Swaminathan, J.M. 2006. Managing new and remanufactured products. *Management Science* 52 (1):15–26.

Ferrer, G. and Swaminathan, J.M. 2010. Managing new and differentiated remanufactured products. *European Journal of Operational Research* 203 (2):370–379.

Ferrer, G. and Whybark, D.C. 2001. Material planning for a remanufacturing facility. *Production & Operations Management* 10 (2):112–124.

de Figueiredo, J.N. and Mayerle, S.F. 2008. Designing minimum-cost recycling collection networks with required throughput. *Transportation Research Part E* 44 (5):731–752.

Fleischmann, M., Beullens, P., Bloemhof-Ruwaard, J.M., and Van Wassenhove, L.N. 2001. The impact of product recovery on logistics network design. *Production & Operations Management* 10 (2):156–173.

Fleischmann, M., Bloemhof-Ruwaard, J.M., Dekker, R., van der Laan, E., van Nunen, J., and Van Wassenhove, L.N. 1997. Quantitative models for reverse logistics: A review. *European Journal of Operational Research* 103 (1):1–17.

Fleischmann, M., Krikke, H.R., Dekker, R., and Flapper, S.D.P. 2000. A characterisation of logistics networks for product recovery. *Omega* 28 (6):653–666.

Fleischmann, M. and Kuik, R. 2003. On optimal inventory control with independent stochastic item returns. *European Journal of Operational Research* 151 (1):25–37.

Fleischmann, M., Kuik, R., and Dekker, R. 2002. Controlling inventories with stochastic item returns: A basic model. *European Journal of Operational Research* 138 (1):63–75.

Franke, C., Basdere, B., Ciupek, M., and Seliger, S. 2006. Remanufacturing of mobile phones—Capacity, program and facility adaptation planning. *Omega* 34 (6):562–570.

Gao, M., Zhou, M.C., and Tang, Y. 2004. Intelligent decision making in disassembly process based on fuzzy reasoning Petri nets. *IEEE Transactions on Systems, Man, and Cybernetics, Part B: Cybernetics* 34 (5):2029–2034.

Gehin, A., Zwolinski, P., and Brissaud, D. 2008. A tool to implement sustainable end-of-life strategies in the product development phase. *Journal of Cleaner Production* 16 (5):566–576.

Georgiadis, P. and Athanasiou, E. 2010. The impact of two-product joint lifecycles on capacity planning of remanufacturing networks. *European Journal of Operational Research* 202 (2):420–433.

Georgiadis, P. and Besiou, M. 2008. Sustainability in electrical and electronic equipment closed-loop supply chains: A system dynamics approach. *Journal of Cleaner Production* 16 (15):1665–1678.

Georgiadis, P. and Vlachos, D. 2004. The effect of environmental parameters on product recovery. *European Journal of Operational Research* 157 (2):449–464.

Georgiadis, P., Vlachos, D., and Tagaras, G. 2006. The impact of product lifecycle on capacity planning of closed-loop supply chains with remanufacturing. *Production & Operations Management* 15 (4):514–527.

Giudice, F. and Fargione, G. 2007. Disassembly planning of mechanical systems for service and recovery: A genetic algorithms based approach. *Journal of Intelligent Manufacturing* 18 (3):313–329.

Golany, B., Yang, J., and Yu, G. 2001. Economic lot-sizing with remanufacturing options. *IIE Transactions* 33 (11):995–1004.

Gonnuru, V.K. 2010. *Radio-Frequency Identification (RFID) Integrated Fuzzy Based Disassembly Planning and Sequencing for End-of-Life Products*. San Antonio, TX: Mechanical Engineering, The University of Texas at San Antonio.

Gonzalez, B. and Adenso-Diaz, B. 2005. A bill of materials-based approach for end-of-life decision making in design for the environment. *International Journal of Production Research* 43 (10):2071–2099.

Gonzalez, B. and Adenso-Diaz, B. 2006. A scatter search approach to the optimum disassembly sequence problem. *Computers & Operations Research* 33 (6):1776–1793.

Gribkovskaia, I., Halskau, O., Laporte, G., and Vlcek, M. 2007. General solutions to the single vehicle routing problem with pickups and deliveries. *European Journal of Operational Research* 180 (2):568–584.

Grochowski, D.E. and Tang, Y. 2009. A machine learning approach for optimal disassembly planning. *International Journal of Computer Integrated Manufacturing* 22 (4):374–383.

Guide, V.D.R. and Jayaraman, V. 2000. Product acquisition management: Current industry practice and a proposed framework. *International Journal of Production Research* 38 (16):3779–3800.

Guide, V.D.R., Jayaraman, V., and Srivastava, R. 1999. Production planning and control for remanufacturing: A state-of-the-art survey. *Robotics and Computer-Integrated Manufacturing* 15 (3):221–230.

Guide, V.D.R., and Pentico, D.W. 2003. A hierarchical decision model for re-manufacturing and re-use. *International Journal of Logistics: Research and Applications* 6 (1–2):29–35.

Guide, V.D.R. and Spencer, M.S. 1997. Rough-cut capacity planning for remanufacturing firms. *Production Planning & Control* 8 (3):237–244.

Guide, V.D.R., Srivastava, R., and Spencer, M.S. 1997. An evaluation of capacity planning techniques in a remanufacturing environment. *International Journal of Production Research* 35 (1):67–82.

Guide, V.D.R., Teunter, R.H., and Van Wassenhove, L.N. 2003. Matching demand and supply to maximize profits from remanufacturing. *Manufacturing & Service Operations Management* 5 (4):303–316.

Guide, V.D.R. and Van Wassenhove, L.N. 2001. Managing product returns for remanufacturing. *Production & Operations Management* 10 (2):142–155.

Gungor, A. and Gupta, S.M. 1997. An evaluation methodology for disassembly processes. *Computers & Industrial Engineering* 33 (1–2):329–332.

Gungor, A. and Gupta, S.M. 1998. Disassembly sequence planning for products with defective parts in product recovery. *Computers & Industrial Engineering* 35 (1–2):161–164.

Gungor, A. and Gupta, S.M. 2001a. Disassembly sequence plan generation using a branch-and-bound algorithm. *International Journal of Production Research* 39 (3):481–509.

Gungor, A. and Gupta, S.M. 2001b. A solution approach to the disassembly line balancing problem in the presence of task failures. *International Journal of Production Research* 39 (7):1427–1467.

Gungor, A. and Gupta, S.M. 2002. Disassembly line in product recovery. *International Journal of Production Research* 40 (11):2569–2589.

Gupta, S.M. and Imtanavanich, P. 2010. Evolutionary computational approach for disassembly sequencing in a multiproduct environment. *International Journal of Biomedical Soft Computing and Human Sciences* 15 (1):73–78.

Gupta, S.M., Imtanavanich, P., and Nakashima, K. 2010. Using neural networks to solve a disassembly-to-order problem. *International Journal of Biomedical Soft Computing and Human Sciences* 15 (1):67–71.

Gupta, S.M. and Taleb, K.N. 1994. Scheduling disassembly. *International Journal of Production Research* 32 (8):1857–1866.

Gupta, S.M. and Veerakamolmal, P. 2001. Aggregate planning for end-of-life products. In *Greener Manufacturing and Operations: From Design to Delivery and Back*, J. Sarkis (ed.). Sheffield, U.K.: Greenleaf Publishing Ltd.

Hasani, A., Zegordi, S.H., and Nikbakhsh, E. 2012. Robust closed-loop supply chain network design for perishable goods in agile manufacturing under uncertainty. *International Journal of Production Research* 50 (16):4649–4669.

Heese, H.S., Cattani, K., Ferrer, G., Gilland, W., and Roth, A.V. 2005. Competitive advantage through take-back of used products. *European Journal of Operational Research* 164 (1):143–157.

Heisig, G. and Fleischmann, M. 2001. Planning stability in a product recovery system. *OR Spectrum* 23 (1):25–50.

Heyman, D.P. 1977. Optimal disposal policies for a single-item inventory system with returns. *Naval Research Logistics* 24 (3):385–405.

Hong, I. H., Assavapokee, T., Ammons, J., Boelkins, C., Gilliam, K., Oudit, D., Realff, M., Vannicola, J. M., and Wongthatsanekorn, W. 2006. Planning the e-scrap reverse production system under uncertainty in the state of Georgia: A case study. *IEEE Transactions on Electronics Packaging Manufacturing* 29: 150–162.

Hong, I.H., Ammons, J.C., and Realff, M.J. 2008. Centralized versus decentralized decision-making for recycled material flows. *Environmental Science & Technology* 42 (4):1172–1177.

Hsin-Hao, H., Wang, M.H., and Johnson, M.R. 2000. Disassembly sequence generation using a neural network approach. *Journal of Manufacturing Systems* 19 (2):73–82.

Hu, T.-L., Sheu, J.-B., and Huang, K.-H. 2002. A reverse logistics cost minimization model for the treatment of hazardous wastes. *Transportation Research Part E* 38 (6):457–473.

Hui, W., Dong, X., and Guanghong, D. 2008. A genetic algorithm for product disassembly sequence planning. *Neurocomputing* 71 (13–15):2720–2726.

Hula, A., Jalali, K., Hamza, K., Skerlos, S.J., and Saitou, K. 2003. Multi-criteria decision-making for optimization of product disassembly under multiple situations. *Environmental Science & Technology* 37 (23):5303–5313.

Iakovou, E., Moussiopoulos, N., Xanthopoulos, A. et al. 2009. A methodological framework for end-of-life management of electronic products. *Resources, Conservation and Recycling* 53 (6):329–339.

Ilgin, M.A. and Gupta, S.M. 2010. Comparison of economic benefits of sensor embedded products and conventional products in a multi-product disassembly line. *Computers & Industrial Engineering* 59 (4):748–763.

Ilgin, M.A. and Gupta, S.M. 2011a. Evaluating the impact of sensor embedded products on the performance of an air conditioner disassembly line. International *Journal of Advanced Manufacturing Technology* 53 (9–12):1199–1216.

Ilgin, M.A. and Gupta, S.M. 2011b. Recovery of sensor embedded washing machines using a multi kanban controlled disassembly line. *Robotics and Computer Integrated Manufacturing* 27 (2):318–334.

Ilgin, M.A. and Gupta, S.M. 2011c. Performance improvement potential of sensor embedded products in environmental supply chains. *Resources, Conservation and Recycling* 55 (6):580–592.

Imtanavanich, P. and Gupta, S.M. 2006. Calculating disassembly yields in a multi-criteria decision-making environment for a disassembly to order system. In *Applications of Management Science: In Productivity, Finance, and Operations*, K.D. Lawrence and R.K. Klimberg (eds). New York: Elsevier Ltd.

Inderfurth, K. 1997. Simple optimal replenishment and disposal policies for a product recovery system with leadtimes. *OR Spectrum* 19 (2):111–122.

Inderfurth, K., de Kok, A.G., and Flapper, S.D.P. 2001. Product recovery in stochastic remanufacturing systems with multiple reuse options. *European Journal of Operational Research* 133 (1):130–152.

Inderfurth, K. and van der Laan, E. 2001. Leadtime effects and policy improvement for stochastic inventory control with remanufacturing. *International Journal of Production Economics* 71 (1–3):381–390.

Inderfurth, K. and Langella, I.M. 2006. Heuristics for solving disassemble-to-order problems with stochastic yields. *OR Spectrum* 28 (1):73–99.

Jaber, M.Y. and El Saadany, A.M.A. 2009. The production, remanufacture and waste disposal model with lost sales. *International Journal of Production Economics* 120 (1):115–124.

Jaber, M.Y. and El Saadany, A.M.A. 2011. An economic production and remanufacturing model with learning effects. *International Journal of Production Economics* 131 (1):115–127.

Jaber, M.Y. and Rosen, M.A. 2008. The economic order quantity repair and waste disposal model with entropy cost. *European Journal of Operational Research* 188 (1):109–120.

Jayant, A., Gupta, P., and Garg, S.K. 2011. Reverse supply chain management (R-SCM): Perspectives, empirical studies and research directions. *International Journal of Business Insights and Transformation* 4 (2):111–125.

Jayaraman, V. 2006. Production planning for closed-loop supply chains with product recovery and reuse: An analytical approach. *International Journal of Production Research* 44 (5):981–998.

Jayaraman, V., Guide, V.D.R., and Srivastava, R. 1999. Closed-loop logistics model for remanufacturing. *Journal of the Operational Research Society* 50 (5):497–508.

Jayaraman, V., Patterson, R.A., and Rolland, E. 2003. The design of reverse distribution networks: Models and solution procedures. *European Journal of Operational Research* 150 (1):128–149.

Johansson, O. and Hellström, D. 2007. The effect of asset visibility on managing returnable transport items. *International Journal of Physical Distribution & Logistics Management* 37 (10):799–815.

Jorjani, S., Leu, J., and Scott, C. 2004. Model for the allocation of electronics components to reuse options. *International Journal of Production Research* 42 (6):1131–1145.

Jun, H.B., Cusin, M., Kiritsis, D., and Xirouchakis, P. 2007. A multi-objective evolutionary algorithm for EOL product recovery optimization: Turbocharger case study. *International Journal of Production Research* 45 (18–19):4573–4594.

Jung, K. and Hwang, H. 2011. Competition and cooperation in a remanufacturing system with take-back requirement. *Journal of Intelligent Manufacturing* 22 (3):427–433.

Kaebernick, H., O'Shea, B., and Grewal, S.S. 2000. A method for sequencing the disassembly of products. *CIRP Annals—Manufacturing Technology* 49 (1):13–16.

Kang, J.G., Lee, D.H., and Xirouchakis, P. 2003. Disassembly sequencing with imprecise data: A case study. *International Journal of Industrial Engineering* 10 (4):407–412.

Kannan, G. 2009. Fuzzy approach for the selection of third party reverse logistics provider. *Asia Pacific Journal of Marketing and Logistics* 21 (3):397–416.

Kannan, G., Haq, A.N., and Sasikumar, P. 2008. An application of the analytical hierarchy process and fuzzy analytical hierarchy process in the selection of collecting centre location for the reverse logistics multicriteria decision-making supply chain model. *International Journal of Management and Decision Making* 9 (4):350–365.

Kannan, G. and Murugesan, P. 2011. Selection of third-party reverse logistics provider using fuzzy extent analysis. *Benchmarking: An International Journal* 18 (1):149–167.

Kannan, G., Murugesan, P., and Haq, A.N. 2009a. 3PRLP's selection using an integrated analytic hierarchy process and linear programming. *International Journal of Services Technology and Management* 12 (1):61–80.

Kannan, G., Noorul Haq, A., and Devika, M. 2009b. Analysis of closed loop supply chain using genetic algorithm and particle swarm optimisation. *International Journal of Production Research* 47 (5):1175–1200.

Kannan, G., Pokharel, S., and Sasi Kumar, P. 2009c. A hybrid approach using ISM and fuzzy TOPSIS for the selection of reverse logistics provider. *Resources, Conservation and Recycling* 54 (1):28–36.

Kannan, G., Sasikumar, P., and Devika, K. 2010. A genetic algorithm approach for solving a closed loop supply chain model: A case of battery recycling. *Applied Mathematical Modelling* 34 (3):655–670.

Kara, S.S. and Onut, S. 2010. A two-stage stochastic and robust programming approach to strategic planning of a reverse supply network: The case of paper recycling. *Expert Systems with Applications* 37 (9):6129–6137.

Kara, S., Rugrungruang, F., and Kaebernick, H. 2007. Simulation modelling of reverse logistics networks. *International Journal of Production Economics* 106 (1):61–69.

Karakayali, I., Emir-Farinas, H., and Akcali, E. 2007. An analysis of decentralized collection and processing of end-of-life products. *Journal of Operations Management* 25 (6):1161–1183.

Kaya, O. 2010. Incentive and production decisions for remanufacturing operations. *European Journal of Operational Research* 201 (2):420–433.

Kazmierczak, K., Mathiassen, S.E., Forsman, M., and Winkel, J. 2005. An integrated analysis of ergonomics and time consumption in Swedish 'craft-type' car disassembly. *Applied Ergonomics* 36 (3):263–273.

Kazmierczak, K., Neumann, W.P., and Winkel, J. 2007. A case study of serial-flow car disassembly: Ergonomics, productivity and potential system performance. *Human Factors and Ergonomics in Manufacturing* 17 (4):331–351.

Kazmierczak, K., Winkel, J., and Westgaard, R.H. 2004. Car disassembly and ergonomics in Sweden: Current situation and future perspectives in light of new environmental legislation. *International Journal of Production Research* 42 (7):1305–1324.

Kelle, P. and Silver, E.A. 1989. Forecasting the returns of reusable containers. *Journal of Operations Management* 8 (1):17–35.

Kiesmüller, G.P. 2003a. A new approach for controlling a hybrid stochastic manufacturing/remanufacturing system with inventories and different leadtimes. *European Journal of Operational Research* 147 (1):62–71.

Kiesmüller, G.P. 2003b. Optimal control of a one product recovery system with leadtimes. *International Journal of Production Economics* 81:333–340.

Kiesmuller, G.P. and Minner, S. 2003. Simple expressions for finding recovery system inventory control parameter values. *Journal of the Operational Research Society* 54 (1):83–88.

Kiesmüller, G.P. and Scherer, C.W. 2003. Computational issues in a stochastic finite horizon one product recovery inventory model. *European Journal of Operational Research* 146 (3):553–579.

Kim, H.J., Chiotellis, S., and Seliger, G. 2009a. Dynamic process planning control of hybrid disassembly systems. *International Journal of Advanced Manufacturing Technology* 40 (9):1016–1023.

Kim, H.J., Harms, R., and Seliger, G. 2007. Automatic control sequence generation for a hybrid disassembly system. *IEEE Transactions on Automation Science and Engineering* 4 (2):194–205.

Kim, K., Jeong, B., and Jeong, S. 2005. Optimization model for remanufacturing system at strategic and operational level. In *Computational Science and Its Applications—ICCSA 2005*, Singapore. Springer, Berlin, Germany, pp. 383–394.

Kim, H.J., Kernbaum, S., and Seliger, G. 2009b. Emulation-based control of a disassembly system for LCD monitors. *International Journal of Advanced Manufacturing Technology* 40 (3):383–392.

Kim, H.J., Lee, D.H., and Xirouchakis, P. 2006a. A Lagrangean heuristic algorithm for disassembly scheduling with capacity constraints. *Journal of the Operational Research Society* 57:1231–1240.

Kim, H.J., Lee, D.H., and Xirouchakis, P. 2006b. Two-phase heuristic for disassembly scheduling with multiple product types and parts commonality. *International Journal of Production Research* 44 (1):195–212.

Kim, H.J., Lee, D.H., Xirouchakis, P., and Kwon, O.K. 2009c. A branch and bound algorithm for disassembly scheduling with assembly product structure. *Journal of the Operational Research Society* 60:419–430.

Kim, H.J., Lee, D.H., Xirouchakis, P., and Zust, R. 2003. Disassembly scheduling with multiple product types. *CIRP Annals—Manufacturing Technology* 52 (1):403–406.

Kim, K., Song, I., Kim, J., and Jeong, B. 2006c. Supply planning model for remanufacturing system in reverse logistics environment. *Computers & Industrial Engineering* 51 (2):279–287.

Kim, H., Yang, J., and Lee, K.-D. 2009d. Vehicle routing in reverse logistics for recycling end-of-life consumer electronic goods in South Korea. *Transportation Research Part D* 14 (5):291–299.

Klausner, M. and Hendrickson, C.T. 2000. Reverse-logistics strategy for product take-back. *Interfaces* 30 (3):156–165.

Kleber, R. 2006. The integral decision on production/remanufacturing technology and investment time in product recovery. *OR Spectrum* 28 (1):21–51.

Kleber, R., Minner, S., and Kiesmüller, G. 2002. A continuous time inventory model for a product recovery system with multiple options. *International Journal of Production Economics* 79 (2):121–141.

Ko, H.J. and Evans, G.W. 2007. A genetic algorithm-based heuristic for the dynamic integrated forward/reverse logistics network for 3PLs. *Computers & Operations Research* 34 (2):346–366.

Koc, A., Sabuncuoglu, I., and Erel, E. 2009. Two exact formulations for disassembly line balancing problems with task precedence diagram construction using an AND/OR graph. *IIE Transactions* 41:866–881.

Koh, S.G., Hwang, H., Sohn, K.I., and Ko, C.S. 2002. An optimal ordering and recovery policy for reusable items. *Computers & Industrial Engineering* 43 (1–2):59–73.

Kongar, E. and Gupta, S.M. 2002. A multi-criteria decision making approach for disassembly-to-order systems. *Journal of Electronics Manufacturing* 11 (2):171–183.

Kongar, E. and Gupta, S.M. 2006a. Disassembly sequencing using genetic algorithm. *International Journal of Advanced Manufacturing Technology* 30 (5):497–506.

Kongar, E. and Gupta, S.M. 2006b. Disassembly to order system under uncertainty. *Omega* 34 (6):550–561.

Kongar, E. and Gupta, S.M. 2009a. A multiple objective tabu search approach for end-of-life product disassembly. *International Journal of Advanced Operations Management* 1 (2–3):177–202.

Kongar, E. and Gupta, S.M. 2009b. Solving the disassembly-to-order problem using linear physical programming. *International Journal of Mathematics in Operational Research* 1 (4):504–531.

Konstantaras, I. and Papachristos, S. 2006. Lot-sizing for a single-product recovery system with backordering. *International Journal of Production Research* 44 (10):2031–2045.

Konstantaras, I. and Papachristos, S. 2007. Optimal policy and holding cost stability regions in a periodic review inventory system with manufacturing and remanufacturing options. *European Journal of Operational Research* 178 (2):433–448.

Kopacek, P. and Kopacek, B. 2006. Intelligent, flexible disassembly. *International Journal of Advanced Manufacturing Technology* 30 (5):554–560.

Korugan, A. and Gupta, S.M. 1998. A multi-echelon inventory system with returns. *Computers & Industrial Engineering* 35 (1–2):145–148.

Krikke, H., le Blanc, I., van Krieken, M., and Fleuren, H. 2008. Low-frequency collection of materials disassembled from end-of-life vehicles: On the value of on-line monitoring in optimizing route planning. *International Journal of Production Economics* 111 (2):209–228.

Krikke, H.R., van Harten, A., and Schuur, P.C. 1998. On a medium term product recovery and disposal strategy for durable assembly products. *International Journal of Production Research* 36 (1):111–140.

Krikke, H.R., van Harten, A., and Schuur, P.C. 1999a. Business case Océ: Reverse logistic network re-design for copiers. *OR Spectrum* 21 (3):381–409.

Krikke, H.R., van Harten, A., and Schuur, P.C. 1999b. Business case Roteb: Recovery strategies for monitors. *Computers & Industrial Engineering* 36 (4):739–757.

Kroll, E. 1996. Application of work-measurement analysis to product disassembly for recycling. *Concurrent Engineering* 4 (2):149–158.

Kumar, V.V. and Chan, F.T.S. 2011. A superiority search and optimisation algorithm to solve RFID and an environmental factor embedded closed loop logistics model. *International Journal of Production Research* 49 (16):4807–4831.

Kumar, S., Kumar, R., Shankar, R., and Tiwari, M.K. 2003. Expert enhanced coloured stochastic Petri net and its application in assembly/disassembly. *International Journal of Production Research* 41 (12):2727–2762.

Kumar, V., Shirodkar, P.S., Camelio, J.A., and Sutherland, J.W. 2007. Value flow characterization during product lifecycle to assist in recovery decisions. *International Journal of Production Research* 45 (18):4555–4572.

Kuo, T.C. 2000. Disassembly sequence and cost analysis for electromechanical products. *Robotics and Computer-Integrated Manufacturing* 16 (1):43–54.

van der Laan, E., Dekker, R., and Salomon, M. 1996a. Product remanufacturing and disposal: A numerical comparison of alternative control strategies. *International Journal of Production Economics* 45 (1–3):489–498.

van der Laan, E., Dekker, R., Salomon, M., and Ridder, A. 1996b. An (s, Q) inventory model with remanufacturing and disposal. *International Journal of Production Economics* 46:339–350.

van der Laan, E. and Salomon, M. 1997. Production planning and inventory control with remanufacturing and disposal. *European Journal of Operational Research* 102 (2):264–278.

van der Laan, E., Salomon, M., and Dekker, R. 1999a. An investigation of lead-time effects in manufacturing/remanufacturing systems under simple PUSH and PULL control strategies. *European Journal of Operational Research* 115 (1):195–214.

van der Laan, E., Salomon, M., Dekker, R., and Wassenhove, L.V. 1999b. Inventory control in hybrid systems with remanufacturing. *Management Science* 45 (5):733–747.

van der Laan, E.A. and Teunter, R.H. 2006. Simple heuristics for push and pull remanufacturing policies. *European Journal of Operational Research* 175 (2):1084–1102.

Lambert, A.J.D. 1997. Optimal disassembly of complex products. *International Journal of Production Research* 35 (9):2509–2524.

Lambert, A.J.D. 1999. Linear programming in disassembly/clustering sequence generation. *Computers & Industrial Engineering* 36 (4):723–738.

Lambert, A.J.D. 2006. Exact methods in optimum disassembly sequence search for problems subject to sequence dependent costs. *Omega* 34 (6):538–549.

Lambert, A.J.D. 2007. Optimizing disassembly processes subjected to sequence-dependent cost. *Computers & Operations Research* 34 (2):536–551.

Lambert, A.J.D. and Gupta, S.M. 2002. Demand-driven disassembly optimization for electronic products. *Journal of Electronics Manufacturing* 11 (2):121–135.

Lambert, A.J.D. and Gupta, S.M. 2005. *Disassembly Modeling for Assembly, Maintenance, Reuse, and Recycling.* Boca Raton, FL: CRC Press.

Lambert, A.J.D. and Gupta, S.M. 2008. Methods for optimum and near optimum disassembly sequencing. *International Journal of Production Research* 46 (11):2845–2865.

Langella, I.M. 2007. Heuristics for demand-driven disassembly planning. *Computers & Operations Research* 34 (2):552–577.

Lee, D.-H., Bian, W., and Dong, M. 2007a. Multiobjective model and solution method for integrated forward and reverse logistics network design for third-party logistics providers. *Transportation Research Record: Journal of the Transportation Research Board* 2032:43–52.

Lee, D.-H., Bian, W., and Dong, M. 2007b. Multiproduct distribution network design of third-party logistics providers with reverse logistics operations. *Transportation Research Record: Journal of the Transportation Research Board* 2008;26–33.

Lee, D.H. and Dong, M. 2008, A heuristic approach to logistics network design for end-of-lease computer products recovery. *Transportation Research Part E* 44 (3):455–474.

Lee, D.-H. and Dong, M. 2009. Dynamic network design for reverse logistics operations under uncertainty. *Transportation Research Part E* 45 (1):61–71.

Lee, D.-H., Dong, M., and Bian, W. 2010. The design of sustainable logistics network under uncertainty. *International Journal of Production Economics* 128 (1):159–166.

Lee, D.-H., Dong, M., Bian, W., and Tseng, Y.-J. 2007c. Design of product recovery networks under uncertainty. *Transportation Research Record: Journal of the Transportation Research Board* 2008:19–25.

Lee, J.-E., Gen, M., and Rhee, K.-G. 2009. Network model and optimization of reverse logistics by hybrid genetic algorithm. *Computers & Industrial Engineering* 56 (3):951–964.

Lee, D.H., Kim, H.J., Choi, G., and Xirouchakis, P. 2004. Disassembly scheduling: Integer programming models. *Proceedings of the Institution of Mechanical Engineers, Part B: Engineering Manufacture* 218 (10):1357–1372.

Lee, S.G., Lye, S.W., and Khoo, M.K. 2001. A multi-objective methodology for evaluating product end-of-life options and disassembly. *International Journal of Advanced Manufacturing Technology* 18 (2):148–156.

Lee, C., Realff, M., and Ammons, J. 2011. Integration of channel decisions in a decentralized reverse production system with retailer collection under deterministic non-stationary demands. *Advanced Engineering Informatics* 25 (1):88–102.

Lee, D.H. and Xirouchakis, P. 2004. A two-stage heuristic for disassembly scheduling with assembly product structure. *Journal of the Operational Research Society* 55 (3):287–297.

Lee, D.H., Xirouchakis, P., and Zust, R. 2002. Disassembly scheduling with capacity constraints. *CIRP Annals—Manufacturing Technology* 51 (1):387–390.

Li, J., Gonzalez, M., and Zhu, Y. 2009. A hybrid simulation optimization method for production planning of dedicated remanufacturing. *International Journal of Production Economics* 117 (2):286–301.

Li, J.R., Khoo, L.P., and Tor, S.B. 2005. An object-oriented intelligent disassembly sequence planner for maintenance. *Computers in Industry* 56 (7):699–718.

Li, J.R., Khoo, L.P., and Tor, S.B. 2006. Generation of possible multiple components disassembly sequence for maintenance using a disassembly constraint graph. *International Journal of Production Economics* 102 (1):51–65.

Liang, Y., Pokharel, S., and Lim, G.H. 2009. Pricing used products for remanufacturing. *European Journal of Operational Research* 193 (2):390–395.

Lieckens, K. and Vandaele, N. 2007. Reverse logistics network design with stochastic lead times. *Computers & Operations Research* 34 (2):395–416.

Lieckens, K. and Vandaele, N. 2012. Multi-Level reverse logistics network design under uncertainty. *International Journal of Production Research* 50 (1):23–40.

Lim, H. and Noble, J. 2006. The impact of facility layout on overall remanufacturing system performance. *International Journal of Industrial and Systems Engineering* 1 (3):357–371.

Linton, J.D. and Yeomans, J.S. 2003. The role of forecasting in sustainability. *Technological Forecasting and Social Change* 70 (1):21–38.

Linton, J.D., Yeomans, J.S., and Yoogalingam, R. 2002. Supply planning for industrial ecology and remanufacturing under uncertainty: A numerical study of leaded-waste recovery from television disposal. *Journal of the Operational Research Society* 53 (11):1185–1196.

Linton, J.D., Yeomans, J.S., and Yoogalingam, R. 2005. Recovery and reclamation of durable goods: A study of television CRTs. *Resources, Conservation and Recycling* 43 (4):337–352.

Listes, O. 2007. A generic stochastic model for supply-and-return network design. *Computers & Operations Research* 34 (2):417–442.

Listes, O. and Dekker, R. 2005. A stochastic approach to a case study for product recovery network design. *European Journal of Operational Research* 160 (1):268–287.

Louwers, D., Kip, B.J., Peters, E., Souren, F., and Flapper, S.D.P. 1999. A facility location allocation model for reusing carpet materials. *Computers & Industrial Engineering* 36 (4):855–869.

Lu, Z. and Bostel, N. 2007. A facility location model for logistics systems including reverse flows: The case of remanufacturing activities. *Computers & Operations Research* 34 (2):299–323.

Lu, Q.I.N., Stuart Williams, J.A., Posner, M., Bonawi-Tan, W., and Qu, X. 2006. Model-based analysis of capacity and service fees for electronics recyclers. *Journal of Manufacturing Systems* 25 (1):45–56.

Ma, Y.-S., Jun, H.-B., Kim, H.-W., and Lee, D.-H. 2011. Disassembly process planning algorithms for end-of-life product recovery and environmentally conscious disposal. *International Journal of Production Research* 49 (23):7007–7027.

Mabini, M.C., Pintelon, L.M., and Gelders, L.F. 1992. EOQ type formulations for controlling repairable inventories. *International Journal of Production Economics* 28 (1):21–33.

Mahadevan, B., Pyke, D.F., and Fleischmann, M. 2003. Periodic review, push inventory policies for remanufacturing. *European Journal of Operational Research* 151 (3):536–551.

Majumder, P. and Groenevelt, H. 2001. Competition in remanufacturing. *Production & Operations Management* 10 (2):125–141.

Mangun, D. and Thurston, D.L. 2002. Incorporating component reuse, remanufacture, and recycle into product portfolio design. *IEEE Transactions on Engineering Management* 49 (4):479–490.

Marx-Gomez, J., Rautenstrauch, C., Nurnberger, A., and Kruse, R. 2002. Neuro-fuzzy approach to forecast returns of scrapped products to recycling and remanufacturing. *Knowledge-Based Systems* 15 (1–2):119–128.

Mascle, C. and Balasoiu, B.A. 2003. Algorithmic selection of a disassembly sequence of a component by a wave propagation method. *Robotics and Computer Integrated Manufacturing* 19 (5):439–448.

McGovern, S.M. and Gupta, S.M. 2006. Ant colony optimization for disassembly sequencing with multiple objectives. *International Journal of Advanced Manufacturing Technology* 30 (5):481–496.

McGovern, S.M. and Gupta, S.M. 2007a. A balancing method and genetic algorithm for disassembly line balancing. *European Journal of Operational Research* 179 (3):692–708.

McGovern, S.M. and Gupta, S.M. 2007b. Combinatorial optimization analysis of the unary NP-complete disassembly line balancing problem. *International Journal of Production Research* 45 (18–19):4485–4511.

Meacham, A., Uzsoy, R., and Venkatadri, U. 1999. Optimal disassembly configurations for single and multiple products. *Journal of Manufacturing Systems* 18 (5):311–322.

Meade, L. and Sarkis, J. 2002. A conceptual model for selecting and evaluating third-party reverse logistics providers. *Supply Chain Management: An International Journal* 7 (5):283–295.

Min, H. and Ko, H.-J. 2008. The dynamic design of a reverse logistics network from the perspective of third-party logistics service providers. *International Journal of Production Economics* 113 (1):176–192.

Min, H., Ko, C.S., and Ko, H.J. 2006a. The spatial and temporal consolidation of returned products in a closed-loop supply chain network. *Computers & Industrial Engineering* 51 (2):309–320.

Min, H., Ko, H.J., and Ko, C.S. 2006b. A genetic algorithm approach to developing the multi-echelon reverse logistics network for product returns. *Omega* 34 (1):56–69.

Minner, S. and Kleber, R. 2001. Optimal control of production and remanufacturing in a simple recovery model with linear cost functions. *OR Spectrum* 23 (1):3–24.

Mitra, S. 2007. Revenue management for remanufactured products. *Omega* 35 (5):553–562.

Mitra, S. 2009. Analysis of a two-echelon inventory system with returns. *Omega* 37 (1):106–115.

Mitra, S. and Webster, S. 2008. Competition in remanufacturing and the effects of government subsidies. *International Journal of Production Economics* 111 (2):287–298.

Moore, K.E., Gungor, A., and Gupta, S.M. 1998. A Petri net approach to disassembly process planning. *Computers & Industrial Engineering* 35 (1–2):165–168.

Moore, K.E., Gungor, A., and Gupta, S.M. 2001. Petri net approach to disassembly process planning for products with complex AND/OR precedence relationships. *European Journal of Operational Research* 135 (2):428–449.

Mostard, J. and Teunter, R. 2006. The newsboy problem with resalable returns: A single period model and case study. *European Journal of Operational Research* 169 (1):81–96.

Mourao, M.C. and Almeida, M.T. 2000. Lower-bounding and heuristic methods for a refuse collection vehicle routing problem. *European Journal of Operational Research* 121 (2):420–434.

Mourao, M.C. and Amado, L. 2005. Heuristic method for a mixed capacitated arc routing problem: A refuse collection application. *European Journal of Operational Research* 160 (1):139–153.

Muckstadt, J.A. and Isaac, M.H. 1981. An analysis of single item inventory systems with returns. *Naval Research Logistics* 28 (2):237–254.

Mukhopadhyay, S.K. and Setaputra, R. 2006. The role of 4PL as the reverse logistics integrator: Optimal pricing and return policies. *International Journal of Physical Distribution & Logistics Management* 36 (9):716–729.

Mukhopadhyay, S.K. and Setoputro, R. 2004. Reverse logistics in e-business. *International Journal of Physical Distribution & Logistics Management* 34 (1):70–88.

Mukhopadhyay, S.K. and Setoputro, R. 2005. Optimal return policy and modular design for build-to-order products. *Journal of Operations Management* 23 (5):496–506.

Mutha, A. and Pokharel, S. 2009. Strategic network design for reverse logistics and remanufacturing using new and old product modules. *Computers & Industrial Engineering* 56 (1):334–346.

Nahmias, S. and Rivera, H. 1979. A deterministic model for a repairable item inventory system with a finite repair rate. *International Journal of Production Research* 17 (3):215–221.

Nakashima, K., Arimitsu, H., Nose, T., and Kuriyama, S. 2002. Analysis of a product recovery system. *International Journal of Production Research* 40 (15):3849–3856.

Nakashima, K., Arimitsu, H., Nose, T., and Kuriyama, S. 2004. Optimal control of a remanufacturing system. *International Journal of Production Research* 42 (17):3619–3625.

Oh, Y.H. and Hwang, H. 2006. Deterministic inventory model for recycling system. *Journal of
Intelligent Manufacturing* 17 (4):423–428.

Olorunniwo, F.O. and Li, X. 2010. Information sharing and collaboration practices in reverse
logistics. *Supply Chain Management: An International Journal* 15 (6):454–462.

Olugu, E.U. and Wong, K.Y. 2011a. An expert fuzzy rule-based system for closed-loop
supply chain performance assessment in the automotive industry. *Expert Systems with
Applications* 39 (1):375–384.

Olugu, E.U. and Wong, K.Y. 2011b. Fuzzy logic evaluation of reverse logistics performance in
the automotive industry. *Scientific Research and Essays* 6 (7):1639–1649.

Olugu, E.U., Wong, K.Y., and Shaharoun, A.M. 2011. Development of key performance measures for the automobile green supply chain. *Resources, Conservation and Recycling*
55 (6):567–579.

Ondemir, O., Ilgin, M.A., and Gupta, S.M. 2012. Optimal End-of-Life Management in
Closed Loop Supply Chains Using RFID & Sensors. *IEEE Transactions on Industrial
Informatics* 8 (3):719–728.

Opalic, M., Vuckovic, K., and Panic, N. 2004. Consumer electronics disassembly line layout.
Polimeri 25 (1–2):20–22.

Ostlin, J., Sundin, E., and Bjorkman, M. 2008. Importance of closed-loop supply chain relationships for product remanufacturing. *International Journal of Production Economics*
115 (2):336–348.

Paksoy, T., Bektas, T., and Özceylan, E. 2011. Operational and environmental performance
measures in a multi-product closed-loop supply chain. *Transportation Research Part E:
Logistics and Transportation Review* 47 (4):532–546.

Pan, L. and Zeid, I. 2001. A knowledge base for indexing and retrieving disassembly plans.
Journal of Intelligent Manufacturing 12 (1):77–94.

Pati, R.K., Vrat, P., and Kumar, P. 2008. A goal programming model for paper recycling system. *Omega* 36 (3):405–417.

Piplani, R. and Saraswat, A. 2012. Robust optimisation approach to the design of service networks for reverse logistics. *International Journal of Production Research*
50 (5):1424–1437.

Pishvaee, M.S., Farahani, R.Z., and Dullaert, W. 2010a. A memetic algorithm for bi-objective
integrated forward/reverse logistics network design. *Computers & Operations Research*
37 (6):1100–1112.

Pishvaee, M.S., Jolai, F., and Razmi, J. 2009. A stochastic optimization model for integrated forward/reverse logistics network design. *Journal of Manufacturing Systems*
28 (4):107–114.

Pishvaee, M., Kianfar, K., and Karimi, B. 2010b. Reverse logistics network design using
simulated annealing. *International Journal of Advanced Manufacturing Technology*
47 (1–4):269–281.

Pishvaee, M.S., Rabbani, M., and Torabi, S.A. 2011. A robust optimization approach to closed-
loop supply chain network design under uncertainty. *Applied Mathematical Modelling*
35 (2):637–649.

Pishvaee, M.S. and Razmi, J. 2012. Environmental supply chain network design using
multi-objective fuzzy mathematical programming. *Applied Mathematical Modelling*
36 (8):3433–3446.

Pishvaee, M.S. and Torabi, S.A. 2010. A possibilistic programming approach for closed-loop supply chain network design under uncertainty. *Fuzzy Sets and Systems* 161 (20):2668–2683.

Pishvaee, M.S., Torabi, S.A., and Razmi, J. 2012. Credibility-based fuzzy mathematical programming model for green logistics design under uncertainty. *Computers & Industrial
Engineering* 62 (2):624–632.

Pochampally, K.K. and Gupta, S.M. 2005. Strategic planning of a reverse supply chain network. *International Journal of Integrated Supply Management* 1 (4):421–441.

Pochampally, K.K. and Gupta, S.M. 2008. A multiphase fuzzy logic approach to strategic planning of a reverse supply chain network. *IEEE Transactions on Electronics Packaging Manufacturing* 31 (1):72–82.

Pochampally, K.K. and Gupta, S.M. 2009. Reverse supply chain design: A neural network approach. In *Web-Based Green Products Life Cycle Management Systems: Reverse Supply Chain Utilization*, H.-F. Wang (ed.). Hershey, PA: IGI Global Publication.

Pochampally, K.K. and Gupta, S.M. 2012. Use of linear physical programming and Bayesian updating for design issues in reverse logistics. *International Journal of Production Research* 50 (5):1349–1359.

Pochampally, K.K., Gupta, S.M., and Govindan, K. 2009a. Metrics for performance measurement of a reverse/closed-loop supply chain. *International Journal of Business Performance and Supply Chain Modelling* 1 (1):8–32.

Pochampally, K.K., Nukala, S., and Gupta, S.M. 2009b. Eco-procurement strategies for environmentally conscious manufacturers. *International Journal of Logistics Systems and Management* 5 (1–2):106–122.

Pochampally, K.K., Nukala, S., and Gupta, S.M. 2009c. *Strategic Planning Models for Reverse and Closed-loop Supply Chains*. Boca Raton, FL: CRC Press.

Pokharel, S. and Mutha, A. 2009. Perspectives in reverse logistics: A review. *Resources, Conservation and Recycling* 53 (4):175–182.

Prahinski, C. and Kocabasoglu, C. 2006. Empirical research opportunities in reverse supply chains. *Omega* 34 (6):519–532.

Presley, A., Meade, L., and Sarkis, J. 2007. A strategic sustainability justification methodology for organizational decisions: A reverse logistics illustration. *International Journal of Production Research* 45 (18):4595–4620.

Qiaolun, G., Jianhua, J., and Tiegang, G. 2008. Pricing management for a closed-loop supply chain. *Journal of Revenue and Pricing Management* 7 (1):45–60.

Qin, Z. and Ji, X. 2010. Logistics network design for product recovery in fuzzy environment. *European Journal of Operational Research* 202 (2):479–490.

Qu, X. and Williams, J.A.S. 2008. An analytical model for reverse automotive production planning and pricing. *European Journal of Operational Research* 190 (3):756–767.

Rahimifard, A., Newman, S.T., and Rahimifard, S. 2004. A web-based information system to support end-of-life product recovery. *Proceedings of the Institution of Mechanical Engineers, Part B: Engineering Manufacture* 218 (9):1047–1057.

Rai, R., Rai, V., Tiwari, M.K., and Allada, V. 2002. Disassembly sequence generation: A Petri net based heuristic approach. *International Journal of Production Research* 40 (13):3183–3198.

Realff, M.J., Ammons, J.C., and Newton, D. 1999. Carpet recycling: Determining the reverse production system design. *Polymer-Plastics Technology and Engineering* 38 (3):547–567.

Realff, M.J., Ammons, J.C., and Newton, D. 2000a. Strategic design of reverse production systems. *Computers & Chemical Engineering* 24 (2–7):991–996.

Realff, M.J., Ammons, J.C., and Newton, D.J. 2004. Robust reverse production system design for carpet recycling. *IIE Transactions* 36 (8):767–776.

Realff, M.J., Newton, D., and Ammons, J.C. 2000b. Modeling and decision-making for reverse production system design for carpet recycling. *Journal of the Textile Institute* 91:168–186.

Ren, C. and Ye, J. 2011. Improved particle swarm optimization algorithm for reverse logistics network design. *Energy Procedia* 13:4591–4600.

Reveliotis, S.A. 2007. Uncertainty management in optimal disassembly planning through learning-based strategies. *IIE Transactions* 39 (6):645–658.

Richter, K. 1996a. The EOQ repair and waste disposal model with variable setup numbers. *European Journal of Operational Research* 95 (2):313–324.

Richter, K. 1996b. The extended EOQ repair and waste disposal model. *International Journal of Production Economics* 45 (1–3):443–447.

Richter, K. 1997. Pure and mixed strategies for the EOQ repair and waste disposal problem. *OR Spectrum* 19 (2):123–129.

Richter, K. and Dobos, I. 1999. Analysis of the EOQ repair and waste disposal problem with integer setup numbers. *International Journal of Production Economics* 59 (1–3):463–467.

Richter, K. and Gobsch, B. 2003. The market-oriented dynamic product recovery model in the just-in-time framework. *International Journal of Production Economics* 81:369–374.

Richter, K. and Sombrutzki, M. 2000. Remanufacturing planning for the reverse Wagner/Whitin models. *European Journal of Operational Research* 121 (2):304–315.

Richter, K. and Weber, J. 2001. The reverse Wagner/Whitin model with variable manufacturing and remanufacturing cost. *International Journal of Production Economics* 71 (1–3):447–456.

Rios, P.J. and Stuart, J.A. 2004. Scheduling selective disassembly for plastics recovery in an electronics recycling center. *IEEE Transactions on Electronics Packaging Manufacturing* 27 (3):187–197.

Ritchey, J.R., Mahmoodi, F., Frascatore, M.R., and Zander, A.K. 2005. A framework to assess the economic viability of remanufacturing. *International Journal of Industrial Engineering* 12 (1):89–100.

Robotis, A., Bhattacharya, S., and Van Wassenhove, L.N. 2005. The effect of remanufacturing on procurement decisions for resellers in secondary markets. *European Journal of Operational Research* 163 (3):688–705.

Rogers, D.S. and Tibben-Lembke, R.S. 1999. *Going Backwards: Reverse Logistics Trends and Practices*. Reno, NV: Reverse Logistics Executive Council, University of Nevada.

Rose, C.M. and Ishii, K. 1999. Product end-of-life strategy categorization design tool. *Journal of Electronics Manufacturing* 9 (1):41–51.

Roy, A., Maity, K., Kar, S., and Maiti, M. 2009. A production-inventory model with remanufacturing for defective and usable items in fuzzy-environment. *Computers & Industrial Engineering* 56 (1):87–96.

Rubio, S. and Corominas, A. 2008. Optimal manufacturing-remanufacturing policies in a lean production environment. *Computers & Industrial Engineering* 55 (1):234–242.

Saen, R.F. 2009. A mathematical model for selecting third-party reverse logistics providers. *International Journal of Procurement Management* 2 (2):180–190.

Saen, R.F. 2010. A new model for selecting third-party reverse logistics providers in the presence of multiple dual-role factors. *International Journal of Advanced Manufacturing Technology* 46 (1–4):405–410.

Saen, R.F. 2011. A decision model for selecting third-party reverse logistics providers in the presence of both dual-role factors and imprecise data. *Asia-Pacific Journal of Operational Research* 28 (2):239–254.

Salema, M.I.G., Barbosa-Povoa, A.P., and Novais, A.Q. 2007. An optimization model for the design of a capacitated multi-product reverse logistics network with uncertainty. *European Journal of Operational Research* 179 (3):1063–1077.

Salema, M.I.G., Barbosa-Povoa, A.P., and Novais, A.Q. 2010. Simultaneous design and planning of supply chains with reverse flows: A generic modelling framework. *European Journal of Operational Research* 203 (2):336–349.

Salema, M.I., Povoa, A.P.B., and Novais, A.Q. 2005. A warehouse-based design model for reverse logistics. *Journal of the Operational Research Society* 57 (6):615–629.

Santochi, M., Dini, G., and Failli, F. 2002. Computer aided disassembly planning: State of the art and perspectives. *CIRP Annals—Manufacturing Technology* 51 (2):507–529.

Sarin, S.C., Sherali, H.D., and Bhootra, A. 2006. A precedence-constrained asymmetric traveling salesman model for disassembly optimization. *IIE Transactions* 38 (3):223–237.

Sasikumar, P., Kannan, G., and Haq, A.N. 2009. A heuristic based approach to vehicle routing model for third party reverse logistics provider. *International Journal of Services Technology and Management* 12 (1):106–125.

Sasikumar, P., Kannan, G., and Haq, A. 2010. A multi-echelon reverse logistics network design for product recovery—A case of truck tire remanufacturing. *International Journal of Advanced Manufacturing Technology* 49 (9):1223–1234.

Savaskan, R.C., Bhattacharya, S., and Van Wassenhove, L.N. 2004. Closed-loop supply chain models with product remanufacturing. *Management Science* 50 (2):239–252.

Savaskan, R.C. and Van Wassenhove, L.N. 2006. Reverse channel design: The case of competing retailers. *Management Science* 52 (1):1–14.

Schrady, D.A. 1967. A deterministic inventory model for repairable items. *Naval Research Logistics Quarterly* 14 (3):391–398.

Schultmann, F., Engels, B., and Rentz, O. 2003. Closed-loop supply chains for spent batteries. *Interfaces* 33 (6):57–71.

Schultmann, F., Zumkeller, M., and Rentz, O. 2006. Modeling reverse logistic tasks within closed-loop supply chains: An example from the automotive industry. *European Journal of Operational Research* 171 (3):1033–1050.

Schulz, T. 2011. A new silver-meal based heuristic for the single-item dynamic lot sizing problem with returns and remanufacturing. *International Journal of Production Research* 49 (9):2519–2533.

Seliger, G., Basdere, B., Keil, T., and Rebafka, U. 2002. Innovative processes and tools for disassembly. *CIRP Annals—Manufacturing Technology* 51 (1):37–40.

Seo, K.K., Park, J.H., and Jang, D.S. 2001. Optimal disassembly sequence using genetic algorithms considering economic and environmental aspects. *International Journal of Advanced Manufacturing Technology* 18 (5):371–380.

Shaik, M.N. and Abdul-Kader, W. 2011. Transportation in reverse logistics enterprise: A comprehensive performance measurement methodology. *Production Planning & Control* (in press).

Sharif, A.M., Irani, Z., Love, P.E.D., and Kamal, M.M. 2012. Evaluating reverse third-party logistics operations using a semi-fuzzy approach. *International Journal of Production Research* 50 (9):2515–2532.

Sharma, M., Ammons, J.C., and Hartman, J.C. 2007. Asset management with reverse product flows and environmental considerations. *Computers & Operations Research* 34 (2):464–486.

Sheu, J.B., Chou, Y.H., and Hu, C.C. 2005. An integrated logistics operational model for green-supply chain management. *Transportation Research Part E* 41 (4):287–313.

Shi, J., Zhang, G., and Sha, J. 2011a. Optimal production and pricing policy for a closed loop system. *Resources, Conservation and Recycling* 55 (6):639–647.

Shi, J., Zhang, G., and Sha, J. 2011b. Optimal production planning for a multi-product closed loop system with uncertain demand and return. *Computers & Operations Research* 38 (3):641–650.

Shih, L.H. 2001. Reverse logistics system planning for recycling electrical appliances and computers in Taiwan. *Resources, Conservation and Recycling* 32 (1):55–72.

Shih, L.-H., Chang, Y.-S., and Lin, Y.-T. 2006. Intelligent evaluation approach for electronic product recycling via case-based reasoning. *Advanced Engineering Informatics* 20 (2):137–145.

Shimizu, Y., Tsuji, K., and Nomura, M. 2007. Optimal disassembly sequence generation using a genetic programming. *International Journal of Production Research* 45 (18):4537–4554.

Sim, E., Jung, S., Kim, H., and Park, J. 2004. A generic network design for a closed-loop supply chain using genetic algorithm. In *Genetic and Evolutionary Computation—GECCO 2004*. Seattle, WA. Springer, Berlin, Germany, pp. 1214–1225.

Simpson, V.P. 1978. Optimum solution structure for a repairable inventory problem. *Operations Research* 26 (2):270–281.

Singh, A.K., Tiwari, M.K., and Mukhopadhyay, S.K. 2003. Modelling and planning of the disassembly processes using an enhanced expert Petri net. *International Journal of Production Research* 41 (16):3761–3792.

Souza, G.C. and Ketzenberg, M.E. 2002. Two-stage make-to-order remanufacturing with service-level constraints. *International Journal of Production Research* 40 (2):477–493.

Souza, G., Ketzenberg, M., and Guide, V.D.R. 2002. Capacitated remanufacturing with service level constraints. *Production & Operations Management* 11 (2):231–248.

Srivastava, S.K. 2008a. Network design for reverse logistics. *Omega* 36 (4):535–548.

Srivastava, S.K. 2008b. Value recovery network design for product returns. *International Journal of Physical Distribution & Logistics Management* 38 (4):311–331.

Staikos, T. and Rahimifard, S. 2007a. A decision-making model for waste management in the footwear industry. *International Journal of Production Research* 45 (18):4403–4422.

Staikos, T. and Rahimifard, S. 2007b. An end-of-life decision support tool for product recovery considerations in the footwear industry. *International Journal of Computer Integrated Manufacturing* 20 (6):602–615.

Stuart, J.A., Bonawi-tan, W., Loehr, S., and Gates, J. 2005. Reducing costs through improved returns processing. *International Journal of Physical Distribution & Logistics Management* 35 (7):468–480.

Stuart, J.A. and Christina, V. 2003. New metrics and scheduling rules for disassembly and bulk recycling. *IEEE Transactions on Electronics Packaging Manufacturing* 26 (2):133–140.

Swarnkar, R. and Harding, J.A. 2009. Modelling and optimisation of a product recovery network. *International Journal of Sustainable Engineering* 2 (1):40–55.

Takahashi, K., Morikawa, K., Myreshka, Takeda, D., and Mizuno, A. 2007. Inventory control for a MARKOVIAN remanufacturing system with stochastic decomposition process. *International Journal of Production Economics* 108 (1–2):416–425.

Takata, S., Isobe, H., and Fujii, H. 2001. Disassembly operation support system with motion monitoring of a human operator. *CIRP Annals—Manufacturing Technology* 50 (1):305–308.

Taleb, K.N. and Gupta, S.M. 1997. Disassembly of multiple product structures. *Computers & Industrial Engineering* 32 (4):949–961.

Taleb, K.N., Gupta, S.M., and Brennan, L. 1997. Disassembly of complex product structures with parts and materials commonality. *Production Planning & Control* 8 (3):255–269.

Tan, A. and Kumar, A. 2008. A decision making model to maximise the value of reverse logistics in the computer industry. *International Journal of Logistics Systems and Management* 4 (3):297–312.

Tan, A.W.K., Yu, W.S., and Arun, K. 2003. Improving the performance of a computer company in supporting its reverse logistics operations in the Asia-Pacific region. *International Journal of Physical Distribution & Logistics Management* 33 (1):59–74.

Tang, O. and Grubbström, R.W. 2005. Considering stochastic lead times in a manufacturing/remanufacturing system with deterministic demands and returns. *International Journal of Production Economics* 93:285–300.

Tang, O. and Teunter, R. 2006. Economic lot scheduling problem with returns. *Production & Operations Management* 15 (4):488–497.

Tang, Y. and Zhou, M.C. 2006. A systematic approach to design and operation of disassembly lines. *IEEE Transactions on Automation Science and Engineering* 3 (3):324–329.

Tang, Y. and Zhou, M. 2008. Human-in-the-loop disassembly modeling and planning. In *Environment Conscious Manufacturing*, S. Gupta and A. Lambert (eds). Boca Raton, FL: CRC Press.

Tang, Y., Zhou, M., and Caudill, R. 2001. An integrated approach to disassembly planning and demanufacturing operation. *IEEE Transactions on Robotics and Automation* 17 (6):773–784.

Tang, Y., Zhou, M., and Gao, M. 2006. Fuzzy-Petri-net based disassembly planning considering human factors. *IEEE Transactions on Systems, Man and Cybernetics* 36 (4):718–726.

Teng, H.-M., Hsu, P.-H., Chiu, Y., and Wee, H.M. 2011. Optimal ordering decisions with returns and excess inventory. *Applied Mathematics and Computation* 217 (22):9009–9018.

Teunter, R.H. 2001. Economic ordering quantities for recoverable item inventory systems. *Naval Research Logistics* 48 (6):484–495.

Teunter, R. 2004. Lot-sizing for inventory systems with product recovery. *Computers & Industrial Engineering* 46 (3):431–441.

Teunter, R.H. 2006. Determining optimal disassembly and recovery strategies. *Omega* 34 (6):533–537.

Teunter, R.H., Bayindir, Z.P., and Heuvel, W.V.D. 2006. Dynamic lot sizing with product returns and remanufacturing. *International Journal of Production Research* 44 (20):4377–4400.

Teunter, R., Kaparis, K., and Tang, O. 2008. Multi-product economic lot scheduling problem with separate production lines for manufacturing and remanufacturing. *European Journal of Operational Research* 191 (3):1241–1253.

Thoroe, L., Melski, A., and Schumann, M. 2009. The impact of RFID on management of returnable containers. *Electronic Markets* 19 (2):115–124.

Thurston, D.L. and De La Torre, J.P. 2007. Leasing and extended producer responsibility for personal computer component reuse. *International Journal of Environment and Pollution* 29 (1):104–126.

Tibben-Lembke, R.S. and Rogers, D.S. 2002. Differences between forward and reverse logistics in a retail environment. *Supply Chain Management: An International Journal* 7 (5):271–282.

Tiwari, M.K., Sinha, N., Kumar, S., Rai, R., and Mukhopadhyay, S.K. 2001. A Petri net based approach to determine the disassembly strategy of a product. *International Journal of Production Research* 40 (5):1113–1129.

Toktay, B., Van Der Laan, E., and De Brito, M.P. 2004. Managing product returns: The role of forecasting. In *Reverse Logistics: Quantitative Models for Closed-Loop Supply Chains*, R. Dekker, M. Fleischmann, K. Inderfurth, and L. V. Wassenhove (eds). Berlin, Germany: Springer.

Toktay, L.B., Wein, L.M., and Zenios, S.A. 2000. Inventory management of remanufacturable products. *Management Science* 46 (11):1412–1426.

Topcu, A., Benneyan, J.C., and Cullinane, T.P. 2008. Facility and storage space design issues in remanufacturing. In *Environment Conscious Manufacturing*, S.M. Gupta and A.J.D. Lambert (eds). Boca Raton, FL: CRC Press.

Torres, F., Gil, P., Puente, S.T., Pomares, J., and Aracil, R. 2004. Automatic PC disassembly for component recovery. *International Journal of Advanced Manufacturing Technology* 23 (1):39–46.

Torres, F., Puente, S.T., and Aracil, R. 2003. Disassembly planning based on precedence relations among assemblies. *International Journal of Advanced Manufacturing Technology* 21 (5):317–327.

Tripathi, M., Agrawal, S., Pandey, M.K., Shankar, R., and Tiwari, M.K. 2009. Real world disassembly modeling and sequencing problem: Optimization by algorithm of self-guided ants (ASGA). *Robotics and Computer-Integrated Manufacturing* 25 (3):483–496.

Tsai, W.H. and Hung, S.-J. 2009. Treatment and recycling system optimisation with activity-based costing in WEEE reverse logistics management: An environmental supply chain perspective. *International Journal of Production Research* 47 (19):5391–5420.

Tseng, Y.-J., Kao, H.-T., and Huang, F.-Y. 2010. Integrated assembly and disassembly sequence planning using a GA approach. *International Journal of Production Research* 48 (20):5991–6013.

Tuzkaya, G., and Gulsun, B. 2008. Evaluating centralized return centers in a reverse logistics network: An integrated fuzzy multi-criteria decision approach. *International Journal of Environmental Science and Technology* 5: 339–352.

Tuzkaya, G., Gülsün, B., and Onsel, S. 2011. A methodology for the strategic design of reverse logistics networks and its application in the Turkish white goods industry. *International Journal of Production Research* 49 (15):4543–4571.

Uster, H., Easwaran, G., Akcali, E., and Cetinkaya, S. 2007. Benders decomposition with alternative multiple cuts for a multi-product closed-loop supply chain network design model. *Naval Research Logistics* 54 (8):890–907.

Vadde, S., Kamarthi, S., and Gupta, S.M. 2007. Optimal pricing of reusable and recyclable components under alternative product acquisition mechanisms. *International Journal of Production Research* 45 (18):4621–4652.

Vadde, S., Kamarthi, S., Gupta, S.M., and Zeid, I. 2008. Product life cycle monitoring via embedded sensors. In *Environment Conscious Manufacturing*, S. M. Gupta and A.J.D. Lambert (eds). Boca Raton, FL: CRC Press.

Vadde, S., Zeid, A., and Kamarthi, S.V. 2010. Pricing decisions in a multi-criteria setting for product recovery facilities. *Omega* 39 (2):186–193.

Veerakamolmal, P. and Gupta, S.M. 1998. Optimal analysis of lot-size balancing for multiproducts selective disassembly. *International Journal of Flexible Automation and Integrated Manufacturing* 6 (3–4):245–269.

Veerakamolmal, P. and Gupta, S.M. 1999. Analysis of design efficiency for the disassembly of modular electronic products. *Journal of Electronics Manufacturing* 9 (1):79–95.

Veerakamolmal, P. and Gupta, S.M. 2002. A case-based reasoning approach for automating disassembly process planning. *Journal of Intelligent Manufacturing* 13 (1):47–60.

Vlachos, D. and Dekker, R. 2003. Return handling options and order quantities for single period products. *European Journal of Operational Research* 151 (1):38–52.

Vlachos, D., Georgiadis, P., and Iakovou, E. 2007. A system dynamics model for dynamic capacity planning of remanufacturing in closed-loop supply chains. *Computers & Operations Research* 34 (2):367–394.

Vorasayan, J. and Ryan, S.M. 2006. Optimal price and quantity of refurbished products. *Production & Operations Management* 15 (3):369–383.

Wadhwa, S., Madaan, J., and Chan, F.T.S. 2009. Flexible decision modeling of reverse logistics system: A value adding MCDM approach for alternative selection. *Robotics and Computer-Integrated Manufacturing* 25 (2):460–469.

Walther, G., Schmid, E., and Spengler, T.S. 2008. Negotiation-based coordination in product recovery networks. *International Journal of Production Economics* 111 (2):334–350.

Wang, Z. and Bai, H. 2010. Reverse logistics network: A review. *Proceedings of the 2010 IEEE International Conference on Industrial Engineering and Engineering Management*, December 7–10, Macau, 1139–1143.

Wang, H.-F. and Hsu, H.-W. 2009. A closed-loop logistic model with a spanning-tree based genetic algorithm. *Computers & Operations Research* 37 (2):376–389.

Wang, I.L. and Yang, W.C. 2007. Fast heuristics for designing integrated e-waste reverse logistics networks. *IEEE Transactions on Electronics Packaging Manufacturing* 30 (2):147–154.

Wang, Z., Yao, D.-Q., and Huang, P. 2007. A new location-inventory policy with reverse logistics applied to B2C e-markets of China. *International Journal of Production Economics* 107 (2):350–363.

Webster, S. and Mitra, S. 2007. Competitive strategy in remanufacturing and the impact of take-back laws. *Journal of Operations Management* 25 (6):1123–1140.

Wee, H.M., Yu, J.C.P., Su, L., and Wu, T. 2006. Optimal ordering and recovery policy for reusable items with shortages. *Lecture Notes in Computer Science* 4247:284–291.

Wei, J. and Zhao, J. 2011. Pricing decisions with retail competition in a fuzzy closed-loop supply chain. *Expert Systems with Applications* 38 (9):11209–11216.

Weigl-Seitz, A., Hohm, K., Seitz, M., and Tolle, H. 2006. On strategies and solutions for automated disassembly of electronic devices. *International Journal of Advanced Manufacturing Technology* 30 (5):561–573.

Wiendahl, H.P., Scholz-Reiter, B., Bürkner, S., and Scharke, H. 2001. Flexible disassembly systems-layouts and modules for processing obsolete products. *Proceedings of the Institution of Mechanical Engineers, Part B: Engineering Manufacture* 215 (5):723–732.

Willems, B., Dewulf, W., and Duflou, J.R. 2006. Can large-scale disassembly be profitable? A linear programming approach to quantifying the turning point to make disassembly economically viable. *International Journal of Production Research* 44 (6):1125–1146.

Wojanowski, R., Verter, V., and Boyaci, T. 2007. Retail-collection network design under deposit-refund. *Computers & Operations Research* 34 (2):324–345.

Xanthopoulos, A. and Iakovou, E. 2009. On the optimal design of the disassembly and recovery processes. *Waste Management* 29 (5):1702–1711.

Yalabik, B., Petruzzi, N.C., and Chhajed, D. 2005. An integrated product returns model with logistics and marketing coordination. *European Journal of Operational Research* 161 (1):162–182.

Yang, J., Golany, B., and Yu, G. 2005. A concave-cost production planning problem with remanufacturing options, *Naval Research Logistics* 52:443–458.

Yang, G.-F., Wang, Z.-P., and Li, X.-Q. 2009. The optimization of the closed-loop supply chain network. *Transportation Research Part E* 45 (1):16–28.

Yao, Z., Wu, Y., and Lai, K.K. 2005. Demand uncertainty and manufacturer returns policies for style-good retailing competition. *Production Planning & Control* 16 (7):691–700.

Yuan, K.F. and Gao, Y. 2010. Inventory decision-making models for a closed-loop supply chain system. *International Journal of Production Research* 48 (20):6155–6187.

Zanoni, S., Ferretti, I., and Tang, O. 2006. Cost performance and bullwhip effect in a hybrid manufacturing and remanufacturing system with different control policies. *International Journal of Production Research* 44 (18):3847–3862.

Zarandi, M., Sisakht, A., and Davari, S. 2011. Design of a closed-loop supply chain (CLSC) model using an interactive fuzzy goal programming. *International Journal of Advanced Manufacturing Technology* 56 (5):809–821.

Zeid, I., Gupta, S.M., and Bardasz, T. 1997. A case-based reasoning approach to planning for disassembly. *Journal of Intelligent Manufacturing* 8 (2):97–106.

Zha, X.F. and Lim, S.Y.E. 2000. Assembly/disassembly task planning and simulation using expert Petri nets. *International Journal of Production Research* 38 (15):3639–3676.

Zhou, G., Cao, Z., Cao, J., and Meng, Z. 2005. A genetic algorithm approach on reverse logistics optimization for product return distribution network. In *Computational Intelligence and Security*, Y. Hao, I. Liu, Y. Wang, Y.-M. Cheung, H. Yin, L. Jiao, J. Ma, and Y.-C. Jiao (eds.). Berlin, Germany: Springer.

Zhou, W. and Piramuthu, S. 2010. Effective Remanufacturing via RFID. Paper read at *Proceedings of RFID Systech 2010—European Workshop on Smart Objects: Systems, Technologies and Applications*, June 15–16, Ciudad, Spain.

Zuidwijk, R. and Krikke, H. 2008. Strategic response to EEE returns: Product eco-design or new recovery processes? *European Journal of Operational Research* 191 (3):1206–1222.

Zussman, E. and Zhou, M. 1999. A methodology for modeling and adaptive planning of disassembly processes. *IEEE Transactions on Robotics and Automation* 15 (1):190–194.

Zussman, E. and Zhou, M.C. 2000. Design and implementation of an adaptive process planner for disassembly processes. *IEEE Transactions on Robotics and Automation* 16 (2):171–179.

2 Issues and Challenges in Reverse Logistics

Samir K. Srivastava

CONTENTS

Reverse logistics (RL) is the process of moving products from the consumer—the traditional final destination—to the manufacturer, the point of origin. The concept involves taking a long-term view of products from "cradle to grave" including possible "resurrection." It is gaining justifiable popularity among society, governments, and industry. Today, RL is viewed as an area that offers great potential to reduce costs, increase revenues, and generate additional profitability for firms and their supply chains. It is increasingly becoming an area of organizational competitive advantage, making its pursuit a strategic decision. In recent years, a rapid increase of corporate and legislative initiatives as well as academic publications on RL can be observed. Nearly everyone agrees that an RL network that seizes value-creation opportunities offers significant competitive advantages for early adopters and process innovators. At societal level, managing product returns in a more effective and cost-efficient way will help develop sustainable economies in a sound way. In light of the aforementioned, this chapter describes the concept of RL, its basic activities and scope, drivers and barriers, and major issues and challenges. We also describe a few initiatives and suggested frameworks and models for better RL design and practices.

2.1 INTRODUCTION

Collection of product recalls as well as collection and recycling of postconsumer goods is gaining interest in business and societies worldwide. Many organizations are discovering that improving their RL processes can be a value-adding proposition. Growing green concerns and advancement of green supply chain management concepts and practices make effective and efficient RL all the more relevant. Possible cost reductions, more rigid environmental legislations, and increasing environmental concerns of consumers have led to increasing attention to RL in present times. Research shows that RL may be a worthwhile proposition even in the contexts where regulatory and consumer pressures are insignificant. A well-managed RL system can not only provide important cost savings in procurement, recovery, disposal, inventory holding, and transportation, but also help in customer retention which is very important for organizational competitiveness. It shall become vital as service management activities and take-back for products such as automobiles, refrigerators and other white goods, cellular handsets, lead-acid batteries, televisions, computer peripherals, personal computers, laptops, etc., increase in future. These, in turn, depend on advancements in information and communication technologies (ICT) and their utility in supporting data collection, transmission, and processing. Since RL operations and the supply chains they support are significantly more complex than traditional supply chains, an organization that succeeds in meeting the challenges will possess a formidable advantage that cannot be easily duplicated by its competitors. The strategic importance of RL is evident from classification and categorization of the existing Green SCM literature by Srivastava (2007), as shown in Figure 2.1.

Today, consumers expect to trade in an old product when they buy a new one. Consumer demand for clean manufacturing and recycling is increasing, many times leading to legislation as well. Consequently, retailers expect original equipment manufacturers (OEMs) to set up a proper RL systems and networks. Recovery of products for remanufacturing, repair, reconfiguration, and recycling can create profitable business opportunities. The ability to quickly and efficiently handle the return of product for necessary repair can be critical. Similarly, the value of returned products may decrease more rapidly than their new counterparts, so the disposition decisions and their implementations need to be quick and effective. Firms in sectors like consumer durables, automobiles, print, pharmaceuticals, electronics and telecom, firms offering leasing and logistics services, and the firms venturing into green products market specially need to focus on RL. By determining the factors that most influence a firm's RL undertakings, it can concentrate its limited resources in those areas; rest may be outsourced. However, third-party logistics collection costs act as direct costs to supply chain and may not always be the best choice.

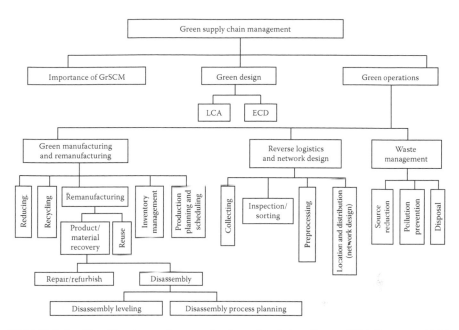

FIGURE 2.1 Classification and categorization of existing Green SCM literature. LCA, life cycle analysis; ECD, environmentally conscious design. (From Srivastava, S.K., *Int. J. Manage. Rev.*, 9, 53, 2007.)

2.2 BASIC REVERSE LOGISTICS ACTIVITIES AND THEIR SCOPE

RL has been used in many applications like photocopiers, cellular telephones, refillable containers, etc. In all these cases, one of the major concerns is to assess whether or not the recovery of used products is economically more attractive than disposal. The added value could be attributed to improved customer service leading to increased customer retention and thereby increased sales. The added value could also be through managing product returns in a more cost-effective manner or due to a new business model. Until recently, RL was not given a great deal of attention in organizations. Many of them are presently in the process of discovering that improving their logistics processes can be a value-adding proposition that can be used to gain a competitive advantage. In fact, implementing RL programs to reduce, reuse, and recycle wastes from distribution and other processes produces tangible and intangible value and can lead to better corporate image. RL is one of the key activities for establishing a reverse supply chain and comprises network design with aspects of product acquisition and remanufacturing. Literature identifies collection, inspection/sorting, preprocessing and logistics, and distribution network design as four important functional aspects in RL. Pure RL networks are generally more complex than pure forward flows because of more uncertainties associated with quality and quantity of returned products. In many cases, RL networks are not set up independently "from scratch" but are intertwined with existing logistics structures. Figure 2.2 shows the basic flow diagram of RL activities where the complexity of operations and the value recovered increase from bottom left to top right.

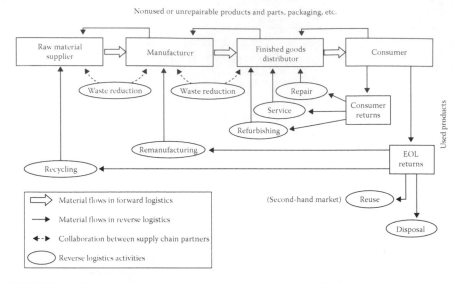

FIGURE 2.2 Basic activities and flows in RL. (Adapted from Srivastava, S.K., *Int. J. Phys. Distrib. Logist. Manage.*, 38, 311, 2008.)

2.3   DRIVERS AND BARRIERS OF REVERSE LOGISTICS

RL has its roots in "environmental management orientation of supply chains" (Srivastava, 2008). Firms have been practicing RL mainly to protect the market, to adopt a green image, and to improve the customer relationships. In literature, different authors mention multiple drivers and barriers of RL, like regulatory, market and societal forces. Three drivers (economic, regulatory, and consumer pressure) drive RL worldwide. The economic driver can be considered as the most important driving force as of now. RL can be economically beneficial for a firm and its supply chain. When production costs and initial purchasing costs decrease, the value of products that are recovered can be incorporated in the product and regained (Fleischmann et al., 1997). Furthermore, RL ensures with the remanufacturing, reuse and recycling of products that less energy is needed to produce products. Regulation refers to the legislation that stipulates that a producer should recover its products or material thereof and take these back. Many industrialized countries have introduced regulations for prevention and management of waste flows related to End-of-life (EOL) vehicles, waste Electrical and electronic equipment, and packaging and packaging waste. Firms can use RL to act in accordance with existing and future regulations and legislation. Walther and Spengler (2005) specifically studied the impact of legislation on supply chains. There are many different names for consumer forces, like corporate social responsibility (CSR), environmental sustainability, and slogans like "going green together." Firms incorporate these aspects in their strategy to express that they respect their environment, society, and nature. They can both be intrinsically or extrinsically motivated. Richey et al.

(2007) offered an interesting perspective on the role of RL in the drive toward sustainable development in emerging economies.

Many barriers can withhold firms from implementing RL. They can be both internal to the firm or external barriers. The most important internal barriers could be lack of awareness (Rogers and Tibben-Lembke, 2001), lack of top management commitment to introduce RL in the firm (Rogers and Tibben-Lembke, 2001), lack of strategic planning (Ravi and Shankar, 2005), financial constraints (Rogers and Tibben-Lembke, 2001), and employees inherently do not like change or are not well educated or trained in economic affairs (Ravi and Shankar, 2005; Rogers and Tibben-Lembke, 2001). Similarly, Erol et al. (2010) found firm policy as the most important reason for not having an efficient RL in electric/electronics industry. They also observe that one of the main barriers to executing RL for all the respondent firms is system inadequacy, which is in line with the findings from Rogers and Tibben-Lembke (2001). The firms face system deficiencies that are partly brought about by inadequate infrastructure such as ICT for developing an efficient reverse supply chain. In the context of external barriers, Lau and Wang (2009) observed that major difficulty in implementing RL in the electronic industry of China is due to the lack of enforceable laws, regulations, or directives to motivate various stakeholders. Furthermore, economic support and preferential tax policies are absent to help manufacturers offset the high investment costs of RL. Erol et al. (2010) too observed lack of economic incentives and legislation as barriers to RL in Turkey. They found that only 28% of the firms in automotive and 25% of the firms in furniture industries have been implementing RL. Even these firms also emphasized that these implementation processes are in a very early stage and continue slowly since the legislations in question have not been enacted yet. Low public awareness of environmental protection and underdevelopment of recycling technologies are some of the other barriers to RL implementation. Separate logistics channels and ill-defined processes with high number of touch points in RL networks may lead to greater errors/deterioration and these too work as barriers. Limited visibility of returns and/or the lack of focus on returns also act as a barrier in many firms and supply chains. Many of these barriers, when overcome, may become potential drivers of RL. Table 2.1 summarizes important drivers and barriers of RL.

2.3.1 MAJOR REVERSE LOGISTICS DECISIONS

Firms and supply chains have to take decisions on issues such as in-house production versus outsourcing, resources allocation, forecasting returns, returns management, RL network design, inventory management, product quality and safety, and marketing and logistics issues. They also need to define clearly the role of ICT and role of various stakeholders. Coordination requirement of two markets (market for products made from virgin materials and market for products made from earlier-used products) and disposition task (further action to be taken on returns) are identified as major challenges in RL literature (Fleischmann et al., 2000). Lambert et al. (2011) identified various RL decisions and categorized them into strategic, tactical, and operational decisions. Table 2.2 lists these major RL decisions.

TABLE 2.1
Important Drivers and Barriers of RL

Drivers	Barriers
Reduction in production/supply chain costs	High costs and lack of supportive economic policies
Improvement of customer service	Lack of awareness and knowledge about RL
Promotion of corporate image	Underdevelopment of appropriate technologies
Support from policies and legislation	Lack of supportive laws and legislation
Fulfillment of environmental obligations	Unpredictability and variability in supply and demand
Top management commitment	Inadequate deployment of resources in RL
Focus on returns management	Resistance to change

TABLE 2.2
Major RL Decisions

Strategic	Tactical	Operational
Whether or not to integrate RL with the forward logistics	Decide transportation means and establish transportation routes	Logistics and operations scheduling
Allocate adequate financial resources	Establish operational policies (production and inventory)	Emphasize cost control
Categorize and define return policies	Define return policies for each item	Return acquisition activities
Determine reasons, stakeholders, and issues related to RL	Define technical support to offer (in-store, subcontractors, etc.)	Consider time value of returns
Evaluate internal expertise in RL and decide about outsourcing a few/all RL activities	Do the RL activities (transportation, warehousing, remanufacturing, etc., in-house or subcontract)	Train personnel on RL concepts and practices
Implement environmental management systems and acquire knowledge of directives, laws, and environmental rules	Develop a planning system for various RL activities and establish quality standards for them	Manage information
Choose activities (repair/rework, reuse, etc.) and identify potential locations	Decide the location and allocation of capacities for RL facilities	Determine level of disassembly
Risk assessment (value of information and uncertainties)	Define performance measures; optimize policies	Analyze returns in order to improve disposition

Source: Adapted from Lambert, S. et al., *Comput. Ind. Eng.*, 61, 561, 2011.

2.4 ISSUES AND CHALLENGES

RL offers unlimited opportunities for firms and supply chains in areas like aftermarkets, EOL vehicles and EOL consumer durables, mobile handsets, refuse collection, e-wastes, hazardous wastes, repair and remanufacturing, and a host of other operations. An important consideration in extracting value from returns is to actively manage their quantity and timing. It is in estimation and control of quantity and timing of returns that firms and other stakeholders face the greatest challenge. Another challenge is related to integrating product design and product take-back. In the case of EOL items, since product usage conditions and lifetimes differ from user to user, there are significant fluctuations in product flows' quantity and quality. The product safety issues and challenges that arise in various industries that are increasingly globalizing their supply chains offer additional RL challenges. Food, pharmaceuticals, medical devices, consumer products, and automobiles are notable industries among these. Large global recalls associated with recent product safety events, for example, the Chinese melamine-adulterated milk contamination in 2008, the adulterated heparin in 2008, and the Toyota recall of 2011, have made the development of tools and technologies for traceability through the reverse supply chain a critical issue in risk control.

Establishing appropriate process controls and deploying appropriate tools and technologies for traceability are therefore important. They can be defined as the formal process of analyzing and tracking returns and measuring returns-related performance criteria aimed at improving the whole RL operation (Rogers et al., 2002). Managing the return flow of product is increasingly recognized as a strategically important activity that involves decisions and actions within and across firms. The issues and challenges in RL may broadly be classified into returns related, process, recovery, and technology related; network design and coordination related; regulatory and sustainability related; and cost-benefit related. We look into each one of these in detail.

2.4.1 RETURNS RELATED

Despite the growing recognition of the importance of RL, many firms are not prepared to meet the challenges involved in handling returns. The rapid growth in the volume of returns often outpaces the abilities of firms to successfully manage the flow of unwanted product coming back from the market. Erol et al. (2010) found that the firms' involvement in the product returns is mainly based on two motives: "national legislative liabilities" and "competitive reasons based on sustainability." However, they miss "capturing value" which generally is the prime driver for RL in most supply chains and businesses. Many firms and supply chains have considered RL as a strategic goal because it is part of the supply chain that offers value. Such value relates to the ability to efficiently and effectively manage "returns."

Managing the return flow of product is increasingly recognized as a strategically important activity that involves decisions and actions within and across firms. Returns in this chapter refer to used products, materials and packaging as well as waste. The dimensions used to characterize RL environments are returns volume, returns timing, returns quality (grade), product complexity, testing and evaluation complexity, and remanufacturing complexity. So, the pattern of quantity, quality, and time of arrival of returns, collection, routing, processing, and resale are of paramount importance.

Various processes are associated with returns management. Return initiation is defined as the process where the customer seeks a return approval from the firm or sends the return directly to the returns center (Rogers et al., 2002). This process relates to the mode of transportation to reach the destination. It is followed by the receiving process which includes verifying, inspecting, and processing the returned product with emphasis on selecting the most efficient disposition option. Quick disposition of returns is the most important part in a successful reverse supply chain. If returns can be disposed in time and processed quickly, profit and service level can be increased. Assigning predisposition codes to the processed return enables fast and accurate determination of disposition options.

Several product characteristics such as composition, deterioration, and use-pattern of products are relevant for the profitability of the RL systems. The main composition characteristics of the products are the homogeneity, disassembility, testability, and standardization. Many components and many materials need to be considered when developing a product (Gungor and Gupta, 1999; Ilgin and Gupta, 2010). Disassembility is another key item in RL. The product should be designed in a way that all the different materials can be easily recycled, which entails an effective disassembly of the product. The testability of the different (hazardous) materials also influences the economics of the RL process.

Product deterioration affects the recovery options strongly. When products are heavily deteriorated recovery is of less economic value in contrast to products that have hardly deteriorated. In order to analyze the deterioration a couple of key items arise; the deterioration sensitivity of different parts and the speed of deterioration in relation to the design cycle. In short their functionality becomes outdated and the product renders obsolete, which makes it more difficult to recover. The next process is crediting the customer/supplier. It involves the charge-back to the buyer's account including credit authorization and potential claim settlements with customers.

2.4.1.1   Returns Policy Issues

Return policies should be properly designed, defined, and communicated to all the relevant stakeholders. They can be used effectively for offering incentives and overcoming hurdles. Fleischmann et al. (2001) suggested that buyback may lead to higher returns leading to economies of scale. Some resolution to customers may be used for this. Offering differentiated take-back prices to consumers based on product model and product quality or charging a return fee is likely to reduce both the number of returns and its variance. Mont et al. (2006) presented a new business model based on leasing prams where the product–service system includes the organization of an RL system with different levels of refurbishment and remanufacturing of prams, partially by retailers. They focus on reducing costs for reconditioning, reduction of time and effort for the same, and finally on environmentally superior solutions.

2.4.2   Process, Recovery, and Technology Related

The literature is rather unanimous on the recovery process and the different recovery options. The process is designed as follows: the product is first of all collected, and this depends on the sort of customer as well as the type of product. When the product has arrived, it is inspected and tested. It becomes clear how much value the product still has and how this value should be recovered. Subsequently, the products that are worth recovering are selected and sorted, and then finally the actual recovery is executed. The recovery can be subdivided by material recovery and added value recovery.

Material recovery boils down to recycling. Recycling denotes material recovery without conserving any product structures. In the case of material recovery, products are usually grinded and their materials are sorted out and grouped according to specifications and quality measures (Fleischmann et al., 1997). Added value recovery can be subdivided in direct recovery and process recovery. Direct recovery stands for putting a product back on the market immediately after its first period of use ended, via resale, reuse, and redistribution. In case this is not possible, but the product can be reprocessed and reworked into something valuable, process recovery can be applied, which stands for the reprocessing of products or parts of it in the production process (Fleischmann et al., 1997).

It has been established that there are three fundamental stages of flow in RL: (1) collection, (2) sort-test, and (3) processing. Barker and Zabinsky (2011) developed a framework for network design decisions, as shown in Figure 2.3. This framework was developed after analysis of 40 case studies to determine the design decisions and associated tradeoff considerations. Each of the three stages of flow in the framework has two decision options and there are eight possible configurations.

Lambert et al. (2011) developed a decisions framework for process mapping and improvements of an RL system. This framework, shown in Figure 2.4, offers flexibility and covers a wide variety of situations that may arise in the practical working environment. The design of an RL system starts at stage 1—the decisions. Once all the decisions (strategic, tactical, operational) have been taken, the selection of performance measures and target setting is undertaken. These first two stages define the RL system to be implemented in stage three. Stage four ensures feedback on the performance of the system while providing a means of returning to previous stages in order to improve the system. A review of the performance measures should be done regularly in order to adjust the objectives to the current market conditions or replace them by better ones. Unless the market has new requirements or the firm has changed its strategic objectives, the program review will be more focused at the operational level.

2.4.3 NETWORK DESIGN AND COORDINATION-RELATED DECISIONS

Strategic planning (also called designing) of a reverse supply chain is a challenging problem due to various crucial issues, such as which EOL products to collect, where to collect them, how to reprocess them, where to reprocess them, etc. (Pochampally and Gupta, 2012). Besides, various stakeholders are involved right form collecting the products from the customers to making sure that these products end up at the manufacturer. The main stakeholders that play a role in the processes are specialized reverse chain players like jobbers, recycling specialists, devoted sector organizations, municipalities, and public or private foundations besides the forward supply chain suppliers, manufacturers, wholesalers, retailers, and sector organizations.

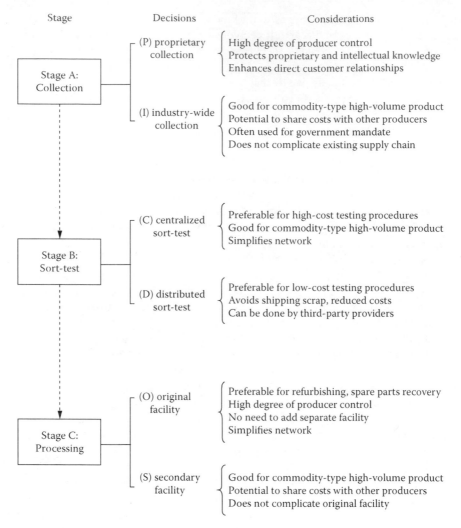

FIGURE 2.3 Framework for network design decisions. (From Barker, T.J. and Zabinsky, Z.B., *Omega*, 39, 558, 2011.)

Common approaches to RL network design are presented by Jayaraman et al. (1999), Fleischmann et al. (2001), and Srivastava (2008) among others.

The products can be recovered via different logistics pathways, or models. Some popular models are as follows:

Model 1: The manufacturer collects the used products directly from the customers. Firms use different methods for different products.

Model 2: The manufacturer contracts the collection of used products to the retailer. The retailer promotes and collects used products in addition to distributing the new products.

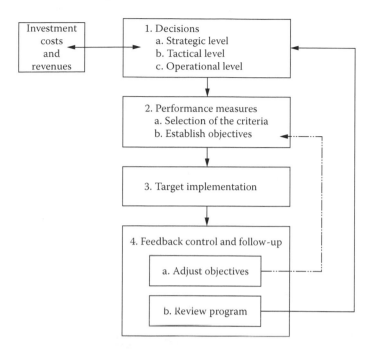

FIGURE 2.4 Framework for process mapping and improvements in RL. (From Lambert, S. et al., *Comput. Ind. Eng.*, 61, 561, 2011.)

Model 3: The manufacturer contracts the collection of used products out to a specialized third party. The third party acts as a broker between the customer and the manufacturer.

Model 4: Different materials are brought back via the manufacturer, the wholesaler and retailer to the supplier who does the actual recovery of its own materials.

Various modeling aspects relevant for designing RL networks such as types of problem formulations, various decision variables and parameters used, data collection and generation techniques, and various solution techniques can be seen in literature. An emerging change of firm objectives in supply chain design from cost minimization only, to simultaneous cost and environmental impact minimization has introduced another dimension of complexity. Most models resemble multilevel warehouse location problems and present deterministic integer programming models to determine the location and capacities of RL facilities. Lee et al. (2010) take hybrid facilities into account and extend the location problem by the decision on the type of depot to install, namely only purely forwarding, returning depots or building hybrid processing facilities for a single period. Srivastava (2008) developed a conceptual model for simultaneous location-allocation of facilities for a cost-effective and efficient RL network covering costs and operations across a wide domain. The proposed RL network consists of collection centers and two types of rework facilities set up by OEMs) or their consortia for a few categories of product returns under various strategic, operational and customer service constraints in the Indian context.

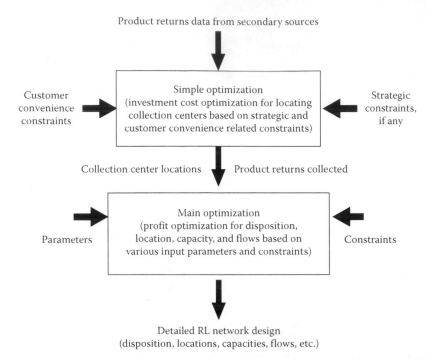

FIGURE 2.5 Multiproduct, multiechelon, profit maximizing value recovery model. (From Srivastava, S.K., *Int. J. Phys. Distrib. Logist. Manage.*, 38, 311, 2008.)

The same is shown in Figure 2.5. The problem has been treated similar to a multistage resource allocation problem. This combinatorial problem resembles a multicommodity network flow problem with a few sequentially dependent decisions for which no special algorithms are applicable apart from decomposition. Various decisions such as the disposition decisions, the sites to be opened, the capacity additions at any period of time as well as the number of products of a particular grade that are to be processed or sold during a particular period of time are decided by the model.

2.4.4 Regulatory and Sustainability Related

Reverse supply chain management has gained increasing popularity in last two decades among researchers, and a rich literature has piled up for various aspects within this important field. One reason of this popularity is the economic value gained back by the recovery processes. Another one is the directives passed by the EU Commission and various governments on environment controls, waste reduction, and product returns. China has issued many laws and regulation such as Cleaner Production Promotion Law, Law on Environmental Impact Assessment (EIA), Renewable Energy Law, Law on the Prevention and Control of Environmental Pollution by Solid Waste, etc. Tianjin promulgated and implemented many laws and regulation in environmental protection, energy saving, and local cleaner production policies and rules. The regulations such as hazardous waste operating license management solution of Tianjin (2004), Tianjin hazardous waste transfer implementation details (2004) issued and implemented gradually. Such legislation efforts have laid a foundation for RL activities according to law.

This is an area where technology originally developed for tracking inventory and assets in the supply chain has proven to be very useful. Traceability systems can bring additional benefits. For instance, Wang et al. (2010) have developed an optimization model that uses traceability data in combination with operations factors to develop an optimal production plan.

2.4.5 Cost-Benefit Analysis Related

Incorporation of returned goods into supply chains account for a significant part of firms' logistics costs and add tremendous complexity. RL can have both a positive and a negative effect on a firm's cash flows. Organizations and supply chains need to understand the financial impact of RL strategies which can generate periodic negative cash flows that are difficult to predict and account for. EOL returns have the potential of generating monetary benefits. Horvath et al. (2005) used a Markov chain approach to model the expectations, risks, and potential shocks associated with cash flows stemming from retail RL activities and actions for avoiding liquidity problems stemming from these activities.

Dowlatshahi (2010) gives a conceptual framework for cost-benefit analysis in RL. This framework shown in Figure 2.6 provides guidelines for managers on how to use and apply cost-benefit analysis for decision-making in RL and addresses two key research questions of what and how. The framework also shows how firms should best pursue their RL strategies at various stages of the development and decision-making process. Figure 2.5 also shows the proper sequence, decision points, and the interaction among subfactors of cost-benefit analysis in a logical and easy to understand way. Managers have found that minimizing decision process unless it adds value, avoiding handling costs unless they add value, keeping processes simple, and linking information tracking/sharing to planning generally yields good results.

2.5 SOME INITIATIVES, OTHER MODELS, AND FRAMEWORKS

In this section, a few initiatives from practice and a few other models and frameworks from recent literature have been presented as they could be useful for the readers.

2.5.1 Reverse Logistics Initiatives in India

Many firms in consumer durables' and automobile sectors in India have introduced exchange offers to tap customers who already own such products mainly from marketing perspective. This has led to increased focus on various R's (reduce, reuse, resell, repair, recycle, refurbish, remanufacture, and RL). Presently, these returned products are either resold directly in the seconds' market or after repair and refurbishment by firm franchisee/local remanufacturers. They are not remanufactured or upgraded by OEMs. The leading car manufacturer and market leader in India,

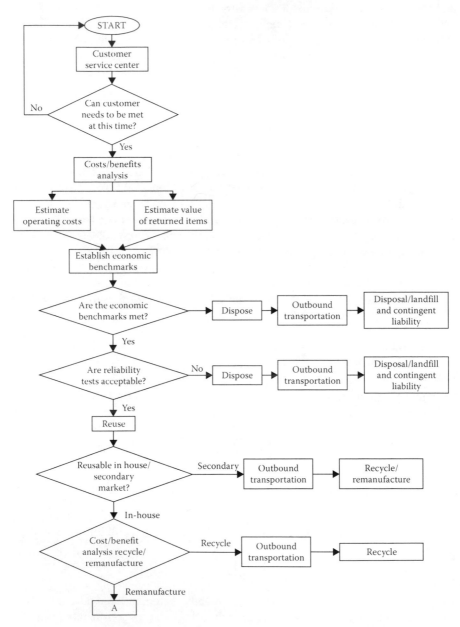

FIGURE 2.6 Conceptual framework of the cost-benefit analysis in RL. (From Dowlatshahi, S., *Int. J. Prod. Res.*, 48, 1361, 2010.)

Maruti Suzuki India Limited, was the first mover with its *True Value* initiative. It has established India's largest certified car dealer network with 358 outlets in 210 cities and is continuously growing. The industry is at a nascent stage but the business potential is considered to be huge. The RL market in India is valued around INR 800 billion currently and is expected to grow rapidly in the future. So far, the most common approach for designing RL networks is the independent design of reverse and forward networks.

2.5.2 GLOBAL ASSET RECOVERY SERVICES AT IBM

IBM has been among the pioneers seeking to unlock the value dormant in product returns. Recognizing the growing importance of RL flows, it assigned the responsibility for managing all product returns worldwide to a dedicated business unit in 1998, named Global Asset Recovery Services (GARS). The main goal of this organization was to manage the dispositioning of returned items and thereby to maximize the total value recovered. To this end, GARS operates some 25 facilities all over the globe where returns are collected, inspected, and assigned to an appropriate recovery option. It assesses which equipment may be remarketable, either "as is" or after a refurbishment process. For this purpose, IBM operates nine refurbishment centers worldwide, each dedicated to a specific product range. Internet auctions, both on IBM's own website and on public sites have become an important sales channel for remanufactured equipment. In particular, dismantling used equipment provides a potential source of spare parts for the service network.

Fleischmann et al. (2002) described integrating Closed-loop Supply Chains and Spare Parts Management at IBM. They emphasize necessity of a holistic perspective while addressing the challenge of integration of used equipment returns as a supply source into spare parts management. They develop an analytic inventory control model and a simulation model and their results show that procurement cost savings largely outweigh RL costs and that information management is the key to an efficient solution. In their analysis they considered two alternative channel designs, denoted as "pull" versus "push" dismantling. In the first case, one builds up a stock of dismantled parts on which test orders can be placed when needed, in analogy with the traditional repair channel. In the second case, dismantled parts are tested as soon as available, after which they are directly added to the serviceable stock. The first option benefits from postponing the investment for testing, which reduces opportunity costs and the risk of testing parts that are no longer needed. On the other hand, the second option avoids stocking defective parts and reduces the throughput time, which may reduce safety stock.

2.5.3 CIRCULAR (SUSTAINABLE) ECONOMY AT TIANJIN

Shen (2008) applies green development index to evaluate the level of RL in Tianjin. Current RL strategies of Tianjin are surveyed in the aspects of practical patterns, policies, legislation, and strategic programming. From 1999, the government adopted *sustainable economy* idea in (1) adjusting industrial structure, shutting down some enterprises that have backward technologies, waste resources, and pollute environment; (2) optimizing energy structure, reducing the proportion of coal in energy, raising the utilization rate of coal, promoting clean coal technologies, and developing renewable energy such as firedamp; and (3) quickening the construction of sewage and garbage treatment facilities to better treat domestic pollution. These include blueness sky project, green water project, quiet project, solid waste pollution prevention project, eco-city and village project, water environmental governance project, and strengthening antiradiation management within Five Year Plans. The government encourages enterprises to engage in circular economy through preferential policies. The enterprises engaged in comprehensive use of resources enjoy tax reduction and exemption policy.

2.5.4 RFID-Based RL System

Lee and Chan (2009) suggested the deployment of an RFID-based RL system, as shown in Figure 2.7.

2.5.5 Implementing JIT Philosophy in RL Systems

Chan et al. (2010) present a framework for implementing just-in-time philosophy to RL systems and the same is reproduced in Figure 2.8. Four key processes that are directly related to RL activities are identified in the Process Model. They are collection, distribution, inventory, and reassembling (or remanufacturing). Collection mainly focuses on the location of collection points and warehouse, whereas distribution covers the transportation planning and route planning, and optimization of the distribution network in terms of cost or efficiency. In addition, it also affects customer satisfaction. Inventory refers to the management and control of stock level. Finally, remanufacturing concentrates on quality control and planning of material requisition for restoring the returned or used products to a usable or resalable condition. The Information System Model comprising MRP, EDI, and other ICT technologies shall primarily capture and process all uncertainty related data and support decision-making in terms of tractability and visibility. As PLC has become shorter, a proper design that takes environmental concerns into consideration (e.g., using green components and reusable material) has become the primary issue for RL. The PLC management model should address these concerns. Finally, reverse logistics structure (RLS) shall be a lean RLS integrating JIT in RL. JIT performance aims at finding out how JIT can help to optimize PM, PLC, and an RL system. It is also important to identify the relationship between an information system and JIT performance in order to derive useful managerial insights. Five main performances are examined in this category. They are cost, speed, flexibility, dependability, and quality, which are also the well-known performance measures of any operations or production related literature.

2.6 CONCLUSIONS AND OUTCOMES

RL is a relatively new concept in the literature and practice, and while it is gaining recognition, it does not have the same foundation as more established streams. It presents a challenge both methodologically and philosophically for further

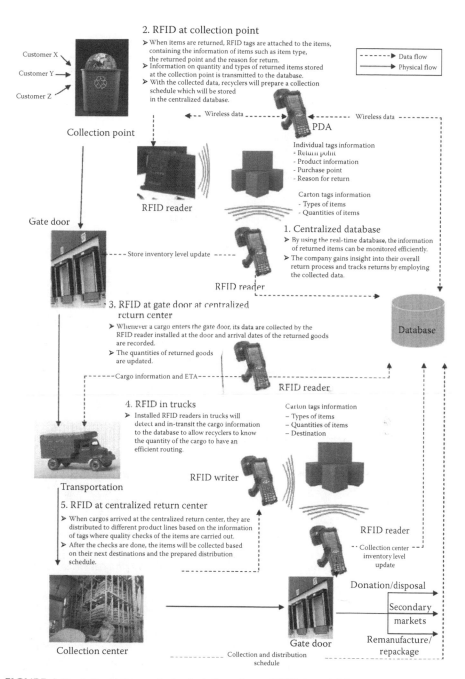

2. RFID at collection point

➤ When items are returned, RFID tags are attached to the items, containing the information of items such as item type, the returned point and the reason for return.
➤ Information on quantity and types of returned items stored at the collection point is transmitted to the database.
➤ With the collected data, recyclers will prepare a collection schedule which will be stored in the centralized database.

Customer X
Customer Y
Customer Z

Collection point

------➤ Data flow
——➤ Physical flow

Wireless data

PDA

Wireless data

Individual tags information
- Return point
- Product information
- Purchase point
- Reason for return

Carton tags information
- Types of items
- Quantities of items

RFID reader

1. Centralized database

➤ By using the real-time database, the information of returned items can be monitored efficiently.
➤ The company gains insight into their overall return process and tracks returns by employing the collected data.

Gate door

Store inventory level update

RFID reader

3. RFID at gate door at centralized return center

➤ Whenever a cargo enters the gate door, its data are collected by the RFID reader installed at the door and arrival dates of the returned goods are recorded.
➤ The quantities of returned goods are updated.

Database

Cargo information and ETA

RFID reader

4. RFID in trucks

➤ Installed RFID readers in trucks will detect and in-transit the cargo information to the database to allow recyclers to know the quantity of the cargo to have an efficient routing.

Carton tags information
– Types of items
– Quantities of items
– Destination

RFID writer

Transportation

5. RFID at centralized return center

➤ When cargos arrived at the centralized return center, they are distributed to different product lines based on the information of tags where quality checks of the items are carried out.
➤ After the checks are done, the items will be collected based on their next destinations and the prepared distribution schedule.

RFID reader

Collection center inventory level update

Donation/disposal

Secondary markets

Collection center

Gate door

Remanufacture/ repackage

Collection and distribution schedule

FIGURE 2.7 Information and physical flows in an RFID-based RL system. (From Lee, C.K.M. and Chan, T.M., *Expert Syst. Appl.*, 36, 9299, 2009.)

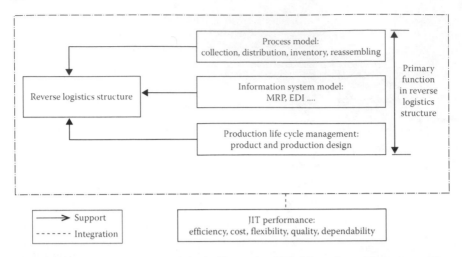

FIGURE 2.8 Proposed framework for implementing JIT philosophy to RL systems. (From Chan, H.K. et al., *Int. J. Prod. Res.*, 48, 6293, 2010.)

advancement. It also offers various opportunities to bridge the gap between sustainability and existing business supply chains. The growth of RL to deal with end-of life products and issues such as product recalls, disposal and reuse options seems very likely to continue, as more firms in different situations begin to face these problems (and develop them into opportunities).

Firms are recognizing the benefits of closed-loop supply chains that integrate product returns into business operations. They are gradually recognizing that reverse supply chain considerations should be a part of their organization's corporate strategy. They are exploring ways and means to use existing and new software for decision-making and workflow management for RL as well. They are eager to adopt better practices and models and are trying to address basic RL implementation issues. Despite developing great sophistication in demand forecasting, few firms to date are collecting any systematic data on product returns. They are generally ignoring the fact that even greater opportunities would come into reach if they use ICT for actively managing their returns rather than accepting them as a given. Advances in ICT, including data logging, radio frequency identification, mobile telephony, and remote sensing provide ever more powerful means for pursuing this road.

Erol et al. (2010) find that given the current uncertainty, many firms are reluctant to invest in infrastructure related to RL which they all consider as a cost driver. There is still a long way to the use of RL system to recover assets as evident from their study in Turkey. Many enterprises have not been able to develop common and key technologies that can help in substantially raising resource efficiency. A number of platform and common technologies should be developed that produce good economic return, consume less resource, and have less pollution, including ICT; better tools and methodologies for managing information during the lifecycle of the product from design through disposal; and technologies for tracing products across the global supply chain and managing recalls. Firms should also explore integration opportunities with 3PL/4PL to facilitate multichannel returns with online visibility using the ERP systems. Industry should work to increase product recyclability, develop life-cycle-analysis capabilities and improve communication among its segments. Efforts should be undertaken to strengthen and expand industry coalitions and link with logistics service providers.

Further, for long-term sustainable development and competitiveness in the global market, the governmental bodies have to set up regulations as soon as possible to promote, control, and standardize RL practices. The technological development for RL should be included in the mid- and long-term scientific development plans of governments. A system and policy environment should be established in favor of the development of RL. Industrial policies should emphasize raising resource efficiency and environment, promoting strategic economic restructuring so that they would be helpful to building a sustainable economy. Policy of logistics should pay more attention to "reverse" industrials. Effective incentive policies, recovery treatment system, and rational pricing mechanism should be designed. Governments can use economic means to build an incentive mechanism for RL. Using market means to promote sustainable economy is an extension of incentive mechanism for environment protection. The tools include taxation, the property right of resources, pricing system and contract energy, etc. Governments should encourage 3PLs/4PLs to invest in RL.

RL opens a number of avenues for experimentations and analysis for firms, researchers and policymakers. Firms may consider under which circumstances should returns be handled, stored, transported, processed jointly with forward flows (integrated logistics), and when should they be treated separately. They may compare cost of remanufacturing with cost of production from virgin materials to decide on proper input mix. A better understanding of the trade-offs inherent in network design decisions is essential for producers and industries to develop efficient RL networks. Integration of RL into the forward logistics operations may provide a potential for competitive differentiation.

As many small firms are likely outsource their RL functions to 3PL/4PL providers initially, it will also be useful for researchers to look at the issues from a different perspective by involving third-party RL service providers in future studies. Another future research direction is to analyze the viability of cost models used in the RL and remanufacturing operations. Many of these models view costs as myopic and isolated from many relevant RL factors and priorities. Efforts should be undertaken to improve the overall effectiveness of cost models. Further, the joint life-cycle dynamics and implications of new versus remanufactured products can be explored. This is an important issue given such factors as sales patterns of both new and remanufactured products, the available supply of used products, and the overall capacity of the OEM.

The future research could also explore as to whether firms in the same industry are equally predisposed to embark on RL. Further, the factors that encourage some firms to choose RL and others not to choose RL can be investigated. It should also explore how the nature of industries, products, and issues facing the firms engaged in RL require changes or customization. The mutual impacts of the external factors affecting RL development and the issues involved in collaborative RL management need to be investigated. This will augment current theories and models of RL. Similarly, there is much scope to explore the potential attractiveness of various control and postponement strategies in designing their reverse flows. Another interesting area is to design changes in a firm's RL strategy for a particular product over the course of the product's entire life cycle. Finally, researchers and practitioners should explore ways and means to establish leanagile RL systems.

REFERENCES

Barker, T.J. and Zabinsky, Z.B. A multicriteria decision making model for reverse logistics using analytical hierarchy process. *Omega* 39: 558–573 (2011).

Chan, H.K., Yin, S., and Chan, F.T.S. Implementing just-in-time philosophy to reverse logistics systems: A review. *International Journal of Production Research* 48: 6293–6313 (2010).

Dowlatshahi, S. A cost-benefit analysis for the design and implementation of reverse logistics systems: Case studies approach. *International Journal of Production Research* 48: 1361–1380 (2010).

Erol, I., Velioglu, M.N., Serifoglu, F.S., Buyukozkan, G., Aras, S., Kakar, N.D., and Korugan, A. Exploring reverse supply chain management practices in Turkey. *Supply Chain Management: An International Journal* 15: 43–54 (2010).

Fleischmann, M., Beullens, P., Bloemhof-Ruwaard, J.M., and van Wassenhove, L.N. The impact of product recovery on logistics network design. *Production and Operations Management* 10: 156–173 (2001).

Fleischmann, M., Bloemhof-Ruwaard, J., Dekker, R., van der Laan, E., van Nunen, J., and van Wassenhove, L.N. Quantitative models for reverse logistics: A review. *European Journal of Operational Research* 103: 1–17 (1997).

Fleischmann, M., Krikke, H.R., Dekker, R., and Flapper, S.D.P. A characterization of logistics networks for product recovery. *Omega* 28: 653–666 (2000).

Fleischmann, M., van Nunen, J., and Gräve, B. Integrating closed-loop supply chains and spare parts management at IBM. ERIM Report Series Research in Management, ERS-2002-107-LIS (2002).

Gungor, A. and Gupta, S.M. Issues in environmentally conscious manufacturing and product recovery: A survey. *Computers & Industrial Engineering* 36: 811–853 (1999).

Horvath, P.A., Autry, C.W., and Wilcox, W.E. Liquidity implications of reverse logistics for retailers: A Markov chain approach. *Journal of Retailing* 81: 191–203 (2005).

Ilgin, M.A. and Gupta, S.M. Environmentally conscious manufacturing and product recovery (ECMPRO): A review of the state of the art. *Journal of Environmental Management* 91: 563–591 (2010).

Jayaraman, V., Guide, V.D.R., Jr., and Srivastava, R. A closed-loop logistics model for remanufacturing. *Journal of the Operational Research Society* 50: 497–508 (1999).

Lambert, S., Riopel, D., and Abdul-Kader, W. A reverse logistics decisions conceptual framework. *Computers & Industrial Engineering* 61: 561–581 (2011).

Lau, K.H. and Wang, Y. Reverse logistics in the electronic industry of China: A case study. *Supply Chain Management: An International Journal* 14: 447–465 (2009).

Lee, C.K.M. and Chan, T.M. Development of RFID-based reverse logistics system. *Expert Systems with Applications* 36: 9299–9307 (2009).

Lee, D.-H., Dong, M., and Bian, W. The design of sustainable logistics network under uncertainty. *International Journal of Production Economics* 128: 159–166 (2010).

Mont, O., Dalhammar, C., and Jacobsson, N. A new business model for baby prams based on leasing and product remanufacturing. *Journal of Cleaner Production* 14: 1509–1518 (2006).

Pochampally, K.K. and Gupta, S.M. Use of linear physical programming and Bayesian updating for design issues in reverse logistics. *International Journal of Production Research* 50: 1349–1359 (2012).

Ravi, V. and Shankar, R. Analysis of interactions among the barriers of reverse logistics. *Technological Forecasting and Social Change* 72: 1011–1029 (2005).

Richey, R.G., Jr., Tokman, M., Wright, R.E., and Harvey, M.G. Monitoring reverse logistics programs: A roadmap to sustainable development in emerging markets *Multinational Business Review* 13: 1–25 (2007).

Rogers, D.S., Lambert, D.M., Croxton, K.L., and Garcia-Dastugue, S.J. The returns management process. *International Journal of Logistics Management* 13: 1–17 (2002).

Rogers, D.S. and Tibben-Lembke, R. An examination of reverse logistics practices. *Journal of Business Logistics* 22: 129 148 (2001).

Shen, C. Reverse logistics strategies and implementation: A survey of Tianjin. (2008). Available at: http://www.mendeley.com/research/reverse-logistics-strategies-implementation-survey-tianjin (accessed on November 30, 2011).

Srivastava, S.K. Green supply chain management: A state-of-the-art literature review. *International Journal of Management Reviews* 9: 53–80 (2007).

Srivastava, S.K. Value recovery network design for product returns. *International Journal of Physical Distribution and Logistics Management* 38: 311–331 (2008).

Walther, G. and Spengler, T. Impact of WEEE-directive on reverse logistics in Germany. *International Journal of Physical Distribution and Logistics Management* 35: 337–361 (2005).

Wang, X., Li, D., O'Brien, C., and Li, Y. A production planning model to reduce risk and improve operations management. *International Journal of Production Economics* 124: 463–474 (2010).

3 New-Product Design Metrics for Efficient Reverse Supply Chains

Seamus M. McGovern and Surendra M. Gupta

CONTENTS

3.1 OVERVIEW

As reverse supply chains grow in importance, products are being increasingly disassembled for recycling and remanufacturing at the end of their lifecycle. Just as the assembly line is considered the most efficient way to manufacture large numbers of products, the disassembly line has been successfully used in the reverse manufacturing of end-of-life products. While products are frequently designed for ease of assembly, there is growing need to design new products that are equally efficient at later being disassembled. Disassembly possesses considerations that add to its line's complexity when compared to an assembly line, including treatment of hazardous parts, and a used-part demand that varies between components. In this chapter, metrics are presented for quantitatively comparing competing new-product designs for end-of-life disassembly on a reverse-production line. A case study consisting of three design alternatives—each equally desirable and efficient in terms of assembly—of a notional consumer product is analyzed to illustrate application of the metrics.

The new-product design metrics are shown to lead to better decisions than may have otherwise been made without the metrics.

3.2 PROBLEM INTRODUCTION

Manufacturers are increasingly recycling and remanufacturing their post-consumer products as a result of new, more rigid environmental legislation, increased public awareness, and extended manufacturer responsibility. In addition, the economic attractiveness of reusing products, subassemblies, or parts instead of disposing of them has furthered this effort. This has contributed to widespread adoption of the full spectrum of the reverse supply chain.

Recycling is a process performed to retrieve the material content of used and nonfunctioning products. *Remanufacturing*, on the other hand, is an industrial process in which worn-out products are restored to like-new conditions. Thus, remanufacturing provides the quality standards of new products with used parts. The first crucial step of both of these processes is disassembly. *Disassembly* is defined here as the methodical extraction of valuable parts/subassemblies and materials from discarded products through a series of operations. After disassembly, reusable parts/subassemblies are cleaned, refurbished, tested, and directed to inventory for the remanufacturing portion of the reverse supply chain. The recyclable materials can be sold to raw-material suppliers, while what remains is sent to landfills. The multiobjective nature of disassembly necessitates a solution schedule which provides a feasible disassembly sequence, minimizes the number of workstations and total idle time, and ensures similar idle times at each workstation, as well as addressing other disassembly-specific concerns.

To connect both ends of a product's lifecycle in the reverse supply chain, its design must not only satisfy functional specifications and be easy to assemble but should also lend itself to disassembly while possessing a host of other end-of-life attributes. This has led to the emergence of concepts such as *design for environment*, *planning for disassembly*, and *design for disassembly*. Quantifying the merits of different product designs allows manufacturers to intelligently plan for a wide variety of potential future contingencies.

In this chapter a method to quantitatively evaluate product design alternatives with respect to the disassembly process is proposed. A product design can make a significant difference in the product's retirement strategy. It is not uncommon for a designer to be faced with the dilemma of choosing among two or more competing design alternatives. A product designer may wish to place equal importance on designing products that accommodate disassembly, reuse, and recycling, in addition to the product's appeal and functionality.

3.3 LITERATURE REVIEW

Graedel and Allenby [3] suggest that a product's design has the highest influence on the product's lifecycle, with design being the first priority toward the "greening" of products; in order to assess the environmental impact of a product, their concept of *green design* requires considering each stage of a given product's lifecycle.

Ishii et al. [9] developed a methodology to design a product for retirement using a hierarchical semantic network that consists of components and subassemblies. Navin-Chandra [17] presented an evaluation methodology for design for disassembly and developed software that optimizes the component recovery plan. Subramani and Dewhurst [18] investigated procedures to assess service difficulties and their associated costs at the product design stage. Isaacs and Gupta [8] developed an evaluation methodology that enables an automobile designer to measure disassembly and recycling potential for different designs using goal programming to analyze the trade-off between the profit levels of disassembling and of shredding. Remanufacturing models are visited by Ilgin and Gupta [7]. Johnson and Wang [10] used a disassembly tree in designing products to enhance material recovery opportunities. Vujosevic et al. [20] studied the design of products that can be easily disassembled for maintenance. Brennan et al. [1] and Gupta and Taleb [5] investigated the problems associated with disassembly planning and scheduling. Torres et al. [19] reported a study for nondestructive automatic disassembly of personal computers. The literature also includes several thorough surveys of existing research [4,6,12] as well as comprehensive reviews of green supply chains [21] and disassembly [13,15].

3.4 DESIGN METRICS FOR END-OF-LIFE PROCESSING

3.4.1 DISASSEMBLY MODEL INTRODUCTION

The specific design metrics proposed in this chapter seek to measure the following five objectives:

1. Minimize the number of disassembly workstations and hence minimize the total idle time
2. Ensure the idle times at each workstation are similar (i.e., balance the line)
3. Remove hazardous components early in the disassembly sequence (to prevent damage to, or contamination of, other components)
4. Remove high-demand components before low-demand components
5. Minimize the number of direction (i.e., the product's or subassembly's orientation) changes required for disassembly

A major constraint is the requirement to provide a *feasible* disassembly sequence for the product being investigated. Solutions consist of an ordered sequence (i.e., n-tuple, where n represents the number of parts—including *virtual parts*, i.e., tasks—for removal) of elements. For example, if a disassembly solution consisted of the eight-tuple $\langle 5, 2, 8, 1, 4, 7, 6, 3 \rangle$, then component 5 would be removed first, followed by component 2, then component 8, and so on.

While different researchers use a variety of definitions for the term "balanced" in reference to assembly [2] and disassembly lines, we adopt the following definition [14,16] that considers the total number of workstations NWS and the station times ST_j (i.e., the total processing time requirement in workstation j) on a paced line.

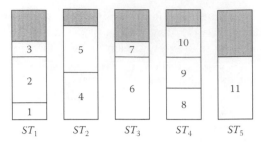

FIGURE 3.1   Line balancing depiction.

Definition 3.1

A paced line is optimally balanced when the fewest possible number of workstations is needed and the variation in idle times between all workstations is minimized, while observing all constraints. This is mathematically described by

$$\text{Minimize } NWS$$

then

$$\text{Minimize } [\max(ST_x) - \min(ST_y)] \quad \forall x, y \in \{1, 2, \ldots, NWS\}$$

Line balancing can be visualized as in Figure 3.1, with the five large boxes representing workstations where the total height of the boxes indicates the cycle time CT (the maximum time available at each workstation). The smaller numbered boxes represent each part (1 through 11 in this example), with each being proportionate in height to its part-removal time, and the gray area being indicative of the idle time.

3.4.2   Metrics

Five design-for-disassembly metrics corresponding to the five objectives detailed at the beginning of this section are developed to quantitatively describe disassembly-related objective functions and performance measures.

The first design-metric is a count of the number of workstations and is obtained by observation once a part-removal sequence is generated. The following provides the formulation of relevant relationships and of the theoretical bounds.

Theorem 3.1

Let PRT_k be the part-removal time for the kth of n parts where CT is the maximum amount of time available to complete all tasks assigned to each workstation.

Then for the most efficient distribution of tasks, the optimal minimum number of workstations NWS^* satisfies

$$NWS^* \geq \left\lceil \frac{\sum_{k=1}^{n} PRT_k}{CT} \right\rceil = NWS_{lower} \qquad (3.1)$$

where NWS_{lower} indicates the theoretical lower bound on the number of workstations.

Proof: If the above inequality is not satisfied, then there must be at least one workstation completing tasks requiring more than CT of time, which is a contradiction.

Subsequent bounds are shown to be true in a similar fashion and are presented without proof. The theoretical upper bound for the number of workstations NWS_{upper} is given by

$$NWS_{upper} = n \qquad (3.2)$$

The balancing metric used here seeks to simultaneously recognize a minimum number of workstations while measuring whether or not idle times at each workstation are similar, though at the expense of the generation of a nonlinear objective function [14,16]. A resulting minimal numerical value is indicative of a more desirable solution, providing both a minimum number of workstations and similar idle times across all workstations.

The balance design-metric F is given by

$$F = \sum_{j=1}^{NWS} (CT - ST_j)^2 \qquad (3.3)$$

The lower bound on F is given by F_{lower} (i.e., simply the square—per Equation 3.3—of the total idle time at the theoretical lower number of workstations divided by the number of workstations; this squared idle time at each workstation is then multiplied by the total number of workstations) and is related to the optimal balance F^* by

$$F^* \geq F_{lower} = \left(\frac{(NWS_{lower} \cdot CT) - \sum_{k=1}^{n} PRT_k}{NWS_{lower}} \right)^2 \cdot NWS_{lower}$$

which reduces to

$$F^* \geq F_{lower} = \frac{\left(NWS_{lower} \cdot CT - \sum_{k=1}^{n} PRT_k \right)^2}{NWS_{lower}} \qquad (3.4)$$

while the upper bound is described by the worst-case balance F_{upper} as

$$F_{upper} = \sum_{k=1}^{n} (CT - PRT_k)^2 \tag{3.5}$$

Note that, in order to make any balance results comparable in magnitude to all subsequent metrics, the effects of squaring portions of Equation 3.3 can be normalized by taking the square root of the final balance metric calculated. For example, solutions having an equal number of workstations (e.g., $NWS = 3$) but differing idle times at each workstation (I_j), resulting in differing balance such as $I_j = \langle 1, 1, 4 \rangle$ and $I_j = \langle 2, 2, 2 \rangle$ (the latter is optimal), would have balance values of 18 and 12, respectively, while the normalized values would stand at 4.24 and 3.46, still not only indicating better balance with the latter solution but also giving a sense of the relative improvement that solution provides, which the metric generated by Equation 3.3 lacks.

A hazard metric H quantifies a design's solution-sequence's performance, with a lower calculated value being more desirable. This metric is based on binary variables that indicate whether a part is considered to contain hazardous material (the binary variable is equal to one if the part is hazardous, else zero) and its position in the sequence. A given design's solution-sequence hazard metric is defined as the sum of hazard binary flags multiplied by their position in the solution-sequence, thereby rewarding the removal of hazardous parts early in the part-removal sequence.

The hazardous-part design-metric is determined using

$$H = \sum_{k=1}^{n} (k \cdot h_{PS_k}), \quad h_{PS_k} = \begin{cases} 1, & \text{hazardous} \\ 0, & \text{otherwise} \end{cases} \tag{3.6}$$

where PS_k identifies the kth part in the solution-sequence PS; i.e., for solution $\langle 3, 1, 2 \rangle$, $PS_2 = 1$. The lower bound on H is given by H_{lower} and is related to the optimal hazard metric H^* by

$$H^* \geq H_{lower} = \sum_{p=1}^{|HP|} p, \quad |HP| = \sum_{k=1}^{n} h_k \tag{3.7}$$

where the set of hazardous parts $HP = \{k: h_k \neq 0 \ \forall k \in P\}$ and where P is the set of n part-removal tasks. For example, a product with three hazardous parts would give an H_{lower} value of $1 + 2 + 3 = 6$. The upper bound on the hazardous-part metric is given by

$$H_{upper} = \sum_{p=n-|HP|+1}^{n} p \tag{3.8}$$

For example, three hazardous parts in a product having a total of twenty would give an H_{upper} value of $18 + 19 + 20 = 57$.

A demand metric D quantifies a design's solution-sequence's performance, with a lower calculated value being more desirable. This metric is based on positive integer values that indicate the quantity required of this part after it is removed—or zero if it is not desired—and its position in the sequence. Any given solution-sequence's demand metric is defined as the sum of each demand value multiplied by its position in the sequence, rewarding the removal of high-demand parts early in the part-removal sequence.

The demand design-metric is calculated using

$$D = \sum_{k=1}^{n} (k \cdot d_{PS_k}), \quad d_{PS_k} \in N, \quad \forall PS_k \tag{3.9}$$

where N represents set of natural numbers, i.e., $\{0, 1, 2, \ldots\}$. The lower bound on the demand metric ($D_{lower} \le D^*$) is given by Equation 3.9 where

$$d_{PS_1} \ge d_{PS_2} \ge \cdots \ge d_{PS_n} \tag{3.10}$$

For example, three parts with demands of 4, 5, and 6, respectively, would give a best-case value of $(1 \cdot 6) + (2 \cdot 5) + (3 \cdot 4) = 28$. The upper bound on the demand metric (D_{upper}) is given by Equation 3.9 where

$$d_{PS_1} \le d_{PS_2} \le \cdots \le d_{PS_n} \tag{3.11}$$

For example, three parts with demands of 4, 5, and 6, respectively, would give a worst-case value of $(1 \cdot 4) + (2 \cdot 5) + (3 \cdot 6) = 32$.

Finally, a direction metric R is developed, with a lower calculated value indicating minimal direction changes in the product's (or subassembly's) orientation during disassembly and, therefore, a more desirable solution. This metric is based on a count of the direction changes. Integer values represent each possible direction (typically $r \in \{+x, -x, +y, -y, +z, -z\}$; in this case $|r| = 6$). These directions are easily expanded to other or different directions in a similar manner.

The direction design-metric is formulated as

$$R = \sum_{k=1}^{n-1} R_k, \quad R_k = \begin{cases} 1, & r_{PS_k} \neq r_{PS_{k+1}} \\ 0, & \text{otherwise} \end{cases} \tag{3.12}$$

The lower bound on the direction metric R is given by R_{lower} and is related to the optimal direction metric R^* by

$$R^* \ge R_{lower} = |r| - 1 \tag{3.13}$$

For example, for a given product containing six parts that are installed/removed in directions $r_k = (-y, +x, -y, -y, +x, +x)$, the resulting best-case value would be $2 - 1 = 1$ (e.g., one possible R_{lower} solution containing the optimal, single-change of product direction would be: $\langle -y, -y, -y, +x, +x, +x \rangle$). In the specific case where the number of unique direction changes is one less than the total number of parts n, the upper bound on the direction metric would be given by

$$R_{upper} = |r|, \quad \text{where } |r| = n - 1 \tag{3.14}$$

Otherwise, the metric varies depending on the number of parts having a given removal direction and the total number of removal directions. It is bounded by

$$|r| \leq R_{upper} \leq n - 1, \quad \text{where } |r| < n - 1 \tag{3.15}$$

For example, six parts installed/removed in directions $r_k = (+x, +x, +x, -y, +x, +x)$ would give an R_{upper} value of 2 as given by the lower bound of Equation 3.15 with a solution-sequence of $\langle +x, +x, -y, +x, +x, +x \rangle$. Six parts installed/removed in directions $r_k = (-y, +x, -y, -y, +x, +x)$ would give an R_{upper} value of $6 - 1 = 5$ as given by the upper bound of Equation 3.15 with a solution-sequence of $\langle -y, +x, -y, +x, -y, +x \rangle$, for example.

In the special case where each part has a unique removal direction, the metrics for R_{lower} and R_{upper} are equal and are given by

$$R_{lower} = R_{upper} = n - 1, \quad \text{where } |r| = n \tag{3.16}$$

The new-product design metrics are therefore given as NWS, F, H, D, and R, where NWS is readily observed from a given sequence while F, H, D, and R are calculated using a given disassembly sequence and Equations 3.3, 3.6, 3.9, and 3.12, respectively.

3.4.3 Metrics as Prototypes

The H, D, and R metrics are also intended as forming the three basic prototypes of any additional end-of-life processing design evaluation criteria. These three different models are then the basis for developing differing or additional metrics. The H metric can be used as the prototype for any binary criteria; for example, a part could be listed according to the categories "valuable" and "not valuable." The D metric can be used as the prototype for any known value (integer or real) criteria; for example, a part can be assigned a D-type metric which contains the part's actual dollar value. The R metric can be used as the prototype for any adjacency or grouping criteria; for example, a part could be categorized as "glass," "metal," or "plastic" if it were desirable to remove parts together in this form of grouping.

3.4.4 Additional Metrics

The primary mathematical evaluation tool developed for comparative quantitative analysis of designs is shown in Equation 3.17 and subsequently referred to as the *efficacy index EI* [14]. The efficacy index is the ratio of the difference between a calculated metric x and its worst-case value x_{worst} to the metric's sample range (i.e., the difference between the best-case metric value x_{best} and the worst-case metric value as given by $\max(X_y) - \min(X_z) \mid y, z \in \{1, 2, \ldots, |X|\}$ from the area of statistical quality control) expressed as a percentage and described by

$$EI_x = \frac{100 \cdot |x_{worst} - x|}{|x_{worst} - x_{best}|}, \quad \text{where } x_{best} \neq x_{worst} \tag{3.17}$$

(with the vertical lines in Equation 3.17 representing absolute value versus cardinality as seen elsewhere in this chapter—while not necessary for the calculations performed in this chapter, use of absolute value provides a more general formulation that allows for application to any future metrics that may make use of values where, unlike each of the metrics developed here, the upper bound indicates the best case). This generates a value between 0% and 100%, indicating the percentage of optimum for any given metric and any given design being evaluated. The caveat that x_{best} should not be equal to x_{worst} protects from a divide by zero; if x_{best} is equal to x_{worst}, the value of 100% would be used by default.

Finally, it should be noted that the values generated using Equation 3.17 can also be calculated using the best-case and worst-case design options instead of the theoretical bounds given by Equations 3.1 through 3.16; this additional type of analysis is demonstrated in Table 3.4 in Section 3.5.3.

3.5   CASE STUDY

3.5.1   PRODUCT DATA

Kongar and Gupta [11] provided the basis for the case study's data. Their instance consists of the data for the disassembly of a notional consumer electronics product, where the objective is to completely disassemble an item that consists of $n = 10$ components and several precedence relationships (e.g., parts 5 and 6 need to be removed prior to part 7) on a paced disassembly line operating at a speed which allows a cycle time of 40 s for each workstation to perform its required disassembly tasks. A slightly modified version of the original instance is seen in Table 3.1.

We consider a simple extension of the Table 3.1 data to clearly illustrate application of the metrics. Using, for example, the assumption that parts 4 and 7 have the same footprints and are completely interchangeable, along with the additional assumption that, alternatively, parts 5 and 8 have the same footprints and are completely interchangeable

---

TABLE 3.1

Knowledge Base of the Consumer Electronics Product Instance

Task	Time	Hazardous	Demand	Direction	Predecessors
1	14	No	No	$+y$	n/a
2	10	No	500	$+x$	1, 8, 9, 10
3	12	No	No	$+x$	1, 8, 9, 10
4	17	No	No	$+y$	n/a
5	23	No	No	$-z$	n/a
6	14	No	750	$-z$	n/a
7	19	Yes	295	$+y$	5, 6
8	36	No	No	$-x$	4, 7
9	14	No	360	$-z$	n/a
10	10	No	No	$-y$	n/a

(as a result, only the precedence is ultimately affected in this example; i.e., the parts still possess their same part-removal times, hazardous-part designations, demand amounts, and removal orientation directions—only their location in the product is changed), three design versions of this product are considered: A, B, and C. Design A is reflected in Table 3.1, design B swaps parts 4 and 7, while design C swaps parts 5 and 8.

3.5.2 NUMERICAL ANALYSIS

Because determining an optimal disassembly sequence is NP-complete [14], a heuristic is applied. The heuristic used here is a greedy search algorithm tailored to the disassembly line balancing problem (DLBP) [14]. A greedy strategy always makes the choice that looks best at the moment. That is, it makes a locally optimal choice in the hope that this choice will lead to a globally optimal solution. The DLBP greedy algorithm was built around first-fit-decreasing (FFD) rules. FFD rules require looking at each element in a list, from largest to smallest and putting that element into the first bin in which it fits.

The DLBP greedy algorithm first sorts the parts. Hazardous parts are put at the front of the list, ranked from largest-to-smallest part-removal times. The same is then done for the nonhazardous parts. Any ties (i.e., two parts with equal hazard typing and equal part-removal times) are not randomly broken, but rather ordered based on the demand for the part, with the higher demand part being placed earlier on the list. Any of these parts also having equal demands is then selected based on their part-removal direction being the same as the previous part on the list (i.e., two parts compared during the sorting that only differ in part-removal direction are swapped if they need to be removed in different directions—the hope being that subsequent parts may have matching part-removal directions).

The DLBP greedy algorithm then places the parts in FFD order while preserving precedence. Each part in the greedy-sorted list is examined from first to last. If the part had not previously been put into the solution-sequence, the part is put into the current workstation if idle time remains to accommodate it and as long as putting it into the sequence at that position will not violate any of its precedence constraints. If the current workstation cannot accommodate it at the given time in the search due to precedence constraints, the part is maintained on the sorted list and the next part on the sorted list is considered. If all parts have been examined for insertion into the current workstation on the greedy solution list, a new workstation is created and the process is repeated, starting from the first part on the greedy-sorted list. Finally, whenever a part is successfully placed in a workstation, the algorithm also returns to the first part on the greedy-sorted list. This process repeats until all parts have been placed.

3.5.3 RESULTS

The greedy algorithm generated the sequence $\langle 5, 4, 6, 7, 8, 9, 1, 10, 3, 2 \rangle$ for design A, $\langle 7, 6, 5, 4, 8, 9, 1, 10, 3, 2 \rangle$ for design B, and $\langle 8, 4, 6, 7, 9, 5, 1, 10, 3, 2 \rangle$ for design C with the associated metrics shown in Table 3.2 (note that the best values from each of the three alternative designs are depicted in bold).

The bound formulations are then used to calculate the case study's upper and lower theoretical bounds as seen in Table 3.3. These values are used along with the

TABLE 3.2

Consumer Electronics Product
Instance Metrics for the Three
Design Alternatives

Case	NWS	$\sqrt{F}$	H	D	R
A	5	19.82	4	10,590	8
B	5	19.82	1	**8,955**	7
C	5	**14.80**	4	10,230	7

TABLE 3.3

Upper and Lower Metric Bounds
for the Consumer Electronics
Product Instance

Bound	NWS	$\sqrt{F}$	H	D	R
Upper	10	76.73	10	16,945	10
Lower	5	13.86	1	4,010	4

values in Table 3.2 to provide the efficacy indices for each design when compared to the best and worst values (provided by the alternative designs), as well as when compared to the theoretical bounds (Table 3.4).

Table 3.2 readily depicts many of the benefits of application of these metrics. The original design (i.e., design A) can be seen to be the best choice only in terms of number of workstations: trivial here as all three options have the same (optimal) number of workstations. Design B can be seen to possess the largest number of best measures; however, design C possesses the desirable trait of having the best balance. Alternatively, the quantitative nature of these metrics is such that a decision maker is not limited to choosing the best alternative, but may see that the benefits of designs B and C in balance, hazardous-part removal, demand, and part-removal direction are

TABLE 3.4

Efficacy Index Metrics Calculated Using the Best and Worst
of the Three Alternatives (Center Column) and the Upper
and Lower Theoretical Bounds (Right Column)

Case	E_{NWS} (%)	$E_{\sqrt{F}}$ (%)	E_H (%)	E_D (%)	E_R (%)	E_{NWS} (%)	$E_{\sqrt{F}}$ (%)	E_H (%)	E_D (%)	E_R (%)
A	100	0	0	0	0	100	90.5	66.7	49.1	33.3
B	100	0	100	100	100	100	90.5	100	61.8	50.0
C	100	100	0	22.0	100	100	98.5	66.7	51.9	50.0

not quantitatively significant enough to warrant selection of these designs over the original design due to some other, less quantitative reason (e.g., aesthetics). Table 3.3 provides this range for the decision maker, culminating in the results depicted in Table 3.4 where each design is positioned in all metric areas as compared to the theoretical best and worst case (rightmost column) as well as the best and worst case in the three (in this example) design options (center column).

Metrics can also be addressed individually. If minimizing the number of workstations is the priority, then any design is equally acceptable. If balancing the workstations is the goal, design C is the preferred option. In the case of removing the hazardous part as quickly as possible and/or removing demanded parts as early as possible, design B is the preference. Where minimizing the number of part-removal direction changes encountered is essential, designs B and C are equally adequate.

While this small example with minimal alternatives (i.e., only differing in the precedence of two parts) is used here to clearly illustrate use of the metrics, products with a greater number of parts, the use of additional metrics (using those described here as prototypes), or a larger number of design options—all of which could be expected in real-world applications—will provide a wider range of metric values, enabling designers to quantitatively measure a variety of end-of-life parameters prior to decision makers committing to a final new-product design that will eventually become part of the reverse supply chain.

3.6   CONCLUSIONS

Application of a thoughtfully designed reverse supply chain is becoming more prevalent for various combinations of regulatory, consumer-driven, and financial reasons. A key component to its success is the efficient disassembly of end-of-life products. Designing products with the expectation of end-of-life disassembly can lead to efficiencies that can minimize future costs and potentially increase future profits. Rather than take an intuitive (e.g., use of a subject matter expert) or qualitative approach to design-for-disassembly, metrics can provide compelling data for the selection of one design over another. This is especially useful when there are multiple and equivalent assembly design options that, due to the multi-criteria nature of disassembly, are not equivalent in terms of disassembly. The metrics proposed here also provide a measure of goodness, showing not only that one design is more efficient during disassembly than another, but in what areas of interest and by how much. This allows a design decision maker to elect trade-offs where one design may be quantitatively preferable, but not by a significant enough margin to justify some other consideration.

REFERENCES

1. Brennan, L., Gupta, S. M., and Taleb, K. N. 1994. Operations planning issues in an assembly/disassembly environment. *International Journal of Operations and Production Management* 14(9): 57–67.
2. Elsayed, E. A. and Boucher, T. O. 1994. *Analysis and Control of Production Systems.* Upper Saddle River, NJ: Prentice Hall.
3. Graedel, T. E. and Allenby, B. R. 1995. *Industrial Ecology.* Englewood Cliffs, NJ: Prentice Hall.

4. Gungor, A. and Gupta, S. M. 1999. Issues in environmentally conscious manufacturing and product recovery: A survey. *Computers and Industrial Engineering* 36(4): 811–853.
5. Gupta, S. M. and Taleb, K. 1994. Scheduling disassembly. *International Journal of Production Research* 32(8): 1857–1866.
6. Ilgin, M. A. and Gupta, S. M. 2010. Environmentally conscious manufacturing and product recovery (ECMPRO): A review of the state of the art. *Journal of Environmental Management* 91(3): 563–591.
7. Ilgin, M. A. and Gupta, S. M. 2012. *Remanufacturing Modeling and Analysis*. Boca Raton, FL: CRC Press.
8. Isaacs, J. A. and Gupta, S. M. 1997. A decision tool to assess the impact of automobile design on disposal strategies. *Journal of Industrial Ecology* 1(4): 19–33.
9. Ishii, K., Eubanks, C. F., and Marco, P. D. 1994. Design for product retirement and material life-cycle. *Materials & Design* 15(4): 225–233.
10. Johnson, M. R. and Wang, M. H. 1995. Planning product disassembly for material recovery opportunities. *International Journal of Production Research* 33(11): 3119–3142.
11. Kongar, E. and Gupta, S. M. 2002. A genetic algorithm for disassembly process planning. *SPIE International Conference on Environmentally Conscious Manufacturing II*, Newton, MA, Vol. 4569, pp. 54–62.
12. Lambert, A. J. D. 2003. Disassembly sequencing: A survey. *International Journal of Production Research* 41(16): 3721–3759.
13. Lambert, A. J. D. and Gupta, S. M. 2005. *Disassembly Modeling for Assembly, Maintenance, Reuse, and Recycling*. Boca Raton, FL: CRC Press.
14. McGovern, S. M. and Gupta, S. M. 2007. Combinatorial optimization analysis of the unary NP-complete disassembly line balancing problem. *International Journal of Production Research* 45(18–19): 4485–4511.
15. McGovern, S. M. and Gupta, S. M. 2011. *The Disassembly Line: Balancing and Modeling*. New York: McGraw-Hill.
16. McGovern, S. M., Gupta, S. M., and Kamarthi, S. V. 2003. Solving disassembly sequence planning problems using combinatorial optimization. *Northeast Decision Sciences Institute Conference*, Providence, RI, pp. 178–180.
17. Navin-Chandra, D. 1994. The recovery problem in product design. *Journal of Engineering Design* 5(1): 65–86.
18. Subramani, A. K. and Dewhurst, P. 1991. Automatic generation of product disassembly sequence. *Annals of the CIRP* 40(1): 115–118.
19. Torres, F., Gil, P., Puente, S. T., Pomares, J., and Aracil, R. 2004. Automatic PC disassembly for component recovery. *International Journal of Advanced Manufacturing Technology* 23(1–2): 39–46.
20. Vujosevic, R., Raskar, T., Yetukuri, N. V., Jothishankar, M. C., and Juang, S. H. 1995. Simulation, animation, and analysis of design assembly for maintainability analysis. *International Journal of Production Research* 33(11): 2999–3022.
21. Wang, H.-F. and Gupta, S. M. 2011. *Green Supply Chain Management: Product Life Cycle Approach*. New York: McGraw Hill.

4 Application of Theory of Constraints' Thinking Processes in a Reverse Logistics Process

Hilmi Yüksel

CONTENTS

4.1 INTRODUCTION

The importance of reverse logistics increases rapidly with the growth of environmental problems. Environmental laws force companies to take responsibility for their products that reach the end of their life cycles. The life spans of the products, especially electronic products, shorten and the products come to the end of their life cycles very rapidly. Besides, an inefficient reverse logistics can be very costly for the company. These recent developments highlight the companies to address concern about their reverse logistics activities. Nowadays, reverse logistics can be a way to sustain the competitiveness of the firms.

The theory of constraints (TOC), which was developed by Goldratt, is a management philosophy that focuses on continuous improvement. The TOC provides the methods and tools to determine and eliminate the constraints that hinder achievement of the goals of the organizations. The first applications of the TOC were implemented in manufacturing firms. Following the development of the TOC thinking processes by Goldratt, they have also been used in service firms.

The TOC thinking processes are used to determine the core problems in manufacturing and service firms and to improve the processes by eliminating these core problems. The questions "what to change?", "what to change to?", and "how to cause the change?" form the framework of the TOC thinking processes.

In this chapter, the TOC thinking processes have been used in order to determine the core problems in the reverse logistics of a firm in the electronics sector. The core problems in reverse logistics have been determined, and the solutions that address the core problems have been developed.

4.2   THEORY OF CONSTRAINTS

The TOC developed by Goldratt is a management philosophy that focuses on continuous improvement process. TOC postulates that there is at least one constraint in every organization that hinders the organization from achieving its goal. The capacity of the organizations to perform is limited within these constraints. Without any physical constraint, if an organization could produce more than it can sell, then the market for the product would itself be the constraint (Cox and Spencer 1998).

The TOC is extremely powerful in forcing us to precisely first determine and define and only then eliminate these constraints, and thereby improve the performance of the system. TOC states that there is always a constraint in the system being analyzed and emphasizes the fact that after eliminating one system constraint other constraints invariably come up. TOC assures continuous improvement by stating that all the constraints must be eliminated.

TOC is a science of management. It applies the methods of science—specifically, the methods of physics—to the general problem of managing in life (McMullen 1998). TOC has an important role for organizations to identify throughput problems, serve as a guide to correct the throughput problems, and to generate considerable improvements in productivity and efficiency (Pegels and Watrous 2005).

TOC has two major components. One of the components is a philosophy that underpins the working principles of TOC and it consists of five steps of ongoing improvement, the drum-buffer-rope scheduling methodology, and the buffer management information system. The second component of TOC, an approach known as thinking processes, is used to search, analyze, and solve the complex problems (Rahman 2002). One of the main assumptions of TOC theory is that every business has the primary goal of "making more money now as well as in the future" without violating certain necessary conditions (Gupta 2003). Related to this goal, TOC prescribes new performance measurements that are quite different from the traditional cost-accounting system. To measure an organization's performance in achieving this goal, two sets of measurements have been prescribed by Goldratt and Cox (1992): global (financial) measurements and operational measurement (Rahman 1998).

The firms can gain important achievements in their performance by applying philosophy and tools of TOC. Mabin and Balderstone (2003) examined the results of TOC applications in the literature. According to this research, 80% mentioned improvements in lead times, cycle times, DDP, and/or inventory, and, of these, over 40% also mentioned improvements in financial performance.

4.3   THEORY OF CONSTRAINTS' THINKING PROCESSES

According to TOC, the organizations are viewed as chains and the production rates of the organizations are controlled by the weakest link of those chains. In order to maximize the production of the organizations, this weakest link must be identified and improved upon to the point where it is no longer the link that limits the chain. The weakest link is the constraint and it is not visible at every case. The approach of thinking processes has been developed by Goldratt to determine the constraints of the systems. The thinking processes present logical tools to provide a road map that gives answers to the questions "what to change?", "what to change to?", and "how to cause the change?" (Goldratt and Cox 1992).

Goldratt's thinking processes are the set of logical tools used solely or used interconnected based on causal relationships (Cox and Spencer 1998). TOC thinking processes allow the decision makers to identify the core problem, to identify and test win–win solutions before the implementation, and to create implementation plans (Walker and Cox 2006). TOC thinking processes and diagrams can be used for strategic planning, policy formulation, process management, project management, day-to-day problem solving, and day-to-day management (McMullen 1998). In a logical extension of the thinking process application tools, several authors have also begun to experiment with the use of the tools for analyzing and formulating strategy (Watson et al. 2007).

The strength of TOC comes from its understanding of cause and effect relationships better than the other strategic planning models (Dettmer 2003). One of the important benefits of TOC thinking processes is its support in determining the barriers that must be eliminated, in addition to defining the problem, introducing the solution, and implementing the solution.

The thinking processes start with the symptoms and ends with a plan that shows the activities required in the application of the solution of the problem. The thinking processes provide five tools organized as cause–effect diagrams. The thinking processes start with the question "what to change?" in the aim of identifying the root problem. In order to determine the current situation of the system, current reality tree (CRT) is used. After determining the root problem, it attempts to find the answer to the question of "what to change to?" At this stage the evaporating cloud (EC) is used and searches for a solution to the problem. By applying EC it is expected that the system will improve according to the changes determined. The last question is "how to cause the change?" At this stage the future reality tree (FRT), the prerequisite tree (PRT), and the transition tree (TT) are used. FRT is a strategic tool to plan the changes and PRT and TT are used for determining the obstacles in the application of the solution of the problem, and for presenting plans in order to eliminate these obstacles. According to the literature review (Kim et al. 2008), EC and CRT are the most used tools of TOC thinking processes.

Rahman (2002) used the TOC thinking processes to identify the success factors of supply chains and to determine the relationships between these factors and develop the supply chain strategies according to the cause and effect relationships. Scoggin et al. (2003) applied the TOC thinking process logic tools in a manufacturing firm which aims to design and produce the quality and quantity of generators and schedule now and in the future. Chaudhari and Mukhopadhyay (2003) demonstrated how the TOC thinking processes can be used to identify and overcome policy constraints in the leading integrated poultry business. Reid and Cormier (2003) illustrated the application of the TOC thinking processes in a service firm to guide and structure the managerial analysis of the first two stages of the change sequence. Choe and Herman (2004) applied the TOC thinking processes to Euripa labs, a research organization, to determine the core problems and to address solutions for the core problems. Shoemaker and Reid (2005) used the TOC thinking processes in a public sector service organization. Taylor et al. (2006) examined the factors that affect employee retention and turnover for the metropolitan police and fire departments and determined the solutions to decrease the loss of public safety personnel at the city under scrutiny with the TOC thinking processes. Walker and Cox (2006) applied CRT to a white-collar working environment as an example of ill-structured problems that had no solution strategy because of limitations of problem-solving tools.

There is a criticism regarding the reliability of the tools due to their reliance on subjective interpretation of perceived reality and the qualitative nature of the subject matter (Watson et al. 2007). According to the literature review (Kim et al. 2008), there are many studies that are essentially descriptive in nature; therefore, further empirical studies would be valuable in order to verify the usefulness of the TOC thinking processes in implementation, and to show the effectiveness of the TOC thinking processes quantitatively (Kim et al. 2008).

4.3.1 What to Change?

The CRT is a cause–effect logic diagram that has been developed by Goldratt (1994) and is designed to help identify the system constraints, root causes, or core problems responsible for a significant majority of the undesirable effects (UDEs) (Scoggin et al. 2003). The purpose of the CRT is to make the connections between a current situation's many symptoms, facts, root causes, and core problems explicitly clear to everyone (McMullen 1998). If the symptoms of a core problem are UDEs, the UDEs are merely symptoms brought on by the core problem itself (Taylor et al. 2006). UDEs reflect poor system performance and are symptomatic of underlying systemic problems (Scoggin et al. 2003).

According to Goldratt, the first stage in TOC thinking processes is determining at least 10–12 UDEs applied to the current problem. The construction of CRT starts with the observed UDEs. All the UDEs are connected according to cause–effect logic until the core problem is determined. The CRT is constructed from top-down by defining the UDEs and presenting probable causes to these effects. However, it is read from bottom-up. After the construction of the CRT, the UDEs can be distinguished at the bottom of the CRT. These UDEs are the core problems of the system.

When the CRT is constructed, it must be analyzed with the aim of determining the core problems. A core problem should be connected to at least 70% of the UDEs (Cox and Spencer 1998). The CRT is a particularly effective tool if the constraint is a policy, as opposed to the physical limitation of the existing system (Kim et al. 2008).

4.3.2 WHAT TO CHANGE TO?

After defining the core problem, the solution can be sought. EC is used to eliminate the core problems. EC aids the decision makers in identifying breakthrough actions that can resolve the problems by underpinning assumptions, in addition to further explaining the dilemma (Mabin et al. 2006). EC verbalizes the inherent conflict, clarifies the assumptions, and provides a mechanism to come up with the ideas, which can be used to resolve the core problem (Gupta 2003). The strength of EC is the focus on the system problem instead of the local problems, and so it can be possible to improve the system's performance according to the desired goal.

The EC starts with an objective that is the opposite of the core problem. From the objective, the requirements are listed. Each requirement will have at least one prerequisite. It is the prerequisite that depicts the conflict. All the requirements and prerequisites are based on the assumptions that keep the people in the conflicted environment (Taylor et al. 2006). EC helps the decision maker search for a solution by challenging the assumptions underlying this conflict (Choe and Herman 2004).

4.3.3 HOW TO CAUSE THE CHANGE?

When the EC is broken, the FRT is built using the injections from the EC. FRT shows that once the injections are implemented, the desirable effects can be accomplished, and assures that all the UDEs would be eliminated using the resolution identified in the EC (Taylor et al. 2006).

The main aim of FRT is to logically research the efficiency of the new ideas and injections before they are applied. FRT helps illustrate that desired effects will take the place of undesired effects with the proposed changes, and it assures that all the undesired effects can be eliminated with the injections defined by EC. If the injections are not sufficient to eliminate the symptoms, or the injections cause new negative results, then the solution is readjusted. The solution process continues until the elimination of the original symptoms without causing new UDEs.

FRT is read from bottom-up using "if ... then" format. FRT shows that the proposed interventions should logically produce a more desirable system future state by eliminating many of its current problems, while the negative branch reservation (NBR) shows some of the potential or unintended negative consequences associated with the planned interventions (Reid and Cormier 2003). NBR is developed to determine the negative effects if the injections could not be evaluated and managed carefully (Scoggin et al. 2003).

The PRT identifies obstacles for the implementation of new ideas and determines intermediate objectives to overcome the obstacles (Gupta 2003). PRT focuses on defining the critical factors and obstacles that hinder achieving the goal. Dettmer (1997) suggests asking two questions to determine whether PRT is needed or not:

- Is the objective a complex condition? If so, a PRT may be needed to sequence the intermediate steps to achieve it.
- Do I already know how to achieve it? If not, then a PRT will help map out the possible obstacles, the steps involved in overcoming them, and the appropriate sequence.

TT identifies the activities required to apply the solution. It helps the decision maker to structure the details of the activity plan with the effect–cause–effect logic and to examine it comprehensively. The goal of TT is to implement the change by providing the implementations of injections developed by EC and FTR. It is an operational or tactical tool. It provides tactical activity plans for strategic plans.

4.4   REVERSE LOGISTICS

Reverse logistics is a process of moving goods from their typical final destination for the purpose of capturing value or proper disposal (Rogers and Tibben 1998). Reverse logistics involves all the activities associated with the collection and either recovery or disposal of used products (Ilgın and Gupta 2010). In addition to comprising the activities of planning, implementing and controlling the inbound flow, inspection and disposition of returned products, reverse logistics also deals with the related information for the purpose of recovering the value (Srivastava 2008). The products can be returned within a supply chain related to different reasons that are (1) rework, (2) commercial returns and outdated products, (3) product recalls, (4) warranty returns, (5) repairs, (6) end of use returns, and (7) end of life returns (Dekker and Vander Laan 2003). Although each type of return requires a reverse logistics appropriate to the characteristics of the returned products to optimize value recovery (Guide et al. 2003), collection, grading, reprocessing, and redistribution are the four main activities of all reverse logistics (Fleischmann 2003).

Reverse logistics activities become more important with the shortening of the products' lifecycles and with the growth of concerns for environmental problems. Environmental legislations force firms to take the responsibility for their products that have come to their end of life cycles. Meanwhile, original equipment manufacturers should add value to the used products. Otherwise, there would be no incentive to design a reverse logistics system (Mutha and Pokharel 2008). Economics is seen as the driving force to reverse logistics relating to all recovery options, where the company receives both direct as well as indirect economic benefits. A reverse logistics program can bring cost benefits to the companies by emphasizing resource reduction, adding value from the recovery of products, or from reducing the disposal costs (Ravi et al. 2005). The value generated by the reverse logistic activities may materialize either in the form of cost reductions, by substituting original forward logistics inputs, or in the form of revenue increases, by opening new markets (Fleischmann et al. 2004). Out of all the cases of reverse logistics, one of the main concerns is to assess whether or not the recovery of the used products is economically more attractive than their disposal (Srivastava 2008). The success of the reverse logistics activities depends on whether or not there is a market for the remanufactured parts and the quality of the remanufactured materials (Beamon 1999). Successfully marketing remanufactured products involves at least two major activities. The first is developing market awareness, appreciation, and acceptance. The second is supporting these marketing efforts by delivering the expectations created (Ferrer and Whybark 2000).

First of all, there must be products' return. The most obvious driver for acquiring products is their future market value (Fleischmann et al. 2004). For a prospective reverse logistics system to operate efficiently, the designing of that system must involve a selection of collection centers and recovery facilities that have sufficient success potentials (Pochampally and Gupta 2004). Appropriate reverse channel structure for the collection of the products that are at the end of their life cycle is an important factor for the success of reverse logistics (Ilgın and Gupta 2010). Deposit fee, buy back option, reduced price new, fees, and take back with or without costs for supplier are the economic incentives to stimulate the acquisition of products for recovery (Brito et al. 2003).

Grading and disposition of a process's design also has a significant impact on the performance of reverse logistics (Fleischmann et al. 2004). For the success of reverse logistics, the actions that reduce uncertainty in the timing and quantity of returns, the actions that balance return rates with demand rates, and the actions that make material recovery more predictable should be taken (Ferrer and Ketzenberg 2004). One of the biggest challenges that firms face while dealing with reverse logistics is a lack of information regarding the process (Rogers and Tibben 1998). A high degree of interaction and communication between members of reverse logistic systems leads to a higher efficiency level (Freires and Guedes 2008).

In reverse logistics, the value of returned products may decrease more rapidly than their new counterparts. Accelerating the process of reverse logistics to drive value preservation is critical (Veerakamolmal and Gupta 2001). The returned product is worth only a fraction of its initial value. The longer it waits, the more its value declines (Rogers and Tibben 2001). Therefore, it is very important to extract the value from the returned products as soon as possible. This can be achieved by designing efficient reverse logistics networks. A well-managed reverse logistic network can not only provide important cost savings in procurement, recovery, disposal inventory holding, and transportation but also help in customer retention (Srivastava 2008).

4.5　E-WASTE

E-waste encompasses a broad and growing range of electronic devices such as large household devices that have been discarded by their users (Basel Action Network). Electronic waste refers to thousands of discarded electronic devices such as computers, televisions, cell phones, and printers. Electronic waste streams are growing rapidly related to the growing sales of electronic products and shortening the life spans of electronic products.

According to the WEEE directive, e-waste is stated as including all components, subassemblies and consumables, which are part of the product at the time of discarding (EU WEEE Directive, EU 2002). E-waste contains many valuable, recoverable materials such as aluminum, ferrous metals, copper, gold, and silver. E-waste also contains toxic and hazardous waste materials including mercury, lead, cadmium, beryllium, chromium, antimony, and many other chemicals. Thus, e-waste can be a serious threat to the environment and human health. If the recycle, refurbish, and reuse ratio of the components of electronic products that are at the end of their life cycles can be increased, natural resources will be conserved and pollution of the air and water will decrease. In addition, the energy requirement for producing new electronic products will decrease.

Walther and Spengler (2005) have developed a model for the treatment of electrical and electronic wastes in Germany. This model optimizes the allocation of discarded products, disassembly activities and disassembly fractions to participants of the treatment system. Knemeyer et al. (2002) suggested a qualitative assessment to evaluate the feasibility of a reverse logistics system for computers that are at the end of their life cycles. Their study attempts to demonstrate a process for utilizing qualitative research methods to obtain in-depth information concerning the factors affecting the reverse logistics activities for computers. Ravi et al. (2005) proposed an analytic network process model for the problem of the conduct of reverse logistics for end of life computers in a hierarchical form.

4.6   APPLICATION

In this chapter, it is aimed to determine the problems of reverse logistics in electronic products. For this goal, the factors that decrease the efficiency of the reverse logistics of electronic products were determined, the proper causes of these factors were displayed, and the relationships between the causes and effects were examined.

4.6.1   What to Change?

The first question to be answered is "what to change?" The CRT has been used to answer this question. The first step of the Goldratt's thinking processes is to make a list of UDEs related to the current problem and to determine the CRT. A CRT begins with the identification of several surface problems or UDEs through interviews with the parties involved in the situation (Walker and Cox 2006). In this chapter, UDEs for the reverse logistics of electronic products were determined with interviews with a company providing services in the field of recovery of waste of electric and electronic equipment and with one of the biggest company in electronics sector in Turkey.

UDEs of reverse logistics for electronic products were determined as follows:

- Not able to recycle and remanufacture the products without giving harm to environment
- Resistance toward the activities related with reverse logistics
- Domination of the scrap sector instead of the recycling sector
- Lack of ability to collect
- Cost driver

In the process of the construction of a CRT, the focus is on the cause–effect–cause relationships of the UDEs. The main point is to understand the fact that UDEs are not the real problems, but they are the visible effects of the core problems. At this stage, cause–effect relationships that relate to the UDEs should be mapped. The factor at the bottom of this map states the core problem. After determining the cause–effect–cause relationships between the UDEs, the core problem, which can be controlled by the system, can be determined. With the effect–cause–effect relationship analysis between the UDEs the CRT has been constructed, as shown in Figure 4.1.

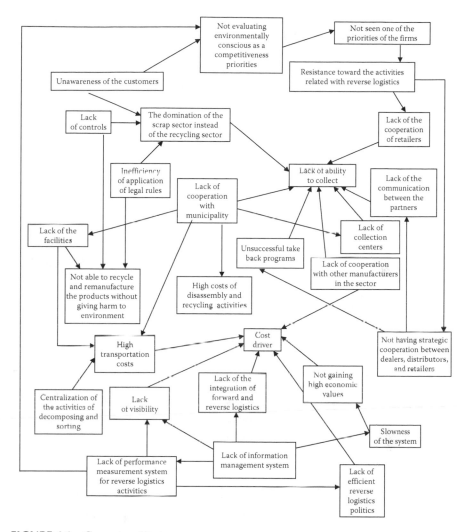

FIGURE 4.1 Current reality tree.

An entity that does not have an arrow entering means that the entity is not caused by some other entity. So these entities can be referred as root causes. The root causes should be determined according to the ability of the systems to control them. If the root causes can be controlled by the system, then they can be stated as root problems. Otherwise they are core drivers not core problems (Walker and Cox 2006).

According to the CRT, core drivers are as follows:

- Unawareness of the customers
- Lack of controls
- Lack of cooperation with municipality
- Lack of cooperation with other manufacturers in the sector
- Inefficiency application of legal rules

A core problem should be connected to at least 70% of the UDEs. According to the CRT, the core problem can be stated as

- Lack of an information management system

4.6.2 WHAT TO CHANGE TO?

The first question that should be asked is "why the root problem occurs?" There must be conflicts surrounding the root problem. After determining the root conflict, the injections can be determined to solve these conflicts. EC is used to solve the conflict. The EC starts with an opposite goal of the root problem. An example of using EC is shown Figure 4.2.

The assumptions that keep people in the conflicted environment can be stated as the following:

- Being environmentally conscious and active in reverse logistics are cost drivers and cannot be evaluated as competitiveness priorities or be used in a way to increase the profits.
- The cooperation between the partners in the reverse logistics is very difficult, the partners are not willing to share information and "win–win" thought is not possible.

The injections can be determined as follows in order to challenge these assumptions underlying the conflicts:

- Increase customer awareness for the products of the firms that are conscious of the environment and are responsible for their products that reach the end of their life cycles. Related to the increase of the customers' demands for these products, the firms can gain profits in addition to preventing pollution and minimizing environmental impacts with their investments in reverse logistics.
- Increase the support of the municipalities by providing collection centers. Thus, it can be possible for the firms to decrease their costs related to the reverse logistics, and can pave the way to gain profits from the investments in them.

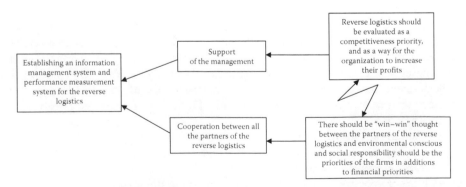

FIGURE 4.2 Evaporating cloud.

- To find new ways to cooperate with other firms in the sector and with the partners of reverse logistics in order to increase the efficiency of the reverse logistics, and establish a "win–win" situation at the same time.
- To establish a system based on RFID in order to increase the visualization.

The implementation of the suggestions would provide a solution for the inefficiency of the reverse logistics of electronic products. With the EC constructed according to the goal related to the core problem, the injections were determined for eliminating conflicts. After constructing the EC, the FRT can be easily constructed.

4.7 DISCUSSION

The first point that the managers should evaluate when they are faced with a complex problem that they want to solve is what to change. In this chapter, the factors that affect the efficiency of reverse logistics and the relationships between the factors have been determined. With the application of CRT, the relationships between UDEs and the symptoms of the problem can be determined and the core problem can be identified. According to the cause–effect relationships in the reverse logistics, the strategies for reverse logistics can be determined.

After determining the suggestions for the core problems with the EC, FRT can be easily constructed. With FRT, the improvements can be realized and it is assured that the UDEs can be eliminated with the injections determined with the EC.

According to the CRT constructed in this chapter, the core drivers for the reverse logistics are inefficiency of applications of legal rules, lack of the pressure of the customers for being environmentally consciousness, lack of cooperation with municipality, and unawareness of the customers. According to Mulder et al. (1999), a system in which municipalities take the responsibility for the collection process seems to be both relatively cheap and yield higher returns than other systems, and a successful collection scheme would be best met by municipal collection systems with the voluntary participation of distributors. Thus, for enhancing the efficiency of the reverse logistics, the municipalities have many responsibilities. If it is possible to cooperate with the municipality, especially at the collection stage, one of the core drivers can be eliminated.

The core problem is a lack of an information management system. If the firms can establish a good information management system for their reverse logistics, it can be possible to integrate forward and reverse logistics, and, thus, visibility will increase. With a performance measurement system for the reverse logistics, the activities can be tracked better and the improvements can be achieved more easily. Savaskan et al. (2004) stated that the optimal results for a product-recovery strategy are achieved when the retailer collects the returned products. Kara et al. (2007) modeled the collection of end-of-life electrical appliances and they suggested local councils as collectors because of the lower costs involved. Collection directly from retailers may provide better quality products. So a good cooperation with retailers and the other partners of the reverse logistics is very significant, and for enhancing this cooperation the information system for the reverse logistics among these partners must be established.

Ravi and Shankar (2005) analyzed the barriers that prevent the application of reverse logistics in automobile industries. According to their study, the barriers of reverse logistics are a lack of information and technological systems, problems with the product quality, company policies, resistance to change for activities related to reverse logistics, a lack of appropriate performance metrics, a lack of training related to reverse logistics, financial constraints, a lack of commitment by top management, a lack of awareness about reverse logistics, a lack of strategic planning, and the reluctance of support from dealers, distributors, and retailers. According to our study, one of these barriers was determined as the core problem for reverse logistics. Rogers and Tibben (1998) stated one of the most significant problems faced by the firms in their reverse logistics as the lack of a good information system. The result of this chapter is consistent with the findings of Rogers and Tibben (1998). Information technology devices have the promise to reduce the uncertainty regarding the condition of the returned products and even in reducing the timing uncertainty in product returns (Guide and Wassenhove 2003). The success of reverse logistics is strongly related with the network design (Ilgın and Gupta 2012). For effective reverse logistics networks, information systems and data management must be redesigned or expanded to accommodate returns (Richey et al. 2005). A high degree of interaction and communication among members of reverse logistic systems leads to a higher efficiency level (Freires and Guedes 2008). This can also be achieved with an efficient information system.

Design of reverse logistic networks involves a high degree of uncertainty associated with quality and quantity of returns (Ilgın and Gupta 2012). Guide (2000) stated that the ability to forecast and control timing, quantity, and quality of products return is very important to the success of reverse logistics. For a successful reverse logistics, a good information system must be enhanced. By the information system, visibility of the reverse logistics activities can be improved. The return rates, inventory levels, etc., in reverse logistics are all measured and tracked by a good information system. Visibility of the reverse logistics is related to the ability to forecast the product's return, ability to see the inventory level at all parties in the reverse logistics, the level of information about the return process, and the use of technologies like bar coding and RFID.

According to the literature, there are many papers to measure the performance of forward logistics. However, the research about measuring the performance of reverse logistics is so limited. There are papers discussing the factors that affect the performance of reverse logistics. Besides there is no a framework suggested for measuring the performance of reverse logistics. For managing reverse logistics efficiently, measuring its performance is also very significant. A performance measurement system for reverse logistics should be developed for managing it.

4.8   CONCLUSION

With the application of TOC thinking processes in their processes in determining and eliminating the core problems, many manufacturing and service firms improved the efficiency of their processes. In this chapter, the TOC thinking processes have been used to determine and eliminate the core problems of the reverse logistics of a firm in the electronics sector. With the application of CRT, the core problem in the reverse logistics of the firm has been determined as the "lack of an information system." After constructing CRT, EC was applied, and the assumptions and injections have been stated in order to determine solutions for the core problem. After constructing CRT and EC, FRT can also be constructed easily and the strategies for the reverse logistics can be stated.

The firms can determine and eliminate core problems of their reverse logistics with TOC thinking processes. This chapter shows the application of TOC thinking processes in a reverse logistics of a firm in the electronics sector. TOC thinking processes can also be applied to different reverse logistics, and the core problems depending on different reverse logistics can be compared.

REFERENCES

Beamon B.M., 1999. Designing the green supply chain, *Logistic Information Management*, 12(4), 332–342.

Brito M.P., Dekker R., and Flapper S.D.P., 2003. Reverse logistics: A review of case studies, ERIM Report Series Research in Management, ERS-2003-012-LIS, pp. 1–29.

Chaudhari C.V. and Mukhopadhyay S.K., 2003. Application of theory of constraints in an integrated poultry industry, *International Journal of Production Research*, 41(4), 799–817.

Choe K. and Herman S., 2004. Using theory of constraints tools to manage organizational change: A case study of Euripa labs, *International Journal of Management and Organizational Behavior*, 8(6), 540–558.

Cox J.F. and Spencer M.S., 1998. *The Constraints Management Handbook*, APICS Series on Constraint Management, CRC Press LLC, Boca Raton, FL, 58, pp. 283–307.

Dekker R. and Van Der Laan E.A., 2003. Inventory control in reverse logistics, in *Business Aspects of Closed-Loop Supply Chains*, eds. V.D.R. Guide, Jr. and L.N. Van Wassenhove. Carnegie Mellon University Press, Pittsburgh, PA, pp. 175–199.

Dettmer H.W., 1997. *Goldratt's Theory of Constraints a System Approach to Continuous Improvement*, ASQ Quality Press, Milwaukee, WI.

Dettmer H.W., 2003. *Strategic Navigation: A Systems Approach to Business Strategy*, ASQ Quality Press, Milwaukee, WI, pp. 33–56.

Ferrer G. and Ketzenberg M.E., 2004. Value of information in remanufacturing complex products, *IEE Transactions*, 36(3), 265–277.

Ferrer G. and Whybark C.D., 2000. From garbage to goods: Successful remanufacturing systems and skills, *Business Horizons*, 43(6), 55–64.

Fleischmann M., 2003. Reverse logistics network structures and design, in *Business Aspects of Closed-Loop Supply Chains*, eds. V.D.R. Guide, Jr. and L.N. Van Wassenhove, Carnegie Mellon University Press, Pittsburgh, PA, pp. 117–148.

Fleischmann M., Nunen J.V., Gröve B., and Gapp R., 2004. Reverse logistics-capturing value in the extended supply chain, ERIM Report Series Research in Management, ERS-2004-091-LIS, pp. 1–17.

Freires F.G.M. and Guedes A.P.S., 2008. Power and trust in reverse logistics systems for scrap tires and its impact on performance, *Flagship Research Journal of International Conference of the Production and Operations Management Society*, 1(1), 57–65.

Goldratt E.M., 1994. *It's Not Luck*, North River Press, Great Barrington, MA.

Goldratt E.M. and Cox J., 1992. *The Goal: A Process of Ongoing Improvement*, North River Press, Great Barrington, MA.

Guide V.D.R., 2000. Production planning and control for remanufacturing; industry practice and research needs, *Journal of Operations Management*, 18(4), 467–483.

Guide V.D.R., Harrison T.P., and Wassenhove V.L.N., 2003. The challenge of closed-loop supply chains, *Interfaces*, 33(6), 3–6.

Guide V.D.R. and Wassenhove V.L.N., 2003. Business aspects of closed-loop supply chains, in *Business Aspects of Closed-Loop Supply Chains*, eds. V.D.R. Guide, Jr. and L.N. Van Wassenhove, Carnegie Mellon University Press, Pittsburgh, PA, pp. 17–42.

Gupta M., 2003. Constraints management—Recent advances and practices, *International Journal of Production Research*, 41(4), 647–659.

Ilgın M.A. and Gupta S.M., 2010. Environmentally conscious manufacturing and product recovery (ECMPRO): A review of the state of the art, *Journal of Environmental Management*, 91(3), 563–591.

Ilgın M.A. and Gupta S.M., 2012. *Remanufacturing Modeling and Analysis*, CRC Press, Boca Raton, FL, pp. 73–93.

Kara S., Rugrungruang F., and Kaebernick H., 2007. Simulation modeling of reverse logistics networks, *International Journal of Production Economics*, 106(1), 61–69.

Kim S., Mabin V., and Davies J., 2008. The theory of constraints thinking processes: Retrospect and prospect, *International Journal of Operations and Production Management*, 28(2), 155–184.

Knemeyer A.M., Ponzurick T.G., and Logar C.M., 2002. A qualitative examination of factors affecting reverse logistics systems for end-of life computers, *International Journal of Physical Distribution and Logistics Management*, 32(6), 455–479.

Mabin V. and Balderstone S.J., 2003. The performance of the theory of constraints methodology analysis and discussion of successful TOC applications, *International Journal of Operations and Production Management*, 23(6), 568–595.

Mabin V.J., Davies J., and Cox F., 2006. Using the theory of constraints thinking processes to complement system dynamics' causal loop diagrams in developing fundamental solutions, *International Transactions in Operations Research*, 13(1), 33–57.

McMullen T.B., 1998. *Introduction to the Theory of Constraints Management System*, St. Lucie Press, APICS Series on Constraints Management, CRS Press LLC, Boca Raton, FL, pp. 15, 26–29, 55–61.

Mulder L., Schneidt L.G., and Scneider A., 1999. Collecting electronic waste in Europe: A Sony View, *Proceedings of the 1999 IEEE International Symposium on Electronics and Environment*, Danvers, MA, pp. 244–250.

Mutha A. and Pokharel S., 2008. Strategic network design for reverse logistics and remanufacturing using new and old product modules, *Computers and Industrial Engineering*, 56(1), 334–346.

Pegels C.C. and Watrous C., 2005. Application of the theory of constraints to a bottleneck operation in a manufacturing plant, *Journal of Manufacturing Technology Management*, 16(3), 302–311.

Pochampally K. and Gupta S.M., 2004. A business mapping approach to multi-criteria group selection of collection centers and recovery facilities, *Proceedings of the 2004 IEEE International Symposium on Electronics and Environment*, Phoenix, AZ, pp. 249–254.

Rahman S., 1998. Theory of constraints: A review of the philosophy and its applications, *International Journal of Operations and Production Management*, 18(4), 336–355.

Rahman S., 2002. The theory of constraints' thinking process approach to developing strategies in supply chains, *International Journal of Physical Distribution and Logistics Management*, 32(10), 809–828.

Ravi V. and Shankar R., 2005. Analysis of interactions among the barriers of reverse logistics, *Technological Forecasting and Social Change*, 72(8), 1011–1029.

Ravi V., Shankar R., and Tiwari M.K., 2005. Analyzing alternatives in reverse logistics for end-of-life computers: ANP and balanced scorecard approach, *Computers and Industrial Engineering*, 48(2), 327–356.

Reid R.A. and Cormier J.R., 2003. Applying the TOC TP: A case study in the service sector, *Managing Service Quality*, 13(5), 349–369.

Richey R.G., Chen H., Genchev S.E., and Daugherty P.J., 2005. Developing effective reverse logistics programs, *Industrial Marketing Management*, 34(8), 830–840.

Rogers D.S. and Tibben L.R.S., 1998. *Going Backwards: Reverse Logistics Trends and Practices*, Reverse Logistics Executive Council, Reno Center for Logistics Management, Reno, NV, pp. 2, 9, 33.

Rogers D.S. and Tibben L.R.S., 2001. An examination of reverse logistics practices, *Journal of Business Logistics*, 22(2), 129–148.

Savaskan R.C., Bhattacharya S., and van Wassenhove L.N., 2004. Closed-loop supply chain models with product remanufacturing, *Management Science*, 50(2), 239–252.

Scoggin J.M., Segelhorst R.J., and Reid R.A., 2003. Applying the TOC thinking process in manufacturing: A case study, *International Journal of Production Research*, 41(4), 767–797.

Shoemaker T.E. and Reid R.A., 2005. Applying the TOC thinking process: A case study in the government sector, *Human Systems Management*, 24(1), 21–37.

Srivastava S.K., 2008. Network design for reverse logistics, *Omega*, 36(4), 535–548.

Taylor L.J., Murphy B., and Price W., 2006. Goldratt's thinking process applied to employee retention, *Business Process Management Journal*, 12(5), 646–670.

Veerakamolmal P. and Gupta S.M., 2001. Optimizing the supply chain in reverse logistics, *Environmental Conscious Manufacturing Proceedings of SPIE*, ed. S.M. Gupta, Vol. 4193, SPIE, International Society for Optics and Photonics, pp. 157–166.

Walker E.D. and Cox J.F., 2006. Addressing ill-structured problems using Goldratt's thinking processes—A white collar example, *Management Decision*, 44(1), 137–154.

Walther G. and Spengler T., 2005. Impact of WEEE-directive on reverse logistics in Germany, *International Journal of Physical Distribution and Logistics Management*, 35(5), 337–361.

Watson K.J., Blackstone J.H., and Gardiner S.C., 2007. The evolution of management philosophy: The theory of constraints, *Journal of Operations Management*, 25(2), 387–402.

INTERNET RESOURCES

Basel Action Network, http://www.ban.org/index.html (accessed at 2008).

EU Directive 2002/96/EC of the European parliament and of the council of 27 January 2003 on waste electrical and electronic equipment (WEEE)—joint declaration of the European parliament, the council and the commission relating to article 9 Official Journal L037:0024–39 [February 13, 2003; 2002, http://europa.eu.int/eur-lex/en/].

5 Modeling Supplier Selection in Reverse Supply Chains

Kenichi Nakashima and Surendra M. Gupta

CONTENTS

5.1   INTRODUCTION

A reverse supply chain consists of a series of activities required to retrieve end-of-life (EOL) products from consumers and the activities either recover their leftover market values or dispose them off. Even though the systematic retrieval of EOL products is still in its infancy in the United States, it is becoming mandatory in many countries in Europe. Until recently, environmental regulations were the primary driving force behind driving the original equipment manufacturers (OEMs) to indulge in the business of reverse supply chains. However, as of late, many OEMs have come to appreciate several other drivers that propel the practice of reverse supply chains such as reduction in the production costs by reusing products or components and enhancing their brand image, apart from environmental regulations.

Supplier selection is one of the key decisions to be made in the strategic planning of supply chains that has far-reaching implications in the subsequent stages of planning and implementation of the supply chain strategies. In traditional/forward supply chain, the problem of supplier selection is not new. First publications on supplier selection in traditional/forward supply chains date back to the early 1960s (Wang et al., 2004). Contrary to a traditional/forward supply chain, however, the strategic, tactical, and operational planning issues in reverse supply chains involve decision making under uncertainty (Ilgin and Gupta, 2010). Uncertainty stems from several sources, the quality and timing of availability of the used-products being the major ones. In addition, the relative importance of the different selection criteria varies for each supplier. A typical supplier selection problem involves selecting the suppliers and assigning the order quantities to those suppliers taking into consideration numerous conflicting constraints. Traditionally, in supply chain literature, the supplier selection problem is treated as an optimization problem that requires formulating a single objective function. However, not all supplier selection criteria can be quantified, because of which, only a few quantitative criteria are included in the problem formulation.

In this chapter, we propose a fuzzy mathematical programming approach that utilizes analytic hierarchy process (AHP), Taguchi loss functions, and fuzzy programming techniques to weigh the suppliers qualitatively as well as determine the order quantities under uncertainty. While the Taguchi loss functions quantify the suppliers' attributes to quality loss, the AHP transforms these quality losses into a variable for decision making that can be used in formulating the fuzzy programming objective function to determine the order quantities. We also carry out a sensitivity analysis on how the order quantities of the suppliers vary with the degree of uncertainty. A numerical example is considered to illustrate the methodology.

5.2 METHODOLOGY

5.2.1 Nomenclature Used in the Methodology

B_j	budget allocated for supplier j
c_j	unit purchasing cost of product from supplier j
d_k	demand for product k
g	goal index
j	supplier index, $j = 1, 2, ..., s$
$Loss_j$	total loss of supplier j for all the critical evaluation criteria
p_j	probability of breakage of products purchased from supplier j
p_{max}	maximum allowable probability of breakage
Q_j	decision variable representing the purchasing quantity from supplier j
r_j	capacity of supplier j
s	number of alternate suppliers available
w_i	weight of criterion i calculated by the AHP
X_{ij}	Taguchi loss of criterion i of supplier j

5.2.2 Taguchi Loss Functions

According to Taguchi's quality philosophy, any deviation from a characteristic's target value results in a loss that can be measured by a quadratic loss function (Ross, 1996). Taguchi et al. (2005) proposed three types of loss functions: (1) Nominal value is the better, used when there is a finite target point to achieve; (2) Smaller-is-better, used where it is desired to minimize the result, with the ideal target being zero; and (3) Larger-is-better, used where it is desired to maximize the result, the ideal target being infinity. The three loss functions are shown in Equations 5.1 through 5.3, respectively, and also Figures 5.1 through 5.3 (Wei and Low, 2006).

$$L(y) = k(y - m)^2 \tag{5.1}$$

$$L(y) = k(y)^2 \tag{5.2}$$

$$L(y) = \frac{k}{y^2} \tag{5.3}$$

where
   $L(y)$ is the loss associated with a particular value of quality characteristic y
   m is the nominal value
   k is the loss coefficient

The quality losses of all the critical criteria for all the suppliers are calculated using the aforementioned loss functions.

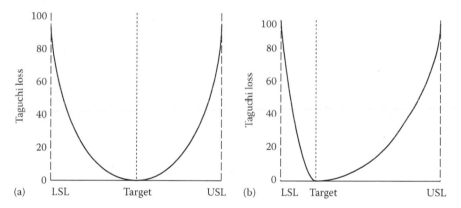

FIGURE 5.1   (a) Nominal-the-better (equal specification). (b) Nominal-the-better (unequal specification).

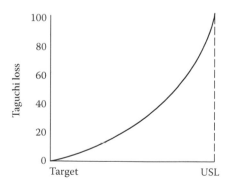

FIGURE 5.2   Smaller-the-better.

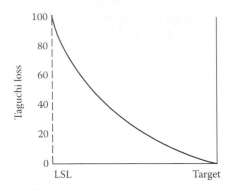

FIGURE 5.3   Larger-the-better.

5.2.3   ANALYTIC HIERARCHY PROCESS

AHP is a tool, supported by simple mathematics, which enables decision-makers to explicitly weigh tangible and intangible criteria against each other for evaluating different alternatives. The process has been formalized by Saaty (1980) and is used in a wide variety of problem areas. In a large number of cases, the tangible and intangible criteria (for evaluation) are considered independent of each other, i.e., those criteria do not in turn depend upon subcriteria and so on. The AHP in such cases is conducted in two steps: (1) Weigh independent criteria using pairwise judgments, (2) Compute the relative ranks of alternatives using pairwise judgments with respect to each independent criterion. Pairwise comparisons express the relative importance of one item versus another in meeting a goal. Table 5.1 shows Saaty's scale for pairwise judgments.

5.2.4   RANKING THE SUPPLIERS

Once the quality losses of all the critical criteria for all the suppliers are calculated using the aforementioned Taguchi loss functions and the weights of all the

TABLE 5.1
Saaty's Scale of Pairwise Judgments

Comparative Importance	Definition
1	Equally important
3	Moderately more important
5	Strongly important
7	Very strongly more important
9	Extremely more important
2, 4, 6, 8	Intermediate judgment values

decision criteria are obtained by AHP, the total loss of all the criteria to each supplier can be calculated as follows:

$$Loss_j = \sum_{i=1}^{n} W_i X_{ij} \qquad (5.4)$$

where

$Loss_j$ is the total loss of supplier j for all the critical evaluation criteria
W_i is the weight of criterion i calculated by AHP
X_{ij} is the Taguchi loss of criterion i of supplier j

Suppliers can be ranked based on the smallest to the largest loss; the best supplier is the one with the smallest loss (Wei and Low, 2006).

5.2.5 FUZZY PROGRAMMING

In real-life situations for a supplier selection problem, much of the input information is uncertain. At the time of selecting a supplier, values of many criteria are expressed in terms of imprecise terms like "approximately more than" or "approximately less than," or "somewhere between," etc. Such vagueness in critical information cannot be captured by deterministic models; hence, the optimal solutions derived from deterministic formulations may not serve the purpose in real-life situations. Therefore, such a problem needs to be modeled as a fuzzy model, in which the overall aspiration level is maximized rather than strictly satisfying the constraints (Kumar et al., 2006). Fuzzy mathematical programming has the capability to handle multi-objective problems and vagueness of the linguistic type (Zimmermann, 1978). The multi-objective programming problem with fuzzy goals and constraints can be transformed into a crisp linear programming formulation that can be solved using conventional optimization tools.

A multi-objective integer programming supplier selection problem (MIP-SSP) for three objectives, namely, total loss of profit (TLP), total cost of purchase (TCP), and percentage rejections (PR) and for relevant system constraints can be represented as follows:

Goal 1: Minimize TLP

$$\sum_{j=1}^{s} Loss_j * Q_j = TLP \qquad (5.5)$$

Goal 2: Minimize TCP

$$\sum_{j=1}^{s} c_j * Q_j = TCP \qquad (5.6)$$

Goal 3: Minimize PR

$$\sum_{j=1}^{s} p_j * Q_j = PR \qquad (5.7)$$

Capacity Constraint

$$Q_i \leq r_i \qquad (5.8)$$

Demand Constraint

$$\sum_j Q_j = d_j \qquad (5.9)$$

Budget Allocation Constraint

$$\sum_j c_j * Q_j \leq B_j \qquad (5.10)$$

Non-Negativity Constraint

$$Q_j \geq 0 \qquad (5.11)$$

The fuzzy programming model for J objectives and K constraints is transformed into the following crisp formulation:

Maximize λ

Subject to:

$$\lambda(Z_j^{\max} - Z_j^{\min}) + Z_j(x) \leq Z_j^{\max} \quad \text{for all } j, \quad j = 1, 2, \ldots, J$$

$$\lambda(d_x) + g_k(x) \leq b_k + d_k \qquad \text{for all } k, \quad k = 1, 2, \ldots, K \qquad (5.12)$$

$$Ax \leq b \qquad \text{for all deterministic constraints}$$

$$x \geq 0 \text{ and integer}$$

$$0 \leq \lambda \leq 1$$

where
   λ is the overall degree of satisfaction
   d_k is the tolerance interval

Zimmerman (1978) suggested the use of individual optima as lower bound $(Z_j^{\min})$ and upper bound $(Z_j^{\max})$ of the optimal values for each objective. The lower bound $(Z_j^{\min})$ and upper bounds $(Z_j^{\max})$ of the optimal values are obtained by solving the (MIP-SSP) as a linear programming problem using one objective each time, ignoring all the others.

A complete solution of the (MIP-SSP) problem is obtained through the following steps:

Step 1: Transform the supplier selection problem into the (MIP-SSP) form.

Step 2: Select the first objective and solve it as a linear programming problem with the system constraints; minimizing the objective gives the lower bound and maximizing the objective gives the upper bound of the optimal values of the objective.

Step 3: Use these values as the lower and upper bounds of the optimal values for the crisp formulation of the problem.

Step 4: Formulate and solve the equivalent crisp formulation of the fuzzy optimization problem maximizing the overall satisfaction level.

5.3 SUPPLIER SELECTION METHODOLOGY: A NUMERICAL EXAMPLE

We consider three suppliers for evaluation. For the qualitative evaluation using Taguchi loss functions and AHP, we consider four criteria: (1) quality of the products delivered (smaller defective rate is better); (2) on-time delivery (lesser the delays or early deliveries the better); (3) proximity (closer the better); and (4) cultural and strategic issues (that include flexibility, level of cooperation and information exchange, supplier's green image, and supplier's financial stability/economic performance). Table 5.2 shows the relative weights of the criteria obtained by carrying out the AHP.

Table 5.3 shows the service factor ratings (SFRs) for the subcriteria considered under the cultural and strategic issues criteria for the three suppliers. The ratings are given on a scale of 1–10, the level of importance being directly proportional to the rating.

Table 5.4 shows the decision variables for calculating the Taguchi losses for the suppliers.

TABLE 5.2
Relative Weights of Criteria

Criteria	Relative Weight
Quality	0.384899
On-time delivery	0.137363
Proximity	0.052674
Cultural and strategic issues	0.425064

TABLE 5.3
Service Factor Ratings for Cultural and Strategic Issues

Supplier	Flexibility	Level of Co-op and Info. Exchange	Green Image	Financial Stability and Economic Performance	Average	Average/10 (%)
1	7	6	6	4	5.75	57.5
2	5	7	8	5	6.25	62.5
3	6	5	8	8	6.75	67.5

TABLE 5.4
Decision Variables for Selecting Suppliers

Criteria	Target Value	Range	Specification Limit
Quality	0%	0%–30%	30%
On-time delivery	0	10-0-5	10 days earlier, 5 days delay
Proximity	Closest	0%–40%	40% higher
Cultural and strategic issues	100%	100%–50%	50%

TABLE 5.5

Characteristic and Relative Values of Criteria

Supplier	Quality Value (%)	Quality Relative Value (%)	On-Time Delivery Value	On-Time Delivery Relative Value	Proximity Value	Proximity Relative Value (%)	Cultural and Strategic Issues Value (%)	Cultural and Strategic Issues Relative Value (%)
1	15	15	+3	+3	8	33.33	57.5	57.5
2	20	20	+1	+1	6	0	62.5	62.5
3	10	10	−8	−8	9	50	67.5	67.5

TABLE 5.6

Supplier Characteristic Taguchi Losses

Supplier	Quality	On-Time Delivery	Proximity	Cultural and Strategic Issues
1	24.99	36	69.43	75.61
2	44.44	4	0	64
3	11.11	64	156.25	54.86

To illustrate the calculation of Taguchi losses, consider for example the criteria, quality. The target defect rate/breakage probability is zero where there is no loss to the manufacturer, and the upper specification limit for the defect rate/breakage probability is 30% where there is 100% loss to the manufacturer. Monczka and Trecha (1998) proposed a SFR that includes performance factors difficult to quantify but are decisive in the supplier selection process. In practice, experts rate these performance factors. For a given supplier, these ratings on all factors are summed and averaged to obtain a total service rating. The supplier's service factor percentage is obtained by dividing the total service rating by the total number of points possible. We assume a specification limit of 50% for the service factor percentage, at which the loss will be 100% while there will be no loss incurred at a service factor percentage of 100%. The value of loss coefficient, k, and the Taguchi losses are computed using Equations 5.1, 5.2, or 5.3 using the characteristic relative values of each criterion for the three suppliers as shown in Table 5.5. Table 5.6 shows the Taguchi losses for each criterion calculated from the appropriate loss functions for the individual suppliers.

The weighted Taguchi loss is then calculated using AHP weights from Table 5.2 and Equation 5.4. Table 5.7 shows the weighted Taguchi loss and the normalized losses for the individual suppliers.

5.3.1 DETERMINING THE ORDER QUANTITIES: FUZZY PROGRAMMING

Table 5.8 shows the supplier profiles we considered in our illustrative example.

TABLE 5.7

Weighted Taguchi Losses

Supplier	Weighted Taguchi Loss	Normalized Loss
1	50.36567	0.360148
2	44.86013	0.32078
3	44.62138	0.319072

TABLE 5.8

Supplier Profiles

Supplier	Unit Cost	Probability of Breakage	Capacity	Budget Allocation
1	1	0.03	300	1500
2	1.5	0.02	500	1750
3	2	0.05	600	1500

Net demand – 1250 units.

We consider the net demand to be a deterministic constraint in our illustrative example. The values of the level of uncertainties for all the fuzzy parameters (capacities and budget allocations) are considered as 15% of the deterministic model. Table 5.9 shows the data for the values at the lowest and highest aspiration levels of the membership functions.

TABLE 5.9

Limiting Values in Membership Function for Fuzzy Objectives and Fuzzy Constraints

	$\mu = 1$	$\mu = 0$
TLP	$399.31 \, (= Z_{TLP}^{\min})$	$413.47 \, (= Z_{TLP}^{\max})$
TCP	$1867.5 \, (= Z_{TCP}^{\min})$	$2220 \, (= Z_{TCP}^{\max})$
PR	$38.35 \, (= Z_{PR}^{\min})$	$49.15 \, (= Z_{PR}^{\max})$
Capacity constraints		
Supplier 1	300	345
Supplier 2	500	575
Supplier 3	600	690
Budget allocations		
Supplier 1	1500	1725
Supplier 2	1750	2012.5
Supplier 3	1500	1725

With the aforementioned data, the equivalent *crisp* formulation of the fuzzy optimization problem is formulated as in (5.12) and solved. The *crisp* formulation for the illustrative example is as follows:

Maximize λ

Subject to:

$14.16\lambda + 0.36Q1 + 0.32Q2 + 0.319Q3 \leq 413.47$

$352.5\lambda + Q1 + 1.5Q2 + 2Q3 \leq 2220$

$10.8\lambda + 0.03Q1 + 0.02Q2 + 0.05Q3 \leq 49.15$

$Q1 + Q2 + Q3 = 1250$

$45\lambda + Q1 \leq 345$

$75\lambda + Q2 \leq 575$

$90\lambda + Q3 \leq 690$ $\qquad\qquad\qquad\qquad\qquad$ (5.13)

$225\lambda + Q1 \leq 1725$

$262.5\lambda + 1.5Q2 \leq 2012.5$

$225\lambda + 2Q3 \leq 1725$

$Q1, Q2, Q3 \geq 0$ and integers

$\lambda(d_x) + g_k(x) \leq b_k + d_k \quad$ for all $k, \quad k = 1, 2, \ldots, K$

$Ax \leq b \qquad\qquad\qquad$ for all deterministic constraints

$x \geq 0$ and integer

$0 \leq \lambda \leq 1$

This model was solved using LINGO8 and the maximum degree of overall satisfaction achieved is $\lambda_{max} = 0.48$ for the supplier quantities: $Q1 = 180$, $Q2 = 539$, and $Q3 = 531$. This solution yields a net TLP = 406.6, TCP = 2050.5, and PR = 42.73.

The aforementioned model is tested for varying degrees of uncertainty in the capacities of suppliers. The solutions are obtained at corresponding increased levels of uncertainties, i.e., the values of b_k is kept the same as the deterministic model, but the value of d_k (= tolerance) is increased in steps of 15% of b_k and the fuzzy model is solved for each step that represents increased vagueness in the supplier's capacities.

Figure 5.4 depicts the percentage changes in the supplier quota allocations at different degrees of uncertainties. The quota allocations depend on the coefficients of the different objectives and the uncertainities captured in the supplier capacities. It can be seen from the figure that at higher degrees of uncertainty, supplier 3's quota allocations drastically reduce while that of suppliers 2 and 1 increases. This can be attributed to the fact that though supplier 3 has the highest capacity, supplier 3's

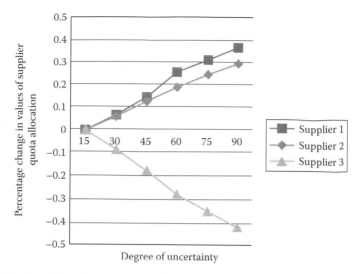

FIGURE 5.4　Supplier allocations at varying degrees of uncertainty.

probability of breakage as well as cost per unit item are higher than those of suppliers 1 and 2. While supplier 2 has the least probability of breakage, supplier 1 has the least cost per unit item. Depending on the decision-makers' relative importance between the two criteria, it can be suggested that allocations be made to suppliers 1 and 2 first and then to supplier 3 if the aggregate demand is not fulfilled by 1 and 2. Apart from the criteria considered in our methodology, there can be several other critical criteria, such as percentage late deliveries, suppliers flexibility, etc., which can be included in the decision making framework.

5.4　CONCLUSIONS

In this chapter, we developed an integrated multi-criteria decision making methodology using Taguchi loss functions, AHP, and fuzzy programming techniques to address the supplier selection problem in a reverse supply chain setting. Our methodology takes into account several qualitative criteria that are hard to quantify, hence ignored in majority of the traditional supplier selection models in the forward supply chains. While the Taguchi loss functions quantify the suppliers' attributes to quality loss, the AHP transforms these quality losses into a variable for decision making that can be used in formulating the fuzzy programming objective function to determine the order quantities. A numerical example was considered to illustrate the proposed methodology.

REFERENCES

Ilgin, M. A. and Gupta, S. M., 2010, Environmentally conscious manufacturing and product recovery (ECMPRO): A review of the state of the art, *Journal of Environmental Management*, **91**(3), 563–591.

Kumar, M., Vrat, P., and Shankar, R., 2006, A fuzzy programming approach for vendor selection problem in a supply chain, *International Journal of Production Economics*, **101**, 273–285.

Monczka, R. M. and Trecha, S. J., 1998, Cost-based supplier performance evaluation, *Journal of Purchasing and Material Management*, **24**(1), 2–7.

Ross, P. J., 1996, *Taguchi Techniques for Quality Engineering*, 2nd edn., McGraw-Hill, New York.

Saaty, T. L., 1980, *The Analytic Hierarchy Process*, McGraw Hill, New York.

Taguchi, G., Chowdhury, S., and Wu, Y., 2005, *Taguchi Quality Engineering Handbook*, John Wiley & Sons, Hoboken, NJ.

Wang, G., Huang, S. H., and Dismukes, J. P., 2004, Product-driven supply chain selection using integrated multi-criteria decision-making methodology, *International Journal of Production Economics*, **91**, 1–15.

Wei, N. P. and Low, C., 2006, Supplier evaluation and selection via Taguchi loss functions and an AHP, *International Journal of Advanced Manufacturing Technology*, **27**, 625–630.

Zimmermann, H. J., 1978, Fuzzy programming and linear programming with several objective functions, *Fuzzy Sets and Systems*, **1**, 45–55.

6 General Modeling Framework for Cost/Benefit Analysis of Remanufacturing

Niloufar Ghoreishi, Mark J. Jakiela,
and Ali Nekouzadeh

CONTENTS

A general modeling framework for cost/benefit analysis of remanufacturing is presented in this chapter. This model consists of three phases: take back, disassembly and reassembly, and resale. The first phase considers the process of buying back the used product from the customers; the second phase focuses on modeling the disassembly of the taken back product into its cores and reassembly of the recovered cores into the remanufactured product; and the last phase models marketing of the remanufactured product. These three phases are modeled separately using the transfer pricing mechanism.

In take back (*tb*) phase, motivating the customers to return their product and other factors that can affect this process like advertisement and transportation are modeled. In disassembly phase, complete and optimum partial disassembly are considered and compared. Common graphical methods to determine the optimum disassembly plan (sequence) are reviewed and a cost model is derived for disassembly process considering all the sources of costs and revenues. In resale phase, a cost model was developed using the conventional method of customer's willingness to include the competition between the remanufactured and new product. Factors that can affect this competition like the warranty plan and the advertisement were included in the model.

6.1   END OF LIFE PLANS

When a product reaches to the point that does not function properly, does not satisfy its owner anymore, or it is out of date and retired, it is considered as an end of life (E.O.L) product. An E.O.L product can be simply disposed. However, for many products and in many situations, although the E.O.L product is no longer suitable for its current application, some or even all of its components may be still in proper working condition. This raises the possibility of reusing the E.O.L product in whole or in parts in production of other products, or for other applications, to recover some of the embodied energy and materials and to save natural resources and reduce waste toward a greener environment. These possibilities have been studied in the more general context of environmentally conscious manufacturing and product recovery (Ilgin and Gupta 2010). Disposal to the landfill, recycling, reuse, refurbishment, and remanufacturing are different plans that may be considered for an E.O.L product. All activities involved in collection and E.O.L treatment of the used products are usually referred to as reverse logistic (Ilgin and Gupta 2010). Because of the mutual impacts between the operations of reverse logistic and forward logistic (e.g., allocating storage space and transportation capacity), sometimes both are studied simultaneously under the concept of closed-loop supply chains, with an emphasis on network design (Amini et al. 2005, Aras et al. 2008, Dehghanian and Mansour 2009, Du and Evans 2008, Kannan et al. 2010, Lee and Dong 2009, Mutha and Pokharel 2009, Pishvaee et al. 2010, Pochampally and Gupta 2008, Pochampally et al. 2009, Qin and Ji 2010, Srivastava 2008, Sutherland et al. 2010, Wang and Hsu 2010, Yang et al. 2009).

The simplest method of treating an E.O.L product is to dispose it to the landfill. This way all its embodied materials and energies are wasted (Ishii et al. 1994). Sometimes disposal of a product to the landfill is discouraged or prohibited by law due to the serious polluting effects of its hazardous materials. In this case product must undergo some treatments before disposal.

Recycling focuses mainly on extracting raw materials from the E.O.L products. In this process raw materials are extracted from the components and parts of the used product. Recycling destroys the value added to the product during fabrication and, economically, is less desirable than reuse and remanufacturing (Klausner and Hendrickson 2000). Recycling is also called material recovery as it recovers the material and reduces the waste compared with disposal.

Another plan for an E.O.L product is to reuse it "as is" in a different application (secondary use). This way the life of the product continues, but not for the primary reason it has been purchased. Secondary use enables the customer who cannot afford a new product to access a second hand one (Heese et al. 2003); therefore, it is an appropriate option for durable used products like cars. Sometimes reuse is defined as the use of a waste product in its original function like refilling discarded container (Asiedu and Gu 1998).

Refurbishment is another possible plan for an E.O.L product to recover some of its materials, energy, and embodied cost. Refurbishment can be divided into two subcategories: repair and reconditioning. In repair, the source of product's defect is determined and repaired. Repair may also include replacement of some minor defected components of the product. Reconditioning is replacement or rebuilding some of the major components of the product that do not function properly.

Finally, in remanufacturing the used product is disassembled into its cores and the functioning and durable cores are used in production of remanufactured product. Currently, the remanufacturing and refurbishment have been implemented successfully for a variety of products, including printer toners and printer cartridges, single-use cameras, photo copiers, cellular phones, electronic components, street lights, vending machines, carpet, and office furniture (Kerr 2000).

Block diagram of Figure 6.1 shows how these processes enter the manufacturing line. Recycling comes in the very early stage of manufacturing line. Remanufacturing, reconditioning, and repair enter in the manufacturing line at later stages. In general, if a process enters in a later stage of production line, it is more productive and has less loss of embodied energy. Also, entering in a later stage of production line means that the used product is less disassembled, and, in general, less work and energy is required to bring it back to the production line.

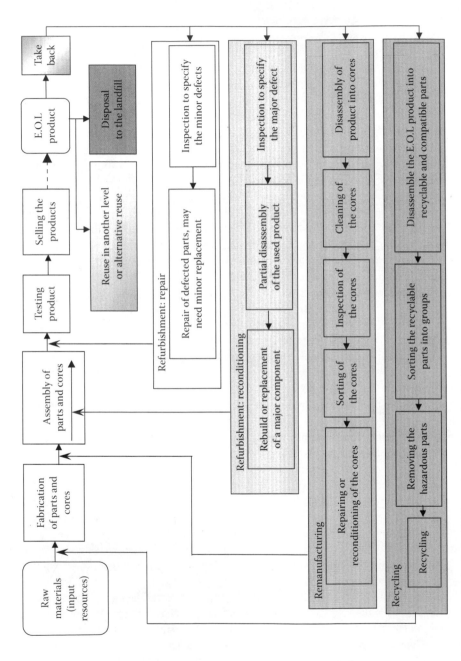

FIGURE 6.1 Bringing back the E.O.L product to the production line.

6.2   REMANUFACTURING

Remanufacturing is defined as an industrial process in which discarded, defective, obsolete, or worn-out durable products are restored to a "like new" condition (Lund 1996). In fact remanufacturing is the entire process of restoring E.O.L products to preserve the added value during the initial design and manufacturing process and to extend the life of the product or its components. The remanufactured product can be the same type as the original product, an upgraded product with a superior performance compared to the original product or another type of product.

Remanufacturing has several environmental and economical benefits. It reduces material and energy consumption as well as pollutant production. The U.S. Environmental Protection Agency (EPA) reported that less energy was used and less waste was produced with remanufacturing activities (U.S. Environmental Protection Agency, EPA 1997). From an economic perspective, remanufacturing generates profit and creates jobs. In the United States there are about 73,000 firms engaged in remanufacturing (Lund 1996). However, regardless of all the benefits of the remanufacturing, it is not likely for a firm to be involved with a remanufacturing process, unless it is profitable; profitability is a strong motivation to initiate a remanufacturing process and to maintain it.

The sale price of a remanufactured product is usually less than the new product as the consumers usually value the remanufactured product less. Also, because it is not necessary to produce the remanufactured product off scratch, its production cost could be less than the production cost of the new product. Therefore, a cost/benefit analysis is required to determine the production cost of the remanufactured product and to decide whether the remanufacturing cost is sufficiently low to compensate for the reduced price of the remanufactured product.

Remanufacturing has been in existence for over 70 years (Parker 1997), but it is not suitable for all types of products. Products with high added value and stable technology and design are appropriate candidates for remanufacturing (Lund 1996). Additionally, profitability of remanufacturing process depends on sufficient numbers of taken back products and convenience of product disassembly (Klausner et al. 1998). A good market acceptance for the remanufactured product is another key issue for a successful remanufacturing process (Klausner et al. 1998). Rate of technology development should also be considered in feasibility study of remanufacturing. Sometimes because of rapid technology change, reuse of product cores is very limited or impossible (Stevels et al. 1999).

Cost/benefit analysis of the remanufactured product may become related to the new product in two stages: in the design stage of the new product and in the marketing stage. The E.O.L plans of the product (including remanufacturing) may be considered in the design stage of the new product to minimize the total cost of E.O.L plan and the environmental impacts. Studying the feasibility of a candidate E.O.L plan at the design stage of the new product is termed design for E.O.L plan (Ishii et al. 1994). If remanufacturing is the E.O.L plan of the product, this may be termed design for remanufacturing. One of the subjects that have been studied frequently in design for E.O.L plan is the convenience and efficiency of disassembly (Hoffmann et al. 2001, Knoth et al. 2001, Li et al. 2001, Shu and Flowers 1995, Wakamatsu et al. 2001).

From the marketing perspective, the remanufactured product should be priced lower than the new product and is usually required to be labeled as remanufactured (Toktay et al. 2000). Therefore, the price of the new product controls the price of the remanufactured product. On the other hand, as the remanufactured product usually becomes available in the market during the life cycle of the new product, it has a cannibalizing effect of the sale dynamics of the new product (Toktay and Wei 2006).

6.3   PROCESSES INVOLVED IN REMANUFACTURING

Remanufacturing consists of several processes including taking back the used product from the customers and transporting it to the remanufacturing site, disassembling the used product into its major parts that are called "cores," inspection of the cores to determine their functioning status, sorting the cores based on their status, repairing or reconditioning of the cores (if needed), cleaning the cores, reassembling the cores into the remanufactured product, testing the remanufactured product, and finally reselling the remanufactured product in the market.

The process of remanufacturing may be divided into three phases (Ghoreishi 2009). The first phase is called the take back phase and includes all the activities that are required to obtain the E.O.L product from the customers and bring it back to the remanufacturing site. This phase is also named reverse logistic (Ferrer and Ketzenberg 2004) and product acquisition (Guide and Van Wassenhove 2001). Take back phase involves activities like informing and motivating the customers (usually by financial Incentives) to return the used product (Guide and Van Wassenhove 2001), collection, sorting and transportation of E.O.L products to a disposition center for processes associated with remanufacturing. In general the remanufacturing firm can control the take back process by setting strategies regarding financial incentives, advertisement and collection/transportation methods (Guide and Srivastava 1988, Guide et al. 1997b). Take back phase requires a market analysis to determine how the customers respond to the financial incentives and other motivating factors to return their used product.

The second phase includes all the engineering and technical activities in the remanufacturing site including test and inspection of the used product, disassembly of the product into the cores, inspection, repair and cleaning of the cores, reassembly of the recovered cores into the remanufactured product, and test and quality control of the remanufactured product. This phase is named the disassembly and reassembly phase.

Finally, the last phase of remanufacturing process is named the resale phase. Resale studies marketing of the remanufactured product and its competition with the new product including its cannibalizing effect on the sale dynamics of the new product.

A transfer pricing mechanism (Edlin and Reichelstein 1995, Vaysman 1988) can be used to study these three phases independently. In a large firm with multiple divisions when the output product of a division is the input product of another division, a price may be considered for the transferring product. This price is considered as revenue for the division that transfers the product and as a cost for the division that receives the product. Transfer pricing mechanism enables manager of a division to optimize the division activities independent of the other divisions. The transfer price is usually set through negotiations between divisions.

For the purpose of this modeling framework there are two transfer prices. One is considered between the *tb* phase and the disassembly and reassembly phase and one is considered between the disassembly and reassembly phase and the resale phase. Value of the taken back product at the remanufacturing site is the transfer price between the *tb* and the disassembly and reassembly. Value of the remanufactured product is the transfer price between the disassembly and reassembly phase and the resale phase. Disassembly and reassembly phase can be divided further into a disassembly phase and a reassembly phase, where the values of the recovered cores are the transfer prices between these two phases. The net profit of remanufacturing does not depend on the values of these transfer prices as they are considered cost in one phase and revenue in another phase; they cancel out in determining the net profit of the entire process. However, their values affect the optimum choice of parameters and consequently the maximum net profits of the phases they connect. As it is not clear for what transfer prices the total net profit of remanufacturing (sum of the net profit of all phases) is maximized, they should be considered as variables in the optimization procedure.

Dividing the remanufacturing process into multiple phases not only simplifies the modeling but also simplifies the optimization process and reduces the computations significantly. For example, assume we have a system with 10 variables and the goal is to determine the optimum values of these variables, computationally. If each variable can assume 10 different values, we need to compute the net profit for 10^{10} different combinations of these variables. If we can divide this system into two subsystems of five variables that are connected by a connecting variable (here the transfer price), then for each value of connecting variable we should compute the net profit of each subsystem for 10^5 different combinations. If the connecting variable assumes 10 different values as well, the total computations are 10^6 which are 4 orders of magnitude smaller than before.

6.4 COST/BENEFIT MODEL FOR TAKE BACK PHASE

In general, the area of *tb* and product acquisition has received limited attention in research and operational level of remanufacturing (Guide et al. 2003b). It is important for the remanufacturing firm to manage the take back process with the right price, quality, and quantity in order to maximize the profit (Guide et al. 2003b). Profitability of remanufacturing in operational level can be affected by the return flow (Guide and Van Wassenhove 2001). Through the process of managing and controlling the quality and quantity of the returned products, we can get a better understanding of the market acceptance and its economic potentials for the remanufactured product. Obtaining the E.O.L products from the customers can be classified into two groups (Guide and Van Wassenhove 2001): waste stream and market driven.

In waste stream, remanufacturer has no policy and strategy to control the quality and quantity of the used products; all the returned products are collected and transferred to the remanufacturing firm; and the remanufacturer can reduce the cost by minimizing the cost of reverse logistic network (Linton and Yeomans 2003,

Linton et al. 2002, 2005). In market driven, customers are motivated to return the end of life product by some type of financial incentive. In this way the remanufacturer can control the quantity and quality of the returned products. This strategy is more applicable to the products that their remanufacturing is profitable. Sometimes there are regulations that obligate the manufacturer to collect the E.O.L products in order to perform treatment and extract dangerous material. In such case, a combination of both strategies may exist. According to Guide (2000), most of the remanufacturing firms in the United States have market-driven strategies, but in Europe, the take back is mostly based on the waste stream.

Market-driven strategy has several advantages over the waste stream strategy including less variability in the quality of the returned products and the quality related costs (e.g., disassembly and repair cost), more manageable inventory control, less failure in operational level of remanufacturing (e.g., disassembly), less operational cost, and less disposal cost. Remanufacturing will be more effective, productive and predictable, and much easier to manage and plan in operational level, if the remanufacturing firm can influence the return flow and control the quality, quantity, rate, and scheduling of the product acquisition (Guide and Srivastava 1988, Guide et al. 1997b).

In general, a firm can control the take back process by setting strategies regarding financial incentives, advertisement, and collection/transportation methods (Guide and Srivastava 1988, Guide et al. 1997b). Usually offering higher incentives (in the form of cash or discount toward purchasing new products) increases the return rate and leads to acquisition of higher quality products. Sometimes a higher incentive can encourage the customers to replace their old products with the new ones earlier (Klausner and Hendrickson 2000). Another way to control the quality of the taken back used products is to have a system to grade them based on their conditions and ages and to pay the financial incentives accordingly (Guide et al. 2003b). Proper advertisement and providing a convenient method for the customers to return their E.O.L products can also increase the return rate (Klausner and Hendrickson 2000).

In the existing models of the take back, all the involved costs are bundled together and called the take back cost; the return rate is modeled as a linear function of the take back cost (Klausner and Hendrickson 2000) or as a linear function (with a threshold) of the financial incentive (Guide et al. 2003b). We developed a market-driven model of take back process by modeling the costs and benefits of the financial incentive, transportation, and advertisement, individually (Ghoreishi et al. 2011). The relation between the financial incentive and the return rate is considered as a market property reflecting the consumers' willingness to return the used product. This model enables operational level decisions over a broader choice of variables and options compared with the existing models.

We considered three important aspects of take back in our model: the financial incentive, the transportation, and the advertisement. Each of these incurs a cost to the process and in return can increase the revenue by increasing the number and the average quality of the returned products. Some of the take back costs are scaled by the number of retuned products, and some are fixed costs independent of the number of returns. Value of a returned product at the recovery site is termed a; a is the price that the recovery firm is willing to pay for the used product at the site. If the take back is performed by the recovery firm then a is a transfer price (Edlin and Reichelstein 1995, Vaysman 1988) which enables the cost/benefit analysis of the take back independent of the rest of the remanufacturing process. We modeled the net profit of the take back during a certain period of time. If the take back process is intended for a limited time, this period could be the entire time of the take back process, and if the take back is intended to be a long lasting process, this period is a time window large enough to average out the stochastic fluctuations in the return rate.

6.4.1 FINANCIAL INCENTIVE

Financial incentive is the cash value that the take back firm offers to the customers to motivate them to return their used products. The financial incentive affects the take back cost, the number of returns, and the average quality of the returned products. Increasing the incentive may increase the net profit by increasing the number of returned products and their average quality or may decrease the net profit by increasing the cost. Therefore, it is an optimizing problem to find the amount of incentive that maximizes the net profit. Number of returns, NR, may be assumed as a function of financial incentive:

$$NR = NR(c) \qquad\qquad (6.1)$$

where c is amount of cash offered to a customer for returning the used product.

Once a customer is motivated to return the used product, the product should be transported to the recovery site. Gathering the used products from the customers that are motivated to return them by the take back firm may impose a significant cost to the take back process. In many situations, it may be possible to reduce the transportation cost by asking the customers to partially or fully contribute to the transportation of their products. This requires the customers to spend some time and energy to return their products, which in average makes the financial incentive less attractive to them. In order to determine the optimum transportation strategy, we should quantify how the transportation methods affect the motivation of the financial incentive and consequently the net profit. One way to include the convenience of the transportation in the model, is to determine (or estimate) how the NR varies with the financial incentive (c) for different transportation methods. Assume i is the index referring to a transportation method, then in general we may write

$$NR = NR_i(c) \qquad\qquad (6.2)$$

Alternatively, a parameter f may be introduced for the convenience of transportation, and the number of returns may be modeled as a function of both c and f:

$$NR = NR(c, f) \qquad\qquad (6.3)$$

The transportation cost can be divided into two costs. All the transportation costs that are scaled by the number of returns are termed the transportation cost per product, tp.

All other transportation costs are bundled together and termed the general cost of transportation, tg. Therefore, the transportation cost can be written as follows:

$$TC = NR \cdot tp + tg \qquad (6.4)$$

6.4.2 ADVERTISEMENT

In this model, advertisement is defined as any action for informing the customers about the take back process. Optimum advertisement strategy depends on many social and psychological factors which are out of the focus of this chapter. Here, we only determine those aspects of advertisement that are important for cost/benefit analysis of the take back process. Advertisement cost can be categorized into two groups: W_1, the cost associated with preparing and designing the advertisement (e.g., flyers, posters, audio clips, or video clips), and W_2, the cost of running and the advertisement (e.g., posting, publishing, distributing, or broadcasting).

Only the customers that are aware of the take back process may return their used product. This means that the number of returns increases by increasing the number of customers that are aware of the take back process. To inform more customers, the take back should be advertised more frequently; this increases the advertisement cost W_2. Therefore, the number of informed customers may be considered as a function of W_2 and we may rewrite the number of returns as

$$NR(c, f, W_2) = N\Omega(W_2) \, \Gamma(c, f) \qquad (6.5)$$

where
  N is the total number of customers having the used product
  Ω is the fraction of customers that are informed by the advertisement
  Γ is the fraction of informed customers that return the used product in response to the motivation effect of the take back process

An estimate can be found for the number of customers that are exposed to the advertisement (Ω function) based on the available information about the advertisement method. Not all the customers can be reached by a specific advertisement method. For example, customers who do not read the newspaper of the advertisement or do not watch or hear the TV or radio program that broadcasts the advertisement will not be exposed to the advertisement, regardless of how frequent the advertisement is posted or broadcasted. The maximum number of the customers that are potential audiences of the advertisement (may see, hear, or watch the advertisement) in frequent runs, is defined as N_{ss}. Also the average fraction of customers that are exposed to the advertisement in each round is defined by λ^*. Both N_{ss} and λ^* are statistical parameters of the advertisement method and are assumed to be known.

Assume N_{ad} is the number of customers (or in general people) that at least one time are exposed to the advertisement (after a number of iterations). The number of customers that have not seen the advertisement, and may be exposed to the advertisement in the next iteration is $N_{ss} - N_{ad}$. Therefore, ΔN_{ad}, the change in N_{ad} after each iteration is

$$\Delta N_{ad} = \lambda^* (N_{ss} - N_{ad}) \tag{6.6}$$

The advertisement cost W_2, is proportional to the number of times the advertisement is broadcasted or published. Let us assume that the cost of running the advertisement is ΔW_2 per each run. We may rewrite Equation 6.6 as

$$\frac{\Delta N_{ad}}{\Delta W_2} = \frac{\lambda^*}{\Delta W_2}(N_{ss} - N_{ad}) = \lambda(N_{ss} - N_{ad}) \tag{6.7}$$

where λ is defined as

$$\lambda = \frac{\lambda^*}{\Delta W_2} \tag{6.8}$$

Although N_{ad} is a discrete function, but when $\lambda \ll 1$ we may approximate it by a continuous function of W_2 and write

$$\frac{dN_{ad}}{dW_2} = \lambda(N_{ss} - N_{ad}) \tag{6.9}$$

and therefore

$$N_{ad}(W_2) = N_{ss}(1 - e^{-\lambda W_2}) = N_{ss}(1 - e^{-W_2/W_{sc}}) \tag{6.10}$$

where W_{sc} is the reciprocal of λ and from a physical point of view is the cost required to inform about 63% $(1 - e^{-1})$ of the potential audiences of an advertisement method. Dividing both sides by N we can find an estimate for Ω

$$\Omega(W_2) = \Omega_{ss}(1 - e^{-W_2/W_{sc}}) \tag{6.11}$$

where
  Ω_{ss} is the maximum fraction of customers that can be informed by an advertisement method
  Ω_{ss} and W_{sc} are different for different advertisement methods

Advertisement, if designed accordingly, can also have a motivating effect by informing the customers about the environmental and global benefits of their contribution.

The response of customers to this motivation effect is assumed to be independent of their response to the financial incentive. Therefore, we modeled this motivation by a constant increase in the motivation effect of financial incentive. g is the parameter that models the motivation effect of advertisement. The model of the number of returns may be rewritten as the following:

$$NR = N\Omega(W_2; \Omega_{ss}, W_{sc}) \; \Gamma(c, g, f) \tag{6.12}$$

As providing a more effective advertisement usually costs more, the motivation effect of advertisement may be considered as a function of W_1:

$$g = g(W_1) \tag{6.13}$$

6.4.3 Cost Model

The cost that is scaled with the number of returns (cost per returned item) consists of the amount of cash incentive, c, and transportation cost, tp. The revenue is generated by the average value of returned product, a, and is also scaled with the number of returns. Advertisement costs, W_1 and W_2, general transportation cost, tg, and any other general cost of take back, termed tbc, are not scaled with the number of returns and are constant costs during the time period of model. Therefore, the net profit of tb, Ψ, can be modeled as the following:

$$\psi = NR \cdot [a - c - t] - W_1 - W_2 - tg - tbc \tag{6.14}$$

The average quality of taken back products is expected to increase by increasing the financial incentive (Guide et al. 2003b). Therefore, we model a as a function of c. Substituting for the number of returns from Equation 6.12, the net profit of take back process is as follows:

$$\psi = N \cdot \Gamma(c, g, f) \cdot \Omega(W_2; \Omega_{ss}, W_{sc}) \cdot [a(c) - c - t] - W_1 - W_2 - tg - tbc \tag{6.15}$$

6.5 COST/BENEFIT MODEL FOR DISASSEMBLY AND REASSEMBLY PHASE

In remanufacturing, the taken back product is broken down into its reusable parts termed cores, in a process termed disassembly. Other recovery processes including refurbishment, repair, and reuse may require replacing one or more cores and therefore may require a partial disassembly. Recycling also involves disassembling of the recyclable parts from the product. Disassembly is a basic phase for all of the recovery processes. Disassembly of a product is not just the reverse of its assembly process. Disassembly modeling is more complicated than the assembly modeling because the final goal is not predetermined; it depends on many factors like quality of the returned products, market demands for the recovered cores or recycled materials, variance in the age and conditions of the cores, and the structure of the returned product. The literature of disassembly is grouped into four categories (Srinivasan et al. 1999, Tang et al. 2002, Yi et al. 2008):

1. Determining the feasible disassembly sequences of the product based on analysis of the product structure and the topology of the cores (Li et al. 2002, Mascle and Balasoiu 2003, Zhang and Kuo 1996, 1997).
2. Disassembly process modeling and planning. This determines to what extent and what cores should be disassembled (Gao et al. 2002, Gungor and Gupta 2001, Hula et al. 2003, Kazmierczak et al. 2004, Lambert 2002, Salomonski and Zussman 1999, Tang and Zhou 2006, Zussman and Zhou 1999).
3. Disassembly at task planning level. Studies about disassembly task planning, scheduling and line balancing are in this group (Gungor and Gupta 2001, Gupta and Taleb 1994, Kazmierczak et al. 2004, Taleb et al. 1997).
4. A fourth category is considered by Lambert (2003) as disassembly concerns at reverse logistic level. The studies of disassembly in the context of industrial ecology, green technology, considering environmental issues, and design for remanufacturing are in this group.

Planning and scheduling in disassembly are the studies of the timing in disassembly process in order to address the demand for different recovered cores. Disassembly planning is studied in the context of material requirement planning (MRP) (Barba-Gutierrez et al. 2008, Depuy et al. 2007, Ferrei and Whybark 2001, Georgiadis et al. 2006, Jayaraman 2006, Li et al. 2009, Lu et al. 2006, Vlachos et al. 2007, Xanthopoulos and Iakovou 2009). MRP in general consists of a set of procedures and timelines to recover the subassemblies and cores of a product, in order to address their expected demands. This includes multistage production and inventory control. An example of a modeling algorithm for scheduling the disassembly process can be found in Gupta and Taleb (1994). Inventory control in remanufacturing process (including disassembly) has been studied from different perspectives as well (Guide et al. 1997a, Li et al. 2006, Nakashima et al. 2004, Teunter 2001, Teunter and van der Laan 2002, Teunter et al. 2000, Toktay et al. 2000, Van der Laan and Salomon 1997, Van der Laan et al. 1999).

Disassembly of a product involves determining all feasible disassembly sequences and determining the optimum sequence among the feasible sequences (groups 1 and 2 of the aforementioned). A disassembly sequence is the order of disassembly operations or steps that should be performed on the product in order to remove the intended cores. Usually the disassembly operations cannot be performed at any arbitrary sequence as removing some joints and connections requires a priori removal of other cores or joints. Therefore, determining the feasible disassembly sequences depends on the assembly structure of the product and in particular the geometrical locations of its cores and their interconnections.

6.5.1   Characterizing the Assembly Structure of a Product

The assembly structure of a product can be quantified and presented by two graphs or alternatively by two matrices associated with these graphs: the components connection graph and the components interference graph (Kuo 2000, Ong and Wong 1999, Tang et al. 2002). Components connection matrix (or graph) represents the names and types of all connections for each component. Components connections can be simple contacts, adhesive, joints, or fastened connection. An interference matrix (or graph) shows the spatial relationships of all the components in the product. It represents the geometrical interferences of one core with the rest of the cores. For a product consisting of n cores, a connection matrix is an $n \times n$ matrix, $E_{n \times n}$, which summarizes the connection graph. It is defined as follows:

$$
E = \begin{bmatrix} e_{11} & e_{12} & e_{1n} \\ e_{21} & e_{22} & e_{2n} \\ e_{n1} & e_{n2} & e_{nn} \end{bmatrix}
\tag{6.16}
$$

where

$$
e_{ij} = \begin{cases} k & \text{if cores } i \text{ and } j \text{ are connected by any connection} \\ 0 & \text{if cores } i \text{ and } j \text{ are not connected} \end{cases}
\tag{6.17}
$$

k can be simply 1 to show existence of a connection between cores or can be a number that represents the type of the connection or the number of joints between the two cores. Connection matrix is a symmetric matrix.

Similarly, for a product consisting of n cores, the interference matrix is an $n \times n$ matrix, $A_{n \times n}$, defined as follows:

$$
A = \begin{bmatrix} a_{11} & a_{12} & a_{1n} \\ a_{21} & a_{22} & a_{2n} \\ a_{n1} & a_{n2} & a_{nn} \end{bmatrix}
\tag{6.18}
$$

where

$$
a_{ij} = \begin{cases} 1 & \text{if core } j \text{ interferes with disassembly of core } i \\ 0 & \text{otherwise} \end{cases}
\tag{6.19}
$$

Interference matrix is sometimes called the disassembly precedence matrix, $DP_{n \times n}$, (Gungor and Gupta 1998, Kuo 2000, Tang et al. 2002) and is introduced in 3D Cartesian coordinate in both positive and negative directions. For example, DP_{x+} denotes the interference (or disassembly precedence) matrix for disassembling the product in $+x$ direction. The interference graph conveys the same information in a graphical manner. In an interference graph, all cores are presented with circles or boxes and their interferences with other cores are presented by arrows. In this graph if core i interferes with removing core j, an arrow is drawn from core j to core i. An example of interference graph is presented in Figure 6.2.

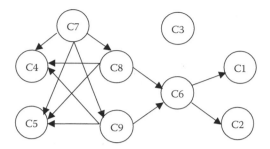

FIGURE 6.2 An example of the interference graph. The cores that interfere with removing a target core can be recognized by following the arrows that leave the target core. For example, in order to remove core C6, cores C1 and C2 should be removed first.

Interference graphs (or matrices) cannot present all the aspects of the topological constraints. For example, sometimes an interfering core may be removed together with some other cores (a subassembly of cores) to make the target core accessible. Or sometimes the target core can be removed within a subassembly of cores with less topological constraints. Also multiple options may exist to make a core accessible for disassembly. Regardless of these limitations, the interference graph or matrix provides a basic model for presenting the topological constraints.

6.5.2 DIFFERENT FORMS OF DISASSEMBLY

Disassembly can be grouped into three categories (Kuo 2000):

1. Targeted (or selective) disassembly
2. Complete (or full) disassembly
3. Optimum partial disassembly

Targeted disassembly is a component-oriented disassembly (Lambert 2003). Sometimes the goal is to disassemble a particular core or a subassembly of cores from the product. This is required in refurbishment, repair, service and maintenance, and sometimes recycling. This type of disassembly is termed selective or targeted disassembly (Garcia et al. 2000, Shyamsundar and Gadh 1996, Srinivasan et al. 1999, Yi et al. 2008). Full disassembly is product oriented, where the goal is to disassemble all of the product cores (Lambert 2003). However, the disassembly process may continue to the extent that is profitable. Where this is the case, the disassembly is termed optimum partial disassembly.

6.5.3 DISASSEMBLY SEQUENCE PLANNING AND OPTIMUM PARTIAL DISASSEMBLY

Usually there are numerous different sequences for disassembling a product. This raises the question of which of these sequences is more efficient and to what extend the disassembly process should be continued. Many studies have been performed on analyzing the disassembly sequences (Bourjault 1984, Gu and Yan 1996, Ko and Lee 1987, Lee 1993, Yokota and Brough 1992, Zussman et al. 1994). To analyze different disassembly sequences, graphical representations of the product cores have been used. These graphs start with the product and consider all the possible disassembly options that break the product into two segments. Then for each resultant subassembly, they consider all the possible disassembly options. This multipath sequence of disassembly steps is continued until the product is broken down to its cores. Note that the precedence relations (Huang and Lee 1989, Lee and Kumara 1992, Rajan and Nof 1996, Wolter et al. 1992) of the product cores are required to draw these graphs. Practically, it is necessary to know whether any core or joint has to be removed in order to perform a particular disassembly step. The following are the four common graphical representations for the disassembly sequences:

1. Connection graph
2. Direct graph
3. And/Or graph
4. Disassembly Petri nets

6.5.3.1   Connection Graph

A connection graph models the structure of the product by showing its cores and noncore parts by boxes or vertices and the physical connections between the cores (and also noncore parts) with lines or edges. A connection graph shows all the connections and joints that should be removed in order to disassemble a core from the product. The connection graph is not intended to show the topological and geometrical constraints. However, this graph may be drawn in a way that represents these geometrical constraints to some extent. Figure 6.3 shows a connection graph of a product that consists of five cores, two noncore parts and eight connections. Assume that in this product disassembling of joint 8 requires removal of core 1 and disassembling of joints 7 requires removal of core 3. These cores may be removed individually or among a subassembly of cores. All other joints can be disassembled independently. A connection graph does not show the disassembly sequences explicitly.

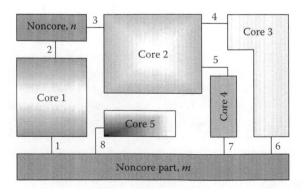

FIGURE 6.3   The connection graph of a product consisting of five cores, two noncore parts, and eight joints.

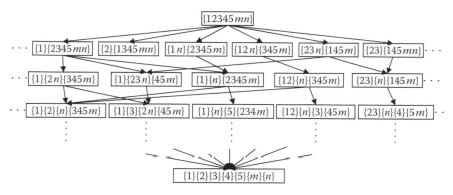

FIGURE 6.4 A small portion of the direct graph of the product shown in Figure 6.3. First line is the complete product. All the possibilities that the product can be broken down into two and three parts are shown in second and third lines, respectively. This process continues until all the seven parts of the product are disassembled in line 7. A disassembly sequence can be determined by following arrows along a possible path from the complete product toward the completely disassembled parts at the bottom of graph.

6.5.3.2 Direct Graph

A direct graph represents all possible sequences of the disassembly. Each node represents a possible partially disassembled state of the product and each edge represents a disassembly task. The nodes are unique, but the edges may repeat at different nodes. This graph will be developed considering the topological constraints and precedence relations. Figure 6.4 shows a small portion of the direct graph for the connection graph presented in Figure 6.3. In this graph each number refers to its associated core. Any combination inside the curly brackets is a subassembly of cores. A disassembly sequence can be determined by following arrows along a path from the complete product toward the completely disassembled parts at the bottom of the graph. As the direct graph considers all the possible combinations of the cores and subassembly of cores, it becomes very large if the product consists of more than a few cores and connections.

6.5.3.3 And/Or Graph

And/Or graph may be interpreted as a reduced version of the direct graph. In direct graph all the edges that exit from a node are considered as (Or); it means that product may go from one state of partially disassembled only through one path to the next state of partially disassembled. In And/Or graph each node shows a subassembly of cores detached in the previous stage of disassembly. For each node (a subassembly of cores), each possible disassembly operation is shown by two edges exiting from the same point of the node to the two resulting subassemblies of cores (And). Other possible disassembly operations (Or) exit from different points of the node. In the And/Or graph the number of nodes is less than the direct graph as the nodes are a possible subassembly of cores rather than a possible disassembly stage of the product. A particular subassembly of cores that appears once in And/Or graph may appear in several

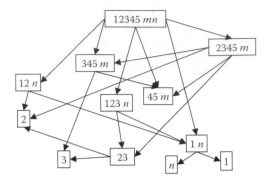

FIGURE 6.5   A portion of the And/Or graph of the product shown in Figure 6.3. Each node represents a possible subassembly of cores. In comparison with Direct graphs, And/Or graphs have less number of nodes.

disassembly stages of the product in direct graph. Figure 6.5 shows a portion of the And/Or graph for the connection graph presented in Figure 6.3. These graphs can be used to determine the optimum disassembly sequence (Penev and deRon 1996).

6.5.3.4   Disassembly Petri Nets

A disassembly Petri net is an alternative representation of the And/Or graph. In comparison to And/Or graph, Petri nets present the disassembly operations with separate units. In And/Or graph disassembly operations are implicit. Explicit presentation of the disassembly units in Petri nets provides a more detailed view of the disassembly process and helps to introduce parameters and decision making criteria more conveniently. Figure 6.6 shows a portion of the disassembly Petri net of the same product.

This net includes a set of places, P, and transitions, t. Each place is a possible subassembly of product cores during the disassembly process. All applicable disassembly operations of place are shown by transitions connected to the place. The product or the subassembly of cores at a place P may go to one of the connected transitions (Or logic between the paths leaving the places) and splits into two (or more) cores or subassembly of cores (And logic between the paths leaving the transitions). Disassembly operations in different transitions could be the same. An algorithm for generating the disassembly Petri nets from the interference matrix has been suggested by Moore et al. (1998). A cell in the Petri net consists of a place and all the connected transitions; each cell of the disassembly Petri net is characterized by the following parameters: $\pi(P)$, the E.O.L value of the place P; $d(P)$, the remanufacturing value of the place P; $d(t)$, the path value defined for each transition connected to the place P; and $\tau(t)$, the transition cost defined for each transition of the cell.

E.O.L value of a place is the value of the product or subassembly of cores "as is" at that place. It is the maximum of reuse value, refurbished value, and recycled value of the subassembly of cores at place P; if none of the aforementioned is an option, it is simply the disposal cost of that subassembly:

$$\pi(P) = \max\left\{\pi_{recycle}(P), \pi_{reuse}(P), \pi_{refurbish}(P)\right\} \quad \text{or} \quad C_{disposal}(P) \qquad (6.20)$$

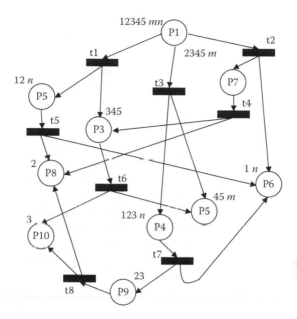

FIGURE 6.6 A portion of the disassembly Petri net of the product shown in Figure 6.3. Each circle shows a place, which is a possible subassembly of the cores. Each solid box shows a transition, which includes one or more disassembly operations. Disassembly operations of the boxes may be similar, but subassemblies in places are always different.

Transition cost, $\tau(t)$, is the disassembly cost at each transition t. The path value, $d(t)$, is the value that can be retrieved from the subassembly at P, if it undergoes the transition t. This value is the sum of the remanufacturing values of the subsequent subassemblies or cores minus the disassembly cost of transition t. For example, in Figure 6.7 the path value for transition t1 is as follows:

$$d(t1) = d(P2) + d(P3) - \tau(t1) \tag{6.21}$$

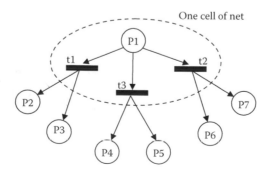

FIGURE 6.7 One cell of the disassembly Petri net. Each cell is defined as one place and all transitions connected to it.

Finally the remanufacturing value of place P, $d(P)$ is defined as the maximum of the E.O.L. value and all of the path values associated with Place P:

$$d(P) = \max\left\{\pi(P), \max\{d(t)\}\right\} \qquad (6.22)$$

The E.O.L values and transition costs are given system parameters, but the path values and remanufacturing values should be calculated using the graph. To calculate these parameters and consequently determining the optimum disassembly path, one should start from the lowest level of graph at places that are associated with one core. At these places the remanufacturing value is the same as the E.O.L value of that core. The paths should be tracked backward from the final cores to find the path values of the transitions. This procedure continues backward until the path values and the remanufacturing values of all cells in the Petri net are calculated. Once all the Petri net parameters are known or calculated, one can find the optimum sequence (or tree) of disassembly by starting from the product in the forward direction. At each place the product or the subassembly of cores should follow the transition that has the maximum path value. The further disassembly of a subassembly of cores should be stopped at a place, if its E.O.L value is more than all the path values.

Petri nets have been widely used in optimizing the disassembly sequence (Kumar et al. 2003, Rai et al. 2002, Salomonski and Zussman 1999, Sarin et al. 2006, Tang et al. 2002, Tiwari et al. 2001, Zussman and Zhou 1999, 2000). Tang et al. (2004) termed the Petri net the "decision tree" approach. They determined the values of the subassemblies of cores at different stages of disassembly (remanufacturing values of places) using a Leontief inverse matrix. This approach is more applicable for assembling a new product and has been modified for disassembly by including large cash outflows to stop the disassembly in certain direction. Johnson and Wang (1998) also proposed a similar Petri net based approach. In their approach there are no remanufacturing values for places and therefore, decision making lacks a quantitative criterion and sometimes the decision for further disassembly is based on personal intuition.

Although disassembly Petri nets are powerful tools to determine the optimum disassembly sequence, their application is limited to products with few cores. Number of places increases exponentially by increasing the number of cores and the Petri net of a product with many cores may become too large to implement. We developed a mechanized nongraphical method based on a new formulation of the disassembly process to overcome these complications (Ghoreishi 2009, Ghoreishi et al. 2012).

A key issue in determining the best disassembly sequence is the uncertainty in the conditions of the product cores, product joints, and disassembly steps. Unlike the new product assembly that all the cores are newly manufactured and tested, in disassembly, the taken back products are used product that may not function appropriately. Therefore, the conditions of the cores are uncertain. Some cores may be good and function appropriately and some may not function. Some of the nonfunctioning cores can be repaired with a reasonable cost and some should be disposed. This makes the values of the product cores stochastic variables. Also, the joints and connections may be deformed or rusty and consequently the disassembly cost is a stochastic variable as well.

Gungor and Gupta (1998) categorized the uncertainty of disassembly process into three groups:

1. Uncertainty in the condition of the cores and joints of the taken back product because of defect or damage.
2. Uncertainty in the product cores because of upgrading or downgrading of the product by the consumers.
3. Uncertainty in the disassembly operations. This includes damaging the cores during disassembly operations.

In some studies, uncertainty of the disassembly model has been considered as an afterthought through sensitivity, analysis, or heuristical adjustment to the solution, when there are significant deviations from the presumed parameters (Erdos et al. 2001, Gungor and Gupta 1998, Lambert 2003, Meacham et al. 1999). Some other studies considered the stochasticity in the parameters of the disassembly model, and assumed that a priori knowledge of these stochastic parameters is available (Geiger and Zussman 1996, Looney 1988, Zussman et al. 1994). However, a priori knowledge of the stochastic distribution of the disassembly parameters may not be available. In such a case, adaptive disassembly models have been suggested (Reveliotis 2007, Zussman and Zhou 1999, 2000). An adaptive model starts with some initial estimates of the stochastic parameters and then trains itself based on the actual data while it is implemented and in use. Therefore, the model parameters and the optimum path and level of disassembly may vary during the accumulation of data.

Zussman and Zhou included the uncertainty of the disassembly operations (steps) in the disassembly Petri net by introducing two pre- and postfiring parameters δ and ρ (Zussman and Zhou 1999, Zussman et al. 1998). ρ is the success rate of a particular disassembly step; it is the ratio between the successful disassembly operations in a disassembly unit to the total number of disassemblies in that unit. δ is the decision value that determines the priority of different paths for disassembling the product or the subassembly of cores at each place. In a typical Petri net the priority (δ) is determined based on the path value of each disassembly unit, $d(t)$. Zussman and Zhou (1999) determined δ based on both the path value and the success rate of each disassembly unit. Sometimes parameters like the high demand for a particular core or obligations to remove hazardous cores also influence setting the value of δ.

The priority of disassembly sequence is not solely dependent on its profitability (Johnson and Wang 1995, 1998, Krikke et al. 1998). Other factors like a high demand for a core or mandated removal of a core (because of its environmental impacts) may affect the priority. Disassembly of multiple used products with some common cores is raised by Kongar and Gupta (2002).

The term reassembly is used for the fabrication process of recovered cores into the remanufactured product. Reassembly is like any other manufacturing line and generally is not a specific subject of the remanufacturing. However, because of uncertainty and stochasticity in the return rate of the used products and disassembly rate of the cores, it has been studied from the perspective of inventory allocation (Kleber et al. 2002).

6.5.4 DISASSEMBLY LINE AND THE CHARACTERISTIC PARAMETERS OF DISASSEMBLY

In each disassembly line, product goes through multiple disassembly units (or disassembly operations) in order to recover some or all of its cores. Before explaining the disassembly line let us define the concepts of core and disassembly unit more specifically.

Core is a durable part of product with a specific function that can be detached from the product with reasonable cost and can be used in the fabrication of a remanufactured product or can be sold in the market. A core does not undergo any further disassembly. It may undergo some limited repairs, though. Life expectancy of the cores (based on both functionality and rate of technology change) should be more than the expected life of the remanufactured product.

Cores and noncore parts of a product are connected to each other by several joints. A disassembly unit disconnects all the joints between two cores or all the joints between a core and a noncore part. Disassembly units may be in the same or different locations and their associated disassembly operations may be performed by the same technician or different technicians. The used product should go through all or some of the disassembly units for disassembling its required cores. In disassembly Petri nets, each transition consists of one or more disassembly units.

Usually disassembly operations are not completely independent from each other; many disassembly operations require prior removal of some cores And/Or joints. Therefore, the disassembly operations in the disassembly sequence should be implemented in a hierarchy order and we may assign a level to each disassembly unit. A disassembly unit that can disassemble a core, a noncore part or a subassembly of cores from the complete product is considered in level one. All the disassembly units that, in order to perform their disassembly operation, require the product to go through at least one disassembly unit in level $k - 1$ are considered in level k.

A schematic presentation of the flow of taken back product through the disassembly line is shown in Figure 6.8. As the return rate is stochastic, an inventory is considered for keeping the taken back products. The taken back product may be inspected before disassembly, to determine the statuses of its cores for an optimum disassembly. In initial inspection four statuses may be assigned to each core: good, repairable, nonrepairable, or undecided (Ghoreishi 2009, Ghoreishi et al. 2012). If a core is in proper working condition, its status is good; if it is not in proper working

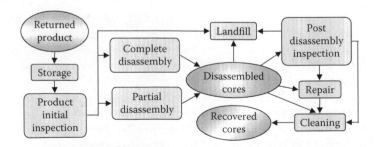

FIGURE 6.8 Schematic presentation of a disassembly line and its major segments and units.

condition, but can be brought back to the proper working condition with a reasonable cost, its status is repairable; if it is defected beyond repair, its status is nonrepairable; finally if the working condition of the core cannot be determined with certainty, its status is undecided.

Based on the condition of the product cores, the product may go for complete or partial disassembly or may be disposed to the landfill. The cores that should be disassembled in partial disassembly depend on the overall statuses of the product cores. In some disassembly lines the initial inspection may not exist. In such lines the product will be disassembled completely. Once the product is disassembled, the non-repairable cores and subassemblies are disposed to the landfill along with noncore parts. The good cores go for cleaning and then are ready to be used in the remanu-factured product. The repairable cores go for repair and after repair for cleaning. The cores that their statuses are not certain from the initial inspection (undecided cores) should be inspected after disassembly to determine if they are good, repairable, or nonrepairable. Once the functionality statuses of these cores are determined they will be sent for cleaning or repair, or will be disposed to the landfill.

In this modeling framework, a disassembly unit is characterized by its level, l_i, its cost, cd_i, and the joints and connections that are removed in that unit. i is the index that refers to different disassembly units. Also, a core is characterized in this modeling framework by its average repair cost (if repairable), $crCr_j$, its landfill cost, $clCr_j$, its cleaning cost, $ccCr_j$, its post disassembly inspection cost, $ciCr_j$, its value after being recovered, VCr_j, and list of all disassembly units required for its disas-sembly, D_{mj}. j is the index that refers to different cores. In addition to these param-eters, two sets of probabilities are required for cost/benefit analysis (or feasibility study) of disassembly. One set is the probabilities of the cores statuses in the initial inspection and other set is the probabilities of the cores statuses in post disassem-bly inspection for undecided cores. Probabilities of the core j is in good, repair-able, nonrepairable, and undecided condition after initial inspection are termed Pg_j, Pr_j, Pnr_j, and Pud_j, respectively. Probabilities of an undecided core turns out good, repairable, or nonrepairable after the post disassembly inspection are termed PgU_j, PrU_j, and $PnrU_j$, respectively.

6.5.5 OPTIMUM PARTIAL DISASSEMBLY BASED ON INITIAL INSPECTION

Once the statuses of different cores have been examined and determined in the initial inspection, the partial disassembly plan of the product can be determined. The partial disassembly plan is the decision about the cores that should be disassembled from the product, or more precisely, the disassembly units that the product should go through to optimize the disassembly cost and maximize the profit. The decision making trees (graphical methods) explained previously are usually used to determine the optimum disassembly sequence. These trees become enormously large by increasing the num-ber of cores. The number of cores in each tree should be kept as small as possible by eliminating the cores that are not required to be included. To do that, we first intro-duce the concept of an independent core and a group of dependent cores.

If a core does not have any disassembly operation in common with any other core, then the core is an independent core. A group of dependent cores consists of cores that each one has at least one disassembly operation in common with at least one other core in the group. Making decision about disassembly of an independent core depends only on the core status, regardless of the statuses of other cores. Also decision about the disassembling cores of a group of dependent cores depends only on the statuses of the cores in that group. Therefore, instead of drawing a decision tree for all the product cores, we need to draw one decision tree for each group of dependent cores.

The product cores have two final destinies: they are either being recovered or being disposed to the land field (or go for recycling if it is an option). If the profit of recovering an independent core is more than the profit of disposing it to the landfill (or recycling it), then the core should be recovered, otherwise, it should be disposed to the landfill. The profits of recovering the independent core Cr_j is the value of recovered core, VCr_j, minus its disassembly cost, $cdCr_j$, its cleaning cost, $ccCr_j$, and if applicable, its repair cost, $crCr_j$, and its post disassembly inspection cost, $ciCr_j$. The net profit of landfill is minus the landfill cost or if recycling is an option is the net profit of recycling.

For a good core there is no repair cost; if the following condition is satisfied, the core should be disassembled:

$$VCr_j - cdCr_j - ccCr_j > -clCr_j \qquad (6.23)$$

which can be rewritten as

$$f_g(Cr_j) = VCr_j - cdCr_j - ccCr_j + clCr_j > 0 \qquad (6.24)$$

f is defined as the net profit of recovering a core. For repairable cores, the repair cost should be included in the cost function as well. Therefore, if the following condition is satisfied the repairable core should be disassembled:

$$f_r(Cr_j) = VCr_j - cdCr_j - ccCr_j + clCr_j - crCr_j > 0 \qquad (6.25)$$

If a core is nonrepairable it should not be disassembled (unless it has the recycling option which generates a net profit more than disassembly cost).

For undecided cores the decision is based on what is more profitable on average. Once the core is disassembled, its status can be determined in a post disassembly inspection unit. If the core turns out good, the net profit of recovering is

$$f_{ud-g}(Cr_j) = VCr_j - cdCr_j - ccCr_j - ciCr_j + clCr_j \qquad (6.26)$$

If the core turns out repairable, the net profit of recovering is

$$f_{ud-r}(Cr_j) = VCr_j - cdCr_j - ccCr_j - ciCr_j - crCr_j + clCr_j \qquad (6.27)$$

And finally if the core turns out nonrepairable, the net profit of recovering that core is

$$f_{ud-nr}(Cr_j) = -cdCr_j - ciCr_j \qquad (6.28)$$

Note that if the core turns out nonrepairable, it has to be disposed to the landfill. Therefore, unlike the good or repairable cores, saving the landfill cost should not be included in the net profit. Using the probabilities of the undecided core statuses, PgU_j, PrU_j, and $PnrU_j$, the net profit of an undecided (on average) is

$$f_{ud}(Cr_j) = PgU_j \cdot f_{ud-g}(Cr_j) + PrU_j \cdot f_{ud-r}(Cr_j) + PnrU_j \cdot f_{ud-nr}(Cr_j) \qquad (6.29)$$

which can be rewritten as

$$f_{ud}(Cr_j) = (PgU_j + PrCr_j) \cdot VCr_j - ciCr_j + (PgU_j + PrU_j) \cdot clCr_j$$
$$-(PgU_j + PrU_j) \cdot ccCr_j - PrCr_j \cdot crU_j - cdCr_j \qquad (6.30)$$

The undecided cores should be disassembled if

$$f_{ud}(Cr_j) > 0 \qquad (6.31)$$

6.5.6 NET PROFIT OF THE DISASSEMBLY PROCESS

To determine the net profit of the disassembly, we need to know all the costs and revenues of the disassembly process. Values of the recovered cores are the source of revenue in disassembly process. Costs in disassembly process includes, take back cost, disassembly cost, inspection cost, repair cost, cleaning cost, and landfill cost. Both revenue and cost depend on the average statuses of cores in used products. Previously, two sets of probabilities were considered for cores statuses, one for the cores statuses after the initial inspection (Pg_j, Pr_j, Pnr_j, and Pud_j) and one for the statuses of undecided cores after the post disassembly inspection (PgU_j, PrU_j, and $PnrU_j$). The overall probabilities of a core being good, repairable, or nonrepairable are termed fPg_j, fPr_j, and $fPnr_j$:

$$fPg_j = Pg_j + Pud_j \cdot PgU_j \qquad (6.32)$$

$$fPr_j = Pr_j + Pud_j \cdot PrUr_j \qquad (6.33)$$

$$fPnr_j = Pnr_j + Pud_j \cdot PnrU_j \qquad (6.34)$$

In complete disassembly, the net profit of disassembly (NPD) can be calculated as an analytical expression in terms of these probabilities. However, in optimum partial disassembly deriving a closed form expression is too complicated if not impossible. The optimum disassembly plan may alter when the statuses of the product cores change. Both the revenue and the cost depend on the optimum disassembly plan. Therefore, to estimate the NPD, cost and revenue should be calculated for all possible combinations of cores' statuses. For a product with n cores, there are 4^n

possibilities for cores' statuses. This makes it very difficult to derive an analytical expression for the net profit. In optimum disassembly, the net profit may be estimated using computer programs (Ghoreishi 2009, Ghoreishi et al. 2012).

In complete disassembly all the product cores become disassembled and there is no need to the initial inspection. However, this does not mean that this stage should be eliminated. Sometimes, testing the functionality of a product core is less costly when it is assembled within the product (e.g., some computer components). Costs and revenue in complete disassembly are explained in the following.

Cost of taking back the used product—this cost is associated with motivation incentives, advertisement, and transportation (Section 6.4). Here, we consider the entire cost of take back as a transfer price. The transfer price is the average price that the remanufacturing segment should pay to the take back segment (or firm) for each used product; this price is termed ctb per product.

Disassembly cost—as all product cores are disassembled in complete disassembly, the disassembly cost is sum of the disassembly costs of all the disassembly units. Total disassembly cost per product is termed cdT and is defined as follows:

$$cdT = \sum_{\text{All } i} cd_i \tag{6.35}$$

Inspection cost—the inspection cost for each product consists of two parts, the cost of initial inspection termed itc and sum of all the post disassembly inspections costs. The probability that a post disassembly inspection is required for core j is Pud_j. The average post disassembly inspection cost of core j is termed ci_j and the total post disassembly inspection cost is termed ciT. Therefore, the total inspection cost in complete disassembly can be written as follows:

$$ciT = itc + \sum_{\text{All } j} Pud_j ci_j \tag{6.36}$$

Repair cost—repair cost should be considered for repairable cores. However, not every repairable core should be repaired. A repairable core should be repaired if the post disassembly value of the recovered core can justify its repair and cleaning costs. More quantitatively a disassembled repairable core will be repaired if

$$VCr_j - ccCr_j - crCr_j > -clCr_j \tag{6.37}$$

or alternatively

$$crCr_j < VCr_j + clCr_j - ccCr_j \tag{6.38}$$

The repair cost may vary among different repairable cores of the same type. These variations can be quantified via the probability density function or cumulative density function of repair cost for each core. Defining $CR_j(crCr_j)$ as the cumulative density function of repair cost of core j, the probability of the repair cost is less than x is $CR_j(x)$. Therefore, the fraction of repairable cores that are repaired is $CR_j(VCr_j + clCr_j - ccCr_j)$. Considering this, the repair cost, crT, is

$$crT = \sum_{\text{All } j} fPr_j \int_0^{VCr_j + clCr_j - ccCr_j} \xi R_j(\xi)d\xi \qquad (6.39)$$

where R_j is the probability density function of repair cost.

Sometimes estimating the distribution of the repair cost or estimating the repair cost before the repair is not practical. In such cases, we may make the decision based on the average repair cost and use a more simplified model. Here $crCr_j$ represents the average repair cost of core j (is not a random variable anymore). In this case, if $crCr_j < VCr_j + clCr_j - ccCr_j$ all the repairable cores are repaired and if $crCr_j \geq VCr_j + clCr_j - ccCr_j$, none of the repairable cores should be repaired.

Assuming $\mathbf{1}(x)$ denotes the unit step function (zero if $x < 0$ and 1 if $x \geq 0$), the repair cost can be approximated as follows:

$$crT = \sum_{\text{All } j} fPr_j \cdot crCr_j \cdot \mathbf{1}(VCr_j + clCr_j - ccCr_j - crCr_j) \qquad (6.40)$$

Cleaning cost—good cores and the repairable cores that are repaired should go through a cleaning process to become completely recovered. Probability of a good core is fPg_j and a probability of a repairable core is fPr_j. However, as discussed before, the number of repaired cores may be less than the number of repairable cores. The fraction of repairable cores that their repair is justifiable is $CR_j(VCr_j + clCr_j - ccCr_j)$, and so the total cleaning cost is

$$ccT = \sum_{\text{All } j} ccCr_j \left[fPg_j + fPr_j \cdot CR_j(VCr_j + clCr_j - ccCr_j) \right] \qquad (6.41)$$

Similar to before if CR_j is not available, cleaning cost may be approximated as follows:

$$ccT = \sum_{\text{All } j} ccCr_j \left[fPg_j + fPr_j \cdot \mathbf{1}(VCr_j + clCr_j - ccCr_j - crCr_j) \right] \qquad (6.42)$$

Landfill cost—the noncore parts of the used product, the nonrepairable cores and the repairable cores that their repairs are not justified, should be disposed to the landfill. The landfill cost of the noncore parts is the same for all used products and is termed, $clNCr$. The total landfill cost, clT, is sum of the landfill costs of the noncore parts, the nonrepairable cores and the repairable cores that are not repaired. Therefore, the landfill cost is

$$clT = clNCr + \sum_{\text{All } j} clCr_j \left[fPnr_j + fPr_j \cdot (1 - CR_j(VCr_j + clCr_j - ccCr_j)) \right] \qquad (6.43)$$

and if CR_j is not available, it may be approximated as follows:

$$clT = clNCr + \sum_{All\ j} clCr_j \left[fPnr_j + fPr_j \cdot (1 - 1(VCr_j + clCr_j - ccCr_j - crCr_j)) \right] \quad (6.44)$$

Revenue of the recovered cores—to calculate the NPD, we assumed that all the recovered cores are used in the later stages of remanufacturing. Therefore, the revenue (benefit) of disassembly, bT, is the value of all recovered cores:

$$bT = \sum_{All\ j} VCr_j [fPg_j + fPr_j \cdot CR_j(VCr_j + clCr_j - ccCr_j)] \quad (6.45)$$

And it can be approximated as follows:

$$bT = \sum_{All\ j} VCr_j [fPg_j + fPr_j \cdot 1(VCr_j + clCr_j - ccCr_j - crCr_j)] \quad (6.46)$$

Net profit of disassembly—the net profit of disassembly, NPD, is the revenue of disassembly minus all the disassembly costs:

$$NPD = NR\ (bT - ctb - ciT - cdT - crT - ccT - clT) - cgd \quad (6.47)$$

where cgd is the cost that is not scaled with the number of products.

6.5.7 REASSEMBLY

Once the cores are recovered from the taken back products they may be sold as is or may be used in manufacturing of the same or different types of products. If these cores are used in the manufacturing of a product, that product should be considered as a remanufactured product. A remanufactured product is a product that all or some of its cores are recovered from the used products. The value of the recovered cores is usually considered less than the new cores and so the total cost of remanufactured product should be less than the new product. This is essential to provide market incentives for the remanufactured product.

To simplify the model, we considered the reassembly process as an independent process (from disassembly); the values of recovered cores are the transfer prices. In this process the inputs are the recovered cores from the disassembly process and the output is the remanufactured product. There is no guarantee that the recovered cores from the disassembly process have the same quantities, therefore, different remanufactured products may have different numbers of recovered cores. Figure 6.9 shows a schematic diagram for the reassembly process. Both the recovered cores and the new cores enter the assembly line and are assembled to produce the remanufactured product. The assembled product will go to a final test and quality control unit, and if assembled

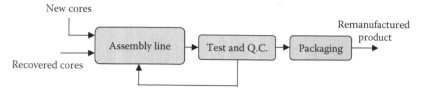

FIGURE 6.9 Block diagram of the reassembly process and its units.

properly, will be sent for packaging and sale. If there is any defect in the assembled product, it will be sent back to the assembly line to fix that defect.

RP_j is defined as the probability of recovering core j from the used product. It can be calculated based on the probabilities of the core statuses as

$$RP_j = (fPg_j) + (fPr_j) \cdot CR_j (VCr_j + clCr_j - ccCr_j) \tag{6.48}$$

Or alternatively, can be approximated as

$$RP_j = (fPg_j) + (fPr_j) \cdot 1(VCr_j + clCr_j - ccCr_j - crCr_j) \tag{6.49}$$

If core m has the maximum recovery percentage of RP_m, the number of remanufactured products is $NR \cdot RP_m$.

Cost of recovered cores—the transfer price that should be paid to the disassembly process for the recovered cores is the cost of recovered cores, $crcT$:

$$crcT = NR \sum_{\text{All } j} RP_j \cdot VCr_j \tag{6.50}$$

Cost of new cores—in order to have $NR\ RP_m$ remanufactured products, some new cores are required to match the number of cores recovered at rates lower than RP_m. The values of new cores are different than the values of recovered cores and are termed $VNCr_j$. The total cost of required new cores, $cncT$, can be calculated as follows:

$$cncT = NR \sum_{\text{All } j} (RP_m - RP_j) \cdot VNCr_j \tag{6.51}$$

Assembly cost—all the costs involved in assembling the cores to the product, test and quality control of the product, and its packaging are bundled together as assembly cost and are shown by cRA. A small fraction of the remanufactured products may not pass the final quality control stage and has to come back to assembly line. For simplicity we define cRA as the assembly cost per successfully remanufactured product. In addition to cRA, the assembly line may have a general cost that is not scaled with the number of remanufactured products. This cost is shown by $cRAg$ in the model. The total cost of reassembly, $cRAT$, is as follows:

$$cRAT = NR \cdot RP_m \cdot cRA + cRAg \qquad (6.52)$$

Revenue of the reassembly—revenue in reassembly is associated with the market value (price) of remanufactured product, termed VRP in this model. VRP is a transfer price that connects the disassembly and reassembly phase to the resale phase. The total revenue of remanufacturing is shown by bRP and can be calculated as follows:

$$bRP = NR \cdot RP_m \cdot VRP$$

Net profit—combining the above costs and revenue, we can determine the net profit of reassembly, NPR, as follows:

$$NPR = bRP - cRAT - cncT - crcT \qquad (6.53)$$

6.6   COST/BENEFIT ANALYSIS OF RESALE PHASE

Once a product is remanufactured in the remanufacturing line, it goes to the marketing stage. The actual revenue of the remanufacturing process is associated with this stage. Although in previous phases we introduced revenues for the taken back and disassembly and reassembly phases, but these revenues are transfer prices introduced to enable independent cost/benefit analysis of each phase.

Resale and marketing of the remanufactured product have been studies from different perspectives including cost and revenue allocation, marketing strategies in remanufacturing, pricing and matching demand and supply, and the dynamics of the joint sale of the new and remanufactured products. In this section, we do not intend to discuss the marketing aspects of remanufacturing. We explain a modeling frame work for resale in the context of existing literature.

Sometimes the remanufacturing is performed by the same firm that manufactures the original (new) product. However, there are usually separate divisions for remanufacturing and manufacturing with separate managers. Examples of these firms are Hewlett Packard (Guide and Van Wassenhove 2002), Bosch (Valenta 2004) and Daimler Chrysler (Driesch et al. 2005). As the same cores that are used in the original new product will be used in the remanufactured product, it is not clear how to allocate the cost of these cores (Toktay and Wei 2005). Also sale of the remanufactured product can adversely affect the sale of the new product, if both marketed simultaneously. Therefore, it is also unclear how to allocate the revenue generated by the remanufactured products.

A partial cost of the new product cores may be allocated to the remanufacturing process as the remanufacturing division uses the new product cores in their second life cycle (Toktay and Wei 2005). However, after the new product is sold to a customer, it is the customer's property and remanufacturing division should buy it back from the customer and bear all the costs associated with the recovering the cores. Therefore, it may be more rational not to allocate any cost of new cores to the remanufacturing process.

The remanufactured product cannibalizes the sale of the new product, and so it may be argued that some of the revenue generated by the remanufactured product should be allocated to the new product division. In our resale model we allocated all the revenue generated by remanufactured product to the remanufacturing process. The rational is that in a free and competitive market, the remanufacturing can be performed by a separate firm and the manufacturing division cannot claim any revenue of the remanufactured product.

6.6.1 Marketing Strategies in Remanufacturing

During the last few decades many industrial firms have gained significant revenues by remanufacturing the used products and showed that there is a big market for remanufacturing. According to Remanufacturing Central in 1997 there were 73,000 remanufacturing firms in the United States with a total sale of $53 billion (Lund 2005). Successful examples of this industry include Kodak, BMW, IBM, and Xerox. The success of remanufacturing process depends highly on marketing the remanufactured product. The market-driven factors of remanufacturing have been discussed in several studies (Atasu et al. 2008, Ferrer and Whybark 2000, Lund and Hauser 2003, McConocha and Speh 1991, Subramanian and Subramanyam 2008, Walle 1988).

Ferrer and Whybark (2000) considered three motivations for the remanufacturing: legislation, prolonging economic life, and strategic initiatives. When there are safety or environmental concerns, the manufacturer may be forced by law to take back the used product and recycle or remanufacture it. An example of this driving force is remanufacturing the x-ray equipment. Automobile or home appliances may be subjected to similar legislations in the future. Subway cars, machine tools, and conveyers are practical examples of equipment that are remanufactured to prolong their life cycle. Many other remanufacturing lines are implemented in large companies as a strategic plan. Xerox copiers, Kodak single-use cameras, and several automobile components are examples of products remanufactured for strategic initiatives.

Market acceptance and supporting the marketing effort are two factors affecting successful marketing (Ferrer and Whybark 2000). The customers' perception of the quality of remanufactured product is generally negative. They have concerns regarding the quality and durability of the used cores within the remanufactured product. Developing the market for the remanufactured products requires educating the customers through advertisement strategies and may be focused on certain segments of the market. After convincing the customers of the benefit of remanufactured products and providing a potential market for remanufactured products, that market should be supported and stabilized by marketing incentives like a lower price of remanufactured products compared with the new product as well as a competitive warranty.

Subramanian and Subramanyam (2008) considered the effect of the following four driving factors for the remanufacturing demand: price advantage, seller reputation, buyer expertise, and quality. They found that price difference between new and remanufactured products is a significant marketing strategy. They concluded that customers expect the remanufactured product to perform as good as original new product or perhaps to have an upgraded performance. They also found that the reputation of the seller is a significant factor for the customers' perception of the quality, and it affects the price difference significantly. Collected data showed that effect of customers' expertise in choosing between new and remanufactured products varies across product categories and on average the buyers of remanufactured products are less experienced.

Knowledge of how the sale of a remanufactured product varies with its price is required for the cost/benefit analysis of remanufacturing. We term the relation between the number of sale and the price of remanufactured product, the demand–price relation. From the factors that can affect this relation, we consider the advertisement and warranty.

6.6.2 Demand–Price Relation

Reducing the price of the remanufactured product may increase the net profit of remanufacturing by increasing the number of sale or may reduce the net profit by reducing the profit per item sold. Therefore, the price of a remanufactured product is a parameter that should be optimized to maximize the total profit of resale phase. Optimal pricing and the demand–price relationship is considered in several studies (Atasu et al. 2008, Celebi 2005, Debo et al. 2005, Guide et al. 2003a, Mitra 2007, Vorasayan and Ryan 2006). The demand price function has been determined by analyzing the willingness of the customers to buy the remanufactured product.

Assume N_{pc} is the number of potential consumers in the market and θ is the willingness of a consumer to pay for the new product. Without loss of generality, we may assume that θ is normalized to P_{xn}, the maximum price a customer may pay for the new product (let say θ for 99% of customers is less than P_{xn}). Therefore, θ varies over the interval [0, 1]. It is usually assumed that variation of θ over [0, 1] is uniform (Atasu et al. 2008, Celebi 2005, Guide et al. 2003a, Vorasayan and Ryan 2006). On average, the willingness of a customer to pay for the remanufactured product is less than the new product. In the model, it is considered that if the customers are willing to pay θ for the new product, they are willing to pay $\delta\theta$ for the remanufactured product. Therefore, willingness of the customers to pay for the remanufactured product varies over the interval [0, δ]. We defined P_r as the price of remanufactured product, d_r as the demand (number of sale) for the remanufactured product, and d_n as the demand for the new product. If the new product and the remanufactured product are being sold in separate markets (no competition effect), their demands are as follows:

$$d_r = \frac{N_{pc}(\delta - P_r)}{\delta} = N_{pc}\left(\frac{1 - P_r}{\delta}\right) \tag{6.54}$$

$$d_n = N_{pc}(1 - P_n) \tag{6.55}$$

When both new product and remanufactured product are being sold in the same market, there will be a competition between them. In this case, consumers purchase the one (new or remanufactured) that is more beneficial for them. To model their competition a utility parameter has been used (Atasu et al. 2008, Debo et al. 2005, Vorasayan and Ryan 2006). Utility, U, is defined as the difference between the willingness of the customer to pay for a product and its price in the market:

$$U = \theta - P \tag{6.56}$$

If utility is greater than zero, the customer purchases the product. When there are several similar products in the market, the customer purchases the product that its utility is the largest. The willingness of a customer to purchase different products (of the same type) depends on many factors including the quality and durability of the products. The perception of a customer for the quality and durability of the remanufactured product is less than the new product (which is included in the model via δ). To simplify the model it is assumed that δ is constant for all customers. U_n and U_r are defined as the utilities of new and remanufactured products, respectively as follows:

$$U_n = \theta - P_n \tag{6.57}$$

$$U_r = \delta\theta - P_r \tag{6.58}$$

If both new and remanufactured products are present in the market, the customer purchases the new product if

$$U_n > U_r \quad \text{and} \quad U_n > 0 \tag{6.59}$$

and purchases the remanufactured product if

$$U_r > U_n \quad \text{and} \quad U_r > 0 \tag{6.60}$$

And finally the customer will not purchase any of them if

$$U_r < 0 \quad \text{and} \quad U_n < 0 \tag{6.61}$$

Depending on the prices of new and remanufactured products, the customers may all purchase the new product or may all purchase the remanufactured product or some buy the new and some buy the remanufactured product. Equation 6.59 can be rewritten as follows:

$$\begin{cases} \text{I)} & \theta > P_n \\ & \text{and} \\ \text{II)} & \theta(1-\delta) + P_r > P_n \end{cases} \tag{6.62}$$

Condition I ensures that for any combination of P_n and P_r there are some customers who purchase the new product. Satisfying condition II, over the range of θ that both remanufactured and new products have positive utilities, ensures that in competition between new and remanufactured product all customers prefer the new product. If $P_n < 1$, then condition I is satisfied for some customers. For any given P_n, utility of both new and remanufactured products are positive if

$$\theta > P_n \quad \text{and} \quad P_r < \theta\delta \tag{6.63}$$

In this region if condition II is satisfied for $\theta = P_n$ (minimum possible), it is satisfied for all θ. Therefore, in the $P_n - P_r$ domain, within the following region all customers purchase the new product:

$$\begin{cases} P_n < 1 \\ \\ P_n(1-\delta) + P_r > P_n \end{cases} \tag{6.64}$$

which can be rewritten as

$$\begin{cases} P_n < 1 \\ \\ P_r > \delta P_n \end{cases} \tag{6.65}$$

Similarly, Equation 6.60 can be rewritten as

$$\begin{cases} \text{I)} \quad \delta\theta > P_r \\ \qquad \text{and} \\ \text{II)} \quad \theta(1-\delta) + P_r < P_n \end{cases} \tag{6.66}$$

To satisfy condition I for some values of θ, P_r should be less than δ. Satisfying condition II for $\theta = 1$ (maximum possible) ensures that, where both remanufactured and new products are competing, all customers purchase remanufactured product. This can be summarized as follows:

$$\begin{cases} P_r < \delta \\ \\ P_n > (1-\delta) + P_r \end{cases} \tag{6.67}$$

Figure 6.10 shows for which combinations of P_n and P_r both products can be sold and for what combinations only new or only remanufactured product will be sold (Vorasayan and Ryan 2006). Knowing these regions, we can estimate the

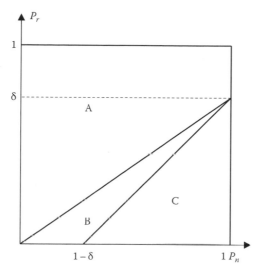

FIGURE 6.10 Different regions in the $P_r - P_n$ price domain, from the perspective of competition between new and remanufactured products. In region A all of the customers purchase the new product, in region C all of the customers purchase the remanufactured product, and in region B some of the customers purchase the new product and some purchase the remanufactured product.

demand price relation of the remanufactured product in the presence of originally manufactured product based on the assumption that the willingness of the customers to purchase the product is distributed uniformly. For a given P_n, we may consider the following situations:

6.6.2.1 $P_n \leq 1 - \delta$

In this case depending on the value of P_r, the price combinations may fall in any of the two regions B and A. In region B, the range of θ over which customers purchase new or remanufactured product can be calculated as follows:

$$0 < P_r < \delta P_n \Rightarrow \begin{cases} \dfrac{P_r}{\delta} < \theta < \dfrac{P_n - P_r}{1 - \delta} & \text{Remanufactured} \\[3mm] \dfrac{P_n - P_r}{1 - \delta} < \theta < 1 & \text{New} \end{cases} \qquad (6.68)$$

In region A no customer purchases the remanufactured product; the range of θ over which customers purchase new product can be calculated as follows:

$$\delta P_n < P_r < 1 \Rightarrow \begin{cases} 0 & \text{Remanufactured} \\[3mm] P_n < \theta < 1 & \text{New} \end{cases} \qquad (6.69)$$

Demand for the new and remanufactured products can be calculated by integrating over the associated ranges of willingness as follows:

$$
d_r = \begin{cases} \dfrac{P_n - P_r}{1 - \delta} - \dfrac{P_r}{\delta} & \text{if } 0 < P_r < \delta P_n \\[4mm] 0 & \text{if } \delta P_n < P_r < 1 \end{cases} \tag{6.70}
$$

$$
d_n = \begin{cases} 1 - \dfrac{P_n - P_r}{1 - \delta} & \text{if } 0 < P_r < \delta P_n \\[4mm] 1 - P_n & \text{if } \delta P_n < P_r < 1 \end{cases} \tag{6.71}
$$

6.6.2.2 $P_n \geq 1 - \delta$

In this case, depending on the value of P_r, price combinations may fall in any of three regions C, B, or A. In region C, a customer purchases the remanufactured product if

$$
0 < P_r < P_n - (1 - \delta) \Rightarrow \begin{cases} \dfrac{P_r}{\delta} < \theta < 1 & \text{Remanufactured} \\[4mm] 0 & \text{New} \end{cases} \tag{6.72}
$$

In region B, the ranges of θ for the new and remanufactured products are

$$
P_n - (1 - \delta) < P_r < \delta P_n \Rightarrow \begin{cases} \dfrac{P_r}{\delta} < \theta < \dfrac{P_n - P_r}{1 - \delta} & \text{Remanufactured} \\[4mm] \dfrac{P_n - P_r}{1 - \delta} < \theta < 1 & \text{New} \end{cases} \tag{6.73}
$$

In region A, a customer purchases the new product if

$$
\delta P_n < P_r < 1 \Rightarrow \begin{cases} 0 & \text{Remanufactured} \\[4mm] P_n < \theta < 1 & \text{New} \end{cases} \tag{6.74}
$$

Demand for the new and remanufactured products can be calculated by integrating over the associated ranges of willingness as follows:

$$d_r = \begin{cases} 1 - \dfrac{P_r}{\delta} & \text{if } 0 < P_r < P_n - (1-\delta) \\\\ \dfrac{P_n - P_r}{1-\delta} - \dfrac{P_r}{\delta} & \text{if } P_n - (1-\delta) < P_r < \delta P_n \\\\ 0 & \text{if } \delta P_n < P_r < 1 \end{cases} \tag{6.75}$$

$$d_n = \begin{cases} 0 & \text{if } 0 < P_r < P_n - (1-\delta) \\\\ 1 - \dfrac{P_n - P_r}{1-\delta} & \text{if } P_n - (1-\delta) < P_r < \delta P_n \\\\ 1 - P_n & \text{if } \delta P_n < P_r < 1 \end{cases} \tag{6.76}$$

Recent advertisement and educational programs on green technology are in favor of consuming the remanufactured products to reduce waste and conserve resources. Therefore, customers may be divided into two groups: regular customers and green customers (Atasu et al. 2008). Green customers value the remanufactured product same as the new product as long as it functions adequately. The willingness of the regular customers for the remanufactured product is reduced by a constant depreciation factor, δ, but the willingness of the green customers for both new and remanufactured product is the same.

This modeling framework does not use multiple depreciation factors for different types of customers, instead, it considers the effect of advertisement and educational green technology programs in the demand–price relation. It is assumed that δ will be affected by two factors: the green advertisement and the warranty of the remanufactured product. Both of these can increase δ and improve the demand–price function toward a higher demand for the same price. Advertisement and warranty programs incur some cost to the resale phase and so there would be a trade off on how much to spend on each to achieve the optimum results and maximize the net profit. In the model, δ is considered as a function of green advertisement cost, C_{ga}, and warranty cost per remanufactured product, C_w:

$$\delta = \delta(C_{ga}, C_w) \tag{6.77}$$

6.6.3 Cost/Benefit Model of Resale

Three marketing conditions are considered in this part: marketing the remanufactured product in a different segment than the new product, marketing the remanufactured product in the same segment of the new product when the remanufacturer is a different firm than the original manufacturing firm (duopoly situation), and marketing the remanufactured product in the same segment of the new product when the remanufacturing and manufacturing is performed by the same firm (monopoly situation).

6.6.3.1 Two Market Segments for Manufacturing and Remanufacturing

In this case marketing the remanufactured product is uncoupled from the originally manufactured product and can be studied independently. Like before, this phase of remanufacturing is connected to the disassembly and reassembly phase by the value of the remanufactured product at the remanufacturing site (the transfer price). This value can be as low as the total cost of remanufacturing, C_{tr}, or as high as the resale price of the remanufactured product, P_r. As in this section the focus of modeling is the resale phase we assumed that all the profit is allocated to the resale phase and therefore, the transfer price is the total cost of remanufacturing including the costs associated with the take back.

Costs: The cost parameters of resale phase are the total cost of remanufacturing per remanufactured product, C_{tr}, the advertisement cost, C_{ga}, and the warranty cost per remanufactured product, C_w.

Revenue: The revenue in this phase is generated by the sale of the remanufactured product and P_r is the associated parameter.

The total number of potential customers in the market is termed, N_{pc}. d_r is defined as the demand normalized to N_{pc}. Therefore, number of sales of the remanufactured product, N_{sr}, is

$$N_{sr} = d_r N_{pc} \qquad (6.78)$$

Number of sales of the remanufactured product should be equal to the number of the products remanufactured in the remanufacturing line:

$$N_{sr} = NR \cdot RP_m \qquad (6.79)$$

Cost/benefit analysis of the resale phase includes choosing the best option for the warranty, deciding how much to spend on advertisement, and determining the optimum price of the remanufactured product. Therefore, C_{ga}, C_w, and P_r are the independent variables of this cost/benefit analysis. The warranty cost is usually a discrete parameter and may take different values for different options of warranty. Duration and coverage of warranty is different in different plans. Warranty is offered for a certain period of time like 1 year, 3 years, or life time; it may be a free replacement of the product or a prorated price for the replacement product. Advertisement may also have multiple options. The frequency of advertisement increases almost continuously by advertisement cost. d_r, in addition to P_r, is a function of C_{ga} and C_w through depreciation factor δ. It is noteworthy that in this case there is no competition with the new product, and it is not necessary to introduce a depreciation factor. The customers' demand for the remanufactured product can be measured directly as a function P_r, C_{ga}, and C_w.

Matching demand and supply in Equation 6.79 makes the remanufacturing cost a function of demand. Note that changing the number of remanufactured products can be achieved by increasing/decreasing the financial incentive, changing the transportation method, and increasing/decreasing the take back advertisement. All these affect the total cost of remanufacturing. Therefore,

$$d_r = d_r(P_r; C_{ga}, C_w); \quad C_{tr} = C_{tr}(d_r) \tag{6.80}$$

By introducing C_{tr} as a function of demand in our modeling framework, we enabled the model to account for the limitation of supply in the remanufacturing process.

The net profit of resale phase for different segment marketing, NPS_{ds}, can be calculated as follows:

$$NPS_{ds} = d_r N_{pc} (P_r - C_w - C_{tr}) - C_{ga}$$

$$d_r = d_r(P_r, C_w, C_{ga}) \tag{6.81}$$

$$C_{tr} = C_{tr}(d_r) = C_{tr}(P_r, C_w, C_{ga})$$

6.6.3.2 Same Market Segment: Duopoly Situation

When both new and remanufactured products are marketed in the same segment, the demand price relation of the new product depends on the price of the remanufactured product, and, similarly, the demand–price relation of remanufactured product depends on the price of the new product. Duopoly is referred to the situation that a local or independent remanufacturer (LR) competes with the original equipment manufacturer (OEM) in the market (Ferguson and Toktay 2006, Ferrer and Swaminathan 2006, Heese et al. 2003, Majumder and Groenevelt 2001).

Cost model of duopoly is similar to the cost model of two market segments, since the objective is to maximize the net profit of remanufacturer. But, the demand–price relation and consequently the optimum values of the parameters (P_r, C_{ga}, and C_w) and the maximum net profit are different. It is noteworthy that when the remanufactured product comes to the market, the OEM may change the price of the new product to compete with the remanufactured product. The net profit in this case is a function of four independent parameters: P_r, C_{ga}, C_w, and P_n. The net profit of resale in duopoly, NPS_{dp}, can be written as follows:

$$NPS_{dp} = d_r N_{pc} (P_r - C_w - C_{tr}) - C_{ga}$$

$$d_r = d_r(P_r, P_n, \delta) = d_r(P_r, P_n, C_w, C_{ga}) \tag{6.82}$$

$$C_{tr} = C_{tr}(d_r) = C_{tr}(P_r, P_n, C_w, C_{ga})$$

Although P_n is a parameter in this cost/benefit analysis, but it cannot be determined by the LR; it is a parameter that the OEM determines its value to maximize its own profit.

6.6.3.3 Same Market Segment: Monopoly Situation

In this case, both the new and the remanufactured products are presented by the same firm to the market and the objective is to maximize the net profit of both new and remanufactured products. This situation is considered as the monopolist manufacturer (Atasu et al. 2008, Debo et al. 2005, 2006, Ferguson and Toktay 2006, Ferrer and Swaminathan 2006). For the case of monopolist manufacturer, the demand–price relations for both new and remanufactured products are similar to the duopoly situation as both the new and the remanufactured products are in the market and competing with each other. But the cost/benefit model is different in two main aspects: first, the goal is to maximize the sum of the net profits of manufacturing and remanufacturing together rather than each separately, and second all four parameters, P_n, C_w, C_{ga}, and P_r, are controlled by the same firm. The net profit of the firm in monopoly is modeled as follows:

$$NPS_{mp} = d_r N_{pc}(P_r - C_w - C_{tr}) - C_{ga} + d_n N_{pc}(P_n - C_{tn})$$

$$d_r = d_r(P_r, P_n, \delta) = d_r(P_r, P_n, C_w, C_{ga})$$

$$d_n = d_n(P_n, P_r, \delta) = d_n(P_n, P_r, C_w, C_{ga})$$ \hfill (6.83)

$$C_{tr} = C_{tr}(d_r) = C_{tr}(P_r, P_n, C_w, C_{ga})$$

where
C_{tn} is the total cost of manufacturing the new product
$d_n N_{pc}(P_n - C_{tn})$ is the net profit of the new product

6.7 PRACTICAL EXAMPLE

Assume a firm that manufactures an industrial monitoring device for power plants and power distribution posts. The firm decides to use some of the components of certain models of used computers in the production of the remanufactured monitoring device. Remanufacturing involves taking back the used computers from the customers of a limited geographical region, recovering the required cores, and using them in the production of the remanufactured monitoring devices. In the following we use the developed modeling framework to optimize this remanufacturing process and maximize its net profit.

6.7.1 CHARACTERISTIC PARAMETERS AND FUNCTIONS OF THE PROBLEM

The net profit of the remanufacturing is modeled over 1 year. The characteristic parameters of the problem are given in the following:

Remanufacturing firm considers three options for the transportation of the used computers to the remanufacturing site:

1. Picking up the used computers from the customers' convenient locations
2. Providing the customers with postage paid boxes
3. Collecting the used computers from the customers at certain locations

It also considers three options for advertising the take back policy:

1. Advertising in local TV channels
2. Advertising in local newspapers
3. Advertising in related retail stores (e.g., stores that are selling or repairing computers)

The characteristic parameters of the transportation and advertisement methods are given in Tables 6.1 and 6.2.

Based on the preliminary study of the customers' behavior, the Γ function is approximated as follows:

$$\Gamma(c,g,f) = \frac{(fc+g)^3}{1.5(fc+g)^3 + 2(fc+g)^2 + 10,000} \tag{6.84}$$

N, the total number of used computers is approximated to be 70,000 and tbc, the general cost of tb is estimated to be $6,000.

The five cores of the taken back computers that are used in the remanufactured monitoring devices are hard drive, CD drive, motherboard, CPU, and RAM. These cores are termed core1 to core 5, respectively. All the five cores are disassembled from all the used computers (complete disassembly). The costs associated with disassembling and recovering these cores are tabulated in Table 6.3. The overall probabilities of the cores that are good, repairable, or nonrepairable are given in Table 6.4. Cost of other components that are required for manufacturing

TABLE 6.1

Parameters of Transportation Options

	t	tg	f
Method 1	15	7,000	1
Method 2	20	1,000	0.7
Method 3	8	30,000	0.6

TABLE 6.2

Parameters of Advertisement Options

	W_1	g	Ω_{ss}	W_{sc}
Method 1	20,000	5	0.9	200,000
Method 2	4,000	1	0.3	20,000
Method 3	5,000	4	0.5	40,000

TABLE 6.3

Characteristic Parameters of Cores

Cores	Inspection Cost ($)	Repair Cost ($)	Cleaning Cost ($)	Landfill Cost ($)	Disassembly Cost ($)	Equivalent New Core ($)
Core 1	2.0	~5.0	0.4	0.5	1.0	30.0
Core 2	2.0	~8.0	1.0	0.5	1.0	25.0
Core 3	3.0	~7.0	0.2	1.5	1.0	45.0
Core 4	1.0	—	1.0	0.1	0.6	40.0
Core 5	1.0	—	0.2	0.1	0.3	10.0

TABLE 6.4

Overall Probabilities of Cores Statuses

Cores	fPg_j	fPr_j	$fPnr_j$
Core 1	0.80	0.10	0.10
Core 2	0.70	0.10	0.20
Core 3	0.90	0.05	0.05
Core 4	0.95	0.00	0.05
Core 5	0.98	0.00	0.02

TABLE 6.5

Other Parameters of Disassembly and Reassembly

General cost of disassembly, cgd	$9,000
Reassembly cost per product, cRA	$125
General cost of reassembly, $cRAg$	$11,000
Value of remanufactured product, VRP	$250

the monitoring device is included in the reassembly and Q.C., cRA. Value of the remanufactured product and other parameters of the disassembly and reassembly phase are given in Table 6.5.

6.7.2 MODELING THE TAKE BACK PHASE

The value of taken back product, a, is the transfer price required for modeling the tb phase. For this problem the average quality of the taken back computers is not expected to vary significantly with the financial incentive; a is considered to be $60 independent of c. To find the maximum net profit, we calculated the net profit using Equation 6.15 as a function of W_2 and c, for all combinations of the advertisement and transportation methods.

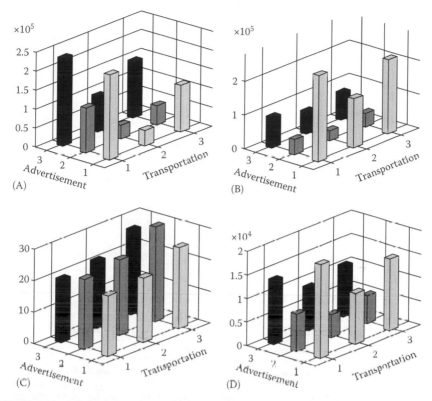

FIGURE 6.11   Maximum net profit (panel A), the associated optimum parameters—optimum advertisement cost and optimum financial incentive (panels B and C, respectively) and the resultant number of returns (panel D), for different combinations of the method of advertisement and method of transportation.

Figure 6.11 compares the maximum net profit and its associated optimum values of c and W_2 for all combinations of the advertisement and transportation methods.

The combination of method 1 of transportation and method 3 of advertisement generates the maximum net profit of $231,000. The optimum advertisement cost, W_2, is $91,000; the optimum financial incentive, c, is $20; the resultant number of returns is 13,600. Method 1 of transportation with method 1 of advertisement generates a profit of $224,000, which is close to the maximum profit (Figure 6.11A) and may be considered as an alternative option. In this case, the optimum financial incentive is almost the same (Figure 6.11C), but the optimum advertisement cost increases substantially to $252,000 (Figure 6.11B). This increased cost is compensated in the total net profit by the increased number of returns (Figure 6.11D). Variations of the net profit, Ψ, and the number of returns, N, by W_2 and c are shown in Figure 6.12 for method 1 of transportation and method 3 of advertisement. Increasing the frequency of advertisement (proportional to advertisement cost) or the financial incentive, initially, increases the net profit by increasing the number of returns; once the maximum point is reached, the net profit decreases because of the increased cost of financial incentive or advertisement.

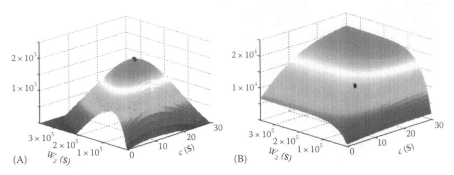

FIGURE 6.12 Net profit of take back ($) (Panel A) and the number of returns (Panel B) as a function of the advertisement cost, W_2, and the financial incentive, c, for method 1 of transportation and method 3 of advertisement. Black circles show the location of the maximum net profit.

6.7.3 MODELING THE DISASSEMBLY AND REASSEMBLY PHASE

The net profits of disassembly and reassembly are modeled by Equations 6.47 and 6.53, respectively. The values of the recovered cores are the transfer prices that connect the two segments of this phase. The values of the recovered cores are chosen equal to 70% of the values of their equivalent new cores. Take back cost, ctb, is equal to a, the value of taken back product at remanufacturing site; it is a transfer price for disassembly phase.

For the given parameters of the disassembly and reassembly phase and the optimum number of returns determined in take back phase, the net profits of disassembly and reassembly are $248,300 and $217,700, respectively, and therefore the NPD and reassembly phase is $465,900. Although, varying the values of recovered cores changes the net profits of both disassembly and reassembly segments individually, it does not change their sum (net profit of the disassembly and reassembly phase).

If the take back phase and the disassembly and reassembly phase are performed by the same firm, the sum of the net profits of both phases should be maximized. The transfer price, a, does not appear in the sum of the net profits, as it is a cost in disassembly and reassembly phase and the revenue in the take back phase. However, the optimum parameters of these phases, and consequently the net profit depend on the value of a. Both the take back phase and the disassembly and reassembly phase are optimized for the range of a between $40 and $140 and the results are shown in Figures 6.13 and 6.14.

The optimum transportation method is always method 1 (Figure 6.13A), and the optimum advertisement method is method 3, if the transfer price is less than $62; it changes to method 1 if the transfer price is larger than $62 (Figure 6.13B). Increasing a increases the net profit of tb (Figure 6.14A) by both increasing the revenue per returned product and increasing the optimum number of returns (Figure 6.14D). At $a = $62, method 3 of advertisement (retail store advertisement) cannot inform sufficient customers for the optimum number of returns and so the optimum method of advertisement alters to method 1 (TV advertisement) that can reach a broader range

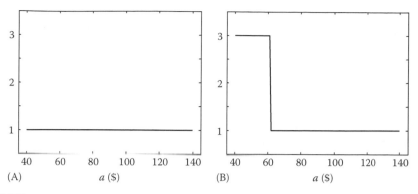

FIGURE 6.13   The optimum transportation method (panel A) and the optimum advertise-ment method (panel B) of the take back phase when the transfer price, a, varies between \$40 and \$140.

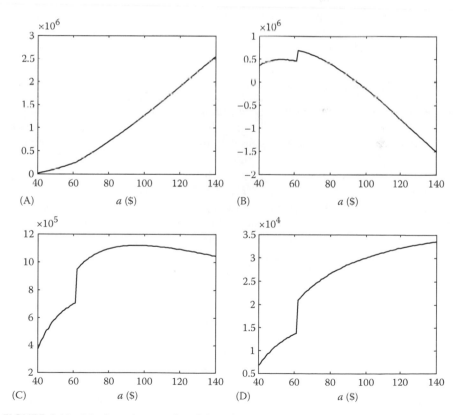

FIGURE 6.14   Maximum net profits of the take back phase (panel A), the disassembly and reassembly phase (panel B), and both phases together (panel C) as functions of transfer price a. The optimum number of returns increases by a (Panel D).

of customers. At this point, although the net profit of take back remains continuous, the take back parameters are discontinuous and follow a different path. Both the optimum advertisement cost and the number of returns increase suddenly to higher values. For these new values, the same net profit is obtained at lower net profit per returned product (because of increased advertisement cost) and higher number of returned products.

A sudden increase in the optimum number of returns at $a = \$62$ causes a sudden increase in the NPD and reassembly phase (Figure 6.14B) and consequently the net profit of both phases together (Figure 6.14C). Figure 6.14C shows that for $a = \$95$, the sum of the net profits of both phases is maximized. The maximum net profit of both phases is \$1,119,700 and the optimum number of returns is 29,400. This maximum profit is obtained by implementing method 1 of advertisement and method 1 of transportation, offering \$26 as financial incentive and spending \$437,500 on advertisement.

In this practical problem, the transfer price that maximizes the net profit of both phases is associated with allocating all the profit to the *tb* phase (no net profit for disassembly and reassembly phase). If the *tb* phase and the disassembly and reassembly phase are performed by different firms, the transfer price will be a negotiated price less than \$95, because the disassembly and reassembly firm has to profit as well. Therefore, from the perspective of operational management, it is more efficient if the disassembly and reassembly firm performs the take back phase as well.

6.8   CONCLUSION

Remanufacturing process can be divided into two marketing phases in the beginning and at the end, and one engineering phase in between. Marketing phases of remanufacturing are buying back the used products from the customers, termed take back phase, and selling the remanufactured product to the customers, termed resale. In marketing phases, an accurate cost/benefit analysis of the process requires an accurate knowledge of the customers' behavior in response to the remanufacturing parameters. In take back phase, the return rate, or more specifically the Γ function, is the response of customers to the financial incentive, advertisement, and transportation method. Γ function is the statistical distribution of the customers' willingness to return their used product in response to the financial incentive. The convenience of transportation method and the motivation effect of advertisement can affect this distribution, and should be considered for a more accurate analysis.

Resale phase is the other marketing phase of remanufacturing. In this phase, the distribution of the customers' willing price for the remanufactured product is required to determine the relation between the price and demand. Also, in competition with the equivalent new product, it is required to know how much the customers depreciate the remanufactured product compared with the new product. A simple method to include this fact in the model is to assume that the willingness of the customers for the remanufactured product is a constant fraction of their willingness for the new product. The customers' perception of the remanufactured product and consequently their willingness to pay for the remanufactured product can be improved by warranty plans, or by promoting the echo-efficiency, quality, and durability of the remanufactured product through green advertisements. These strategic plans affect the cost/benefit analysis of the resale phase and should be included in the analysis of remanufacturing process.

From engineering perspective, disassembly and recovering the used cores are required for the remanufacturing process. Some of the concerns that have to be addressed in the disassembly process are whether a nonfunctioning core should be repaired, recycled, or disposed to the landfill and what cores should be removed from the product and what is the optimum disassembly sequence according to the statuses of a product cores. Statistics of the cores' statuses affect the NPD and also the optimum disassembly plan. These statistics are required for cost/benefit analysis of disassembly. Different phases of remanufacturing can be modeled separately by assigning a value to the product (or its cores) when it transfers from one phase to another. In analyzing the entire remanufacturing process, these transfer prices should be considered as variables and should be optimized along with system parameters to maximize the net profit.

REFERENCES

Amini, M. M., Retzlaff-Roberts, D., and Bienstock, C. C. 2005. Designing a reverse logistics operation for short cycle time repair services. *International Journal of Production Economics*, 96:3, 367–380.

Aras, N., Aksen, D., and Tanugur, A. G. 2008. Locating collection centers for incentive-dependent returns under a pick-up policy with capacitated vehicles. *European Journal of Operational Research*, 191:3, 1223–1240.

Asiedu, Y. and Gu, P. 1998. Product life cycle cost analysis: State of the art review. *International Journal of Production Research*, 36:4, 883–908.

Atasu, A., Sarvary, M., and Van Wassenhove, L. N. 2008. Remanufacturing as a marketing strategy. *Management Science*, 54:10, 1731–1746.

Barba-Gutierrez, Y., Adenso-Diaz, B., and Gupta, S. M. 2008. Lot sizing in reverse MRP for scheduling disassembly. *International Journal of Production Economics*, 111:2, 741–751.

Bourjault, A. 1984. Contribution a une approach methodologique del. Assemblage automatise: Elaboration automatique des sequence operatoires. L.niversite de Franche-Comte, Besançon, France.

Celebi, A. M. 2005. Optimal Multi-period Pricing and Trade-in Rebate Strategy for Ramanufacturable Durable Goods. Master Thesis, Istanbul, Turkey: Bogazici University, Industrial Engineering Department.

Debo, L. G., Toktay, L. B., and Van Wassenhove, L. N. 2005. Market segmentation and product technology selection for remanufacturable products. *Management Science*, 51:8, 1193–1205.

Debo, L. G., Toktay, L. B., and Van Wassenhove, L. N. 2006. Joint life-cycle dynamics of new and remanufactured products. *Production and Operations Management*, 15:4, 498–513.

Dehghanian, F. and Mansour, S. 2009. Designing sustainable recovery network of end-of-life products using genetic algorithm. *Resources Conservation and Recycling*, 53:10, 559–570.

Depuy, G. W., Usher, J. S., Walker, R. L., and Taylor, G. D. 2007. Production planning for remanufactured products. *Production Planning and Control*, 18:7, 573–583.

Driesch, H.-M., Van Oyen, H. E., and Flapper, S. D. P. 2005. Recovery of car engines: The Mercedes-Benz case. In S. D. P. Flapper, J. Van Nunen, and L. N. Van Wassenhove (Eds.) *Managing Closed-Loop Supply Chains*, Berlin, Germany: Springer, pp. 157–166.

Du, F. and Evans, G. W. 2008. A bi-objective reverse logistics network analysis for post-sale service. *Computers & Operations Research*, 35:8, 2617–2634.

Edlin, A. S. and Reichelstein, S. 1995. Specific investment under negotiated transfer pricing: An efficiency result. *Accounting Review*, 70:2, 275–292.

Erdos, G., Kis, T., and Xirouchakis, P. 2001. Modeling and evaluating product end-of-life options. *International Journal of Production Research*, 9, 1203–1220.

Ferguson, M. E. and Toktay, L. B. 2006. The effect of competition on recovery strategies. *Production and Operations Management*, 15:3, 351–368.

Ferrer, G. and Ketzenberg, M. 2004. Value of information in remanufacturing complex products. *IIE Transactions*, 36:3, 265–277.

Ferrer, G. and Swaminathan, J. M. 2006. Managing new and remanufactured products. *Management Science*, 52:1, 15–26.

Ferrer, G. and Whybark, C. D. 2000. From garbage to goods: Successful remanufacturing systems and skills. *Business Horizons*, 43:6, 55–64.

Ferrer, G. and Whybark, D. C. 2001. Material planning for a remanufacturing facility. *Production and Operations Management*, 10:2, 112–124.

Gao, M. M., Zhou, M. C., and Caudill, R. J. 2002. Integration of disassembly leveling and bin assignment for demanufacturing automation. *IEEE Transactions on Robotics and Automation*, 18:6, 867–874.

Garcia, M. A., Larre, A., Lopez, B., and Oller, A. 2000. Reducing the complexity of geometric selective disassembly. *Proceedings of IEEE/RSJ International Conference on Intelligent Robots and Systems*, Takamatsu, Japan, pp. 1474–1479.

Geiger, D. and Zussman, E. 1996. Probabilistic reactive disassembly planning. *Ann CIRP*, 45:1, 49–52.

Georgiadis, P., Vlachos, D., and Tagaras, G. 2006. The impact of product lifecycle on capacity planning of closed-loop supply chains with remanufacturing. *Production and Operations Management*, 15:4, 514–527.

Ghoreishi, N. 2009. A general modeling framework for cost/benefit analysis of remanufacturing. Mechanical Engineering and Materials Science 387, Electronic thesis and dissertations. Saint Louis, MO: Washington University, pp. 1–15.

Ghoreishi, N., Jakiela, M. J., and Nekouzadeh, A. 2011. A cost model for optimizing the take back phase of used product recovery. *Journal of Remanufacturing*, 1:11–15.

Ghoreishi, N., Jakiela, M. J., and Nekouzadeh, A. 2012. A non-graphical method to determine the optimum disassembly plan in remanufacturing. *Journal of Mechanical Design* (in press).

Gu, P. and Yan, X. 1996. A directed automatic assembly sequence planning. *International Journal of Production Research*, 33:11, 3069–3100.

Guide, V. D. R. 2000. Production planning and control for remanufacturing: Industry practice and research needs. *Journal of Operations Management*, 18, 467–483.

Guide, V. D. R., Kraus, M. E., and Srivastava, R. 1997a. Scheduling policies for remanufacturing. *International Journal of Production Economics*, 48, 187–204.

Guide, V. D. R., Ruud, H. T., and Luk, N. V. W. 2003a. Matching demand and supply to maximize profits from remanufacturing. *Manufacturing & Service Operations Management*, 5:4, 303–316.

Guide, V. D. R. and Srivastava, R. 1988. Inventory buffers in recoverable manufacturing. *Journal of Operations Management*, 16:5, 551–568.

Guide, V. D. R., Srivastava, R., and Kraus, M. 1997b. Product structure complexity and scheduling of operations in recoverable manufacturing. *International Journal of Production Research*, 35:11, 3179–3199.

Guide, V. D. R., Teunter, R. H., and Van Wassenhove, L. N. 2003b. Matching demand and supply to maximize profits from remanufacturing. *Manufacturing & Service Operations Management*, 5:4, 303–316.

Guide, V. D. R. and Van Wassenhove, L. N. 2001. Managing product returns for remanufacturing. *Production and Operations Management*, 10:2, 142–155.

Guide, V. D. R. and Van Wassenhove, L. N. 2002. Managing product returns at Hewlett Packard. Teaching Case 05/2002-4940, *INSEAD Case*.

Gungor, A. and Gupta, S. M. 1998. Disassembly sequence planning for products with defective parts in product recovery. *Computers & Industrial Engineering*, 35:1-2, 161-164.

Gungor, A. and Gupta, S. M. 2001. A solution approach to the disassembly line balancing problem in the presence of task failures. *International Journal of Production Research*, 39:7, 1427-1467.

Gupta, S. M. and Taleb, K. N. 1994. Scheduling disassembly. *International Journal of Production Research*, 32:8, 1857-1866.

Heese, H. S., Cattani, K., Ferrer, G., Gilland, W., and Roth, A. V. 2003. Competitive advantage through take-back of used products. *European Journal of Operational Research*, 164:1, 143-157.

Hoffmann, M., Kopacek, B., Kopacek, P., and Knoth, R. 2001. Design for re-use and disassembly. Paper presented at *Proceedings EcoDesign 2001: Second International Symposium on Environmentally Conscious Design and Inverse Manufacturing*, Tokyo, Japan, pp. 378-381.

Huang, Y. F. and Lee, C. S. G. 1989. Precedence knowledge in feature mating operation assembly planning. *Proceedings of 1989 IEEE International Conference on Robotics and Automation*, Scottsdale, AZ, pp. 216-221.

Hula, A., Jalali, K., Hamza, K., Skerlos, S. J., and Saitou, K. 2003. Multi-criteria decision-making for optimization of product disassembly under multiple situations. *Environmental Science and Technology*, 37:23, 5303-5313.

Ilgin, M. A. and Gupta, S. M. 2010. Environmentally conscious manufacturing and product recovery (ECMPRO): A review of the state of the art. *Journal of Environmental Management*, 91:3, 563-591.

Ishii, K., Eubanks, C. F., and DiMarco, P. 1994. Design for product retirement and material life-cycle. *Materials and Design*, 15:4, 225-233.

Jayaraman, V. 2006. Production planning for closed-loop supply chains with product recovery and reuse: An analytical approach. *International Journal of Production Research*, 44:5, 981-998.

Johnson, M. R. and Wang, M. H. 1995. Planning product disassembly for material recovery opportunities. *International Journal of Production Research*, 33:11, 3119-3142.

Johnson, M. R. and Wang, M. H. 1998. Economical evaluation of disassembly operations for recycling, remanufacturing and reuse. *International Journal of Production Research*, 36:12, 3227-3252.

Kannan, G., Sasikumar, P., and Devika, K. 2010. A genetic algorithm approach for solving a closed loop supply chain model: A case of battery recycling. *Applied Mathematical Modelling*, 34:3, 655-670.

Kazmierczak, K., Winkel, J., and Westgaard, R. H. 2004. Car disassembly and ergonomics in Sweden: Current situation and future perspectives in light of new environmental legislation. *International Journal of Production Research*, 42:7, 1305-1324.

Kerr, W. 2000. Remanufacturing and eco-efficiency: A case study of photocopier remanufacturing at Fuji Xerox Australia. IIIEE Communications Report No. 2005:05.

Klausner, M., Grimm, W. M., Hendrickson, C., and Horvath, A. 1998. Sensor-based data recording of use conditions for product takeback. Paper presented at *Proceedings of the 1998 IEEE International Symposium on Electronics and the Environment*, Oak Brook, IL, pp. 138-143.

Klausner, M. and Hendrickson, C. 2000. Reverse-logistic strategy for product take-back. *INTERFACES*, 30:3, 156-165.

Kleber, R., Minner, S., and Kiesmuller, G. 2002. A continuous time inventory model for a product recovery system with multiple options. *International Journal of Production Economics*, 79:2, 121-141.

Knoth, R., Hoffmann, M., Kopacek, B., and Kopacek, P. 2001. Intelligent disassembly of electr(on)ic equipment. Paper presented at *Proceedings EcoDesign 2001: Second International Symposium on Environmentally Conscious Design and Inverse Manufacturing*, Tokyo, Japan, pp. 557–561.

Ko, H. and Lee, K. 1987. Automatic assembling procedure generation from mating conditions. *Computer-Aided Design*, 19:1, 3–10.

Kongar, E. and Gupta, S. M. 2002. A multi-criteria decision making approach for disassembly-to-order systems. *Journal of Electronics Manufacturing*, 11:2, 171–183.

Krikke, H. R., van Harten, A., and Schuur, P. C. 1998. On a medium term product recovery and disposal strategy for durable assembly products. *International Journal of Production Research*, 36:1, 111–139.

Kumar, S., Kumar, R., Shankar, R., and Tiwari, M. K. 2003. Expert enhanced coloured stochastic petri net and its application in assembly/disassembly. *International Journal of Production Research*, 41:12, 2727–2762.

Kuo, T. C. 2000. Disassembly sequence and cost analysis for electromechanical products. *Robotics and Computer-Integrated Manufacturing*, 16:1, 43–54.

Lambert, A. J. D. 2002. Determining optimum disassembly sequences in electronic equipment. *Computers & Industrial Engineering*, 43:3, 553–575.

Lambert, A. J. D. 2003. Disassembly sequencing: A survey. *International Journal of Production Research*, 41:16, 3721–3759.

Lee, Y. Q. 1993. Automated sequence generation and selection for mechanical assembly and part replacement. Doctoral dissertation, Industrial Engineering, Pennsylvania State University, University Park, PA.

Lee, D. H. and Dong, M. 2009. Dynamic network design for reverse logistics operations under uncertainty. *Transportation Research Part E-Logistics and Transportation Review*, 45:1, 61–71.

Lee, Y. Q. and Kumara, S. R. T. 1992. Individual and group disassembly sequence generation through freedom and interference space. *Journal of Design and Manufacturing*, 2, 143–154.

Li, Y., Chen, J., and Cai, X. 2006. Uncapacitated product planning with multiple product types, returned product remanufacturing, and demand substitution. *OR Spectrum*, 28, 101–125.

Li, J., Gonzalez, M., and Zhu, Y. 2009. A hybrid simulation optimization method for production planning of dedicated remanufacturing. *International Journal of Production Economics*, 117:2, 286–301.

Li, J. R., Khoo, L. P., and Tor, S. B. 2002. A novel representation scheme for disassembly sequence planning. *International Journal of Advanced Manufacturing Technology*, 20:8, 621–630.

Li, Y., Saitou, K., Kikuchi, N., Skerlos, S. J., and Papalambros, P. Y. 2001. Design of heat-activated reversible integral attachments for product-embedded disassembly. Paper presented at *Proceedings EcoDesign 2001: Second International Symposium on Environmentally Conscious Design and Inverse Manufacturing*, Tokyo, Japan, pp. 360–365.

Linton, J. D. and Yeomans, J. S. 2003. The role of forecasting in sustainability. *Technological Forecasting and Social Change*, 70:1, 21–38.

Linton, J. D., Yeomans, J. S., and Yoogalingam, R. 2002. Supply planning for industrial ecology and remanufacturing under uncertainty: A numerical study of leaded-waste recovery from television disposal. *Journal of the Operational Research Society*, 53:11, 1185–1196.

Linton, J. D., Yeomans, J. S., and Yoogalingam, R. 2005. Recovery and reclamation of durable goods: A study of television CRTs. *Resources Conservation and Recycling*, 43:4, 337–352.

Looney, C. G. 1988. Fuzzy Petri nets for rule-based decision-making. *IEEE Transactions on Systems, Man, and Cybernetics*, 18:1, 178–183.

Lu, Q., Williams, J. A. S., Posner, M., Bonawi-Tan, W., and Qu, X. L. 2006. Model-based analysis of capacity and service fees for electronics recyclers. *Journal of Manufacturing Systems*, 25:1, 45–57.

Lund, R. T. 1996. *The Remanufacturing Industry: Hidden Giant*. Boston, MA: Boston University.

Lund, R. T. 2005. Remanufacturing central: About the remanufacturing industry. http://www.remancentral.com/about_reman_industry.htm

Lund, R. T. and Hauser, W. 2003. *The Remanufacturing Industry: Anatomy of a Giant*. Boston, MA: Department of Manufacturing Engineering, Boston University, pp. 1–179.

Majumder, P. and Groenevelt, H. 2001. Competition in remanufacturing. *Production and Operations Management*, 10:2, 125–141.

Mascle, C. and Balasoiu, B. A. 2003. Algorithmic selection of a disassembly sequence of a component by a wave propagation method. *Robotics and Computer-Integrated Manufacturing*, 19:5, 439–448.

McConocha, D. M. and Speh, T. W. 1991. Remarketing: Commercialization of remanufacturing technology. *Journal of Business and Industrial Marketing*, 6:1/2, 23–37.

Meacham, A., Uzsoy, R., and Venkatadri, U. 1999. Optimal disassembly configuration for single and multiple products. *Journal of Manufacturing Systems*, 18, 311–322.

Mitra, S. 2007. Revenue management for remanufactured products. *Omega-International Journal of Management Science*, 35:5, 553–562.

Moore, K. E., Gungor, A., and Gupta, S. M. 1998. A Petri net approach to disassembly process planning. *Computers & Industrial Engineering*, 35:1–2, 165–168.

Mutha, A. and Pokharel, S. 2009. Strategic network design for reverse logistics and remanufacturing using new and old product modules. *Computers & Industrial Engineering*, 56:1, 334–346.

Nakashima, K., Arimitsu, H., Nose, T., and Kuriyama, S. 2004. Cost analysis of a remanufacturing system. *Asia Pacific Management Review*, 9:4, 595–602.

Ong, N. S. and Wong, Y. C. 1999. Automatic subassembly detection from a product model for disassembly sequence generation. *International Journal of Advanced Manufacturing Technology*, 15:6, 425–431.

Parker, S. D. 1997. Remanufacturing: The ultimate form of recycling. Paper presented at *Proceedings of Air and Waste Management Association's Annual Meeting and Exhibition*, Toronto, Ontario, Canada, 97-TA44.01.

Penev, K. D. and deRon, A. J. 1996. Determination of a disassembly strategy. *International Journal of Production Research*, 34:2, 495–506.

Pishvaee, M. S., Kianfar, K., and Karimi, B. 2010. Reverse logistics network design using simulated annealing. *International Journal of Advanced Manufacturing Technology*, 47:1–4, 269–281.

Pochampally, K. K. and Gupta, S. M. 2008. A multiphase fuzzy logic approach to strategic planning of a reverse supply chain network. *IEEE Transactions on Electronics Packaging Manufacturing*, 31:1, 72–82.

Pochampally, K. K., Nukala, S., and Gupta, S. M. 2009. *Strategic Planning Models for Reverse and Closed-loop Supply Chains*. Boca Raton, FL: CRC Press.

Qin, Z. F. and Ji, X. Y. 2010. Logistics network design for product recovery in fuzzy environment. *European Journal of Operational Research*, 202:2, 479–490.

Rai, R., Rai, V., Tiwari, M. K., and Allada, V. 2002. Disassembly sequence generation: A Petri net based heuristic approach. *International Journal of Production Research*, 40:13, 3183–3198.

Rajan, V. N. and Nof, S. Y. 1996. Minimal precedence constraints for integrated assembly and execution planning. *IEEE Transactions on Robotics and Automation*, 12:2, 175–186.

Reveliotis, S. A. 2007. Uncertainty management in optimal disassembly planning through learning-based strategies. *IIE Transactions*, 39:6, 645–658.

Salomonski, N. and Zussman, E. 1999. On-line predictive model for disassembly process planning adaptation. *Robotics and Computer-Integrated Manufacturing*, 15:3, 211–220.

Sarin, S. C., Sherali, H. D., and Bhootra, A. 2006. A precedence-constrained asymmetric traveling salesman model for disassembly optimization. *IIE Transactions*, 38:3, 223–237.

Shu, L. H. and Flowers, W. C. 1995. Considering remanufacture and other end-of-life options in selection of fastening and joining methods. Paper presented at *IEEE International Symposium on Electronics the Environment*, Orlando, FL, pp. 75–80.

Shyamsundar, N. and Gadh, R. 1996. Selective disassembly of virtual prototypes. *Proceedings of 1996 IEEE International Conference on Systems, Man, and Cybernetics*, Beijing, China, pp. 3159–3164.

Srinivasan, H., Figueroa, R., and Gadh, R. 1999. Selective disassembly for virtual prototyping as applied to de-manufacturing. *Robotics and Computer-Integrated Manufacturing*, 15:3, 231–245.

Srivastava, S. K. 2008. Network design for reverse logistics. *Omega-International Journal of Management Science*, 36:4, 535–548.

Stevels, A. L. N., Ram, A. A. P., and Deckers, E. 1999. Take-back of discarded consumer electronic products from the perspective of the producer—Conditions for success. *Journal of Cleaner Production*, 7:5, 383–389.

Subramanian, R. and Subramanyam, R. 2008. Key drivers in the market for remanufactured products: Empirical evidence from eBay. SSRN: http://ssrn.com/abstract=1320719

Sutherland, J. W., Jenkins, T. L., and Haapala, K. R. 2010. Development of a cost model and its application in determining optimal size of a diesel engine remanufacturing facility. *CIRP Annals—Manufacturing Technology*, 59:1, 49–52.

Taleb, K. N., Gupta, S. M., and Brennan, L. 1997. Disassembly of complex product structures with parts and materials commonality. *Production Planning and Control*, 8:3, 255–269.

Tang, O., Grubbstrom, R. W., and Zanoni, S. 2004. Economic evaluation of disassembly processes in remanufacturing systems. *International Journal of Production Research*, 42:17, 3603–3617.

Tang, Y. and Zhou, M. C. 2006. A systematic approach to design and operation of disassembly lines. *IEEE Transactions on Automation Science and Engineering*, 3:3, 324–330.

Tang, Y., Zhou, M. C., Zussman, E., and Caudill, R. 2002. Disassembly modeling, planning, and application. *Journal of Manufacturing Systems*, 21:3, 200–217.

Teunter, R. H. 2001. A reverse logistics valuation method for inventory control. *International Journal of Production Research*, 39:9, 2023–2035.

Teunter, R. and van der Laan, E. 2002. On the non-optimality of the average cost approach for inventory models with remanufacturing. *International Journal of Production Economics*, 79:1, 67–73.

Teunter, R. H., van der Laan, E., and Inderfurth, K. 2000. How to set the holding cost rates in average cost inventory models with reverse logistics? *Omega-International Journal of Management Science*, 28:4, 409–415.

Tiwari, M. K., Sinha, N., Kumar, S., Rai, R., and Mukhopadhyay, S. K. 2001. A Petri net based approach to determine the disassembly strategy of a product. *International Journal of Production Research*, 40:5, 1113–1129.

Toktay, L. B. and Wei, D. 2005. Cost allocation in manufacturing-remanufacturing operations. Working Paper. Fontainebleau, France: INSEAD.

Toktay, L. B. and Wei, D. 2006. Cost allocations in manufacturing-remanufacturing operations. Working paper 2005/08/TOM, *INSEAD*.

Toktay, L. B., Wein, L., and Stefanos, Z. 2000. Inventory management of remanufacturable products. *Management Science*, 46, 1412–1426.

Unites States Environmental Protection Agency, WasteWise Update, "Remanufactured products: Good as new," EPA 530-N-97-002, May 1997.

Valenta, R. 2004. Product recovery at Robert Bosch tools, North America. Paper presented at *2004 Closed-Loop Supply Chain Workshop* held at INSEAD, Fontainebleau, France.

Van der Laan, E. A. and Salomon, M. 1997. Production planning and inventory control with remanufacturing and disposal. *European Journal of Operational Research*, 102, 264–278.

Van der Laan, E., Salomon, M., and Dekker, R. 1999. Inventory control in hybrid systems with remanufacturing. *Management Science*, 45, 733–747.

Vaysman, I. 1988. A model of negotiated transfer pricing. *Journal of Accounting and Economics*, 25:3, 349–385.

Vlachos, D., Georgiadis, P., and Iakovou, E. 2007. A system dynamics model for dynamic capacity planning of remanufacturing in closed-loop supply chains. *Computers & Operations Research*, 34.2, 367–394.

Vorasayan, J. and Ryan, S. M. 2006. Optimal price and quantity of refurbished products. *Production and Operations Management*, 15:3, 369–383.

Wakamatsu, H., Tsumaya, A., Shirase, K., and Arai, E. 2001. Development of disassembly support system for mechanical parts and its application to design considering reuse/recycle. Paper presented at *Proceedings EcoDesign 2001: Second International Symposium on Environmentally Conscious Design and Inverse Manufacturing*, Tokyo, Japan, pp. 372–377.

Walle, A. H. 1988. Remanufacturing marketing strategies and developing countries. *Journal of Global Marketing*, 1:4, 75–90.

Wang, H. F. and Hsu, H. W. 2010. A closed-loop logistic model with a spanning-tree based genetic algorithm. *Computers & Operations Research*, 37:2, 376–389.

Wolter, J. D., Chakrabarty, S., and Tsao, J. 1992. Mating constraint languages for assembly sequence planning. *Proceedings of 1992 IEEE International Conference on Robotics and Automation*, Nice, France, pp. 2367–2374.

Xanthopoulos, A. and Iakovou, E. 2009. On the optimal design of the disassembly and recovery processes. *Waste Management*, 29.5, 1702–1711.

Yang, G. F., Wang, Z. P., and Li, X. Q. 2009. The optimization of the closed-loop supply chain network. *Transportation Research Part E: Logistics and Transportation Review*, 45:1, 16–28.

Yi, J. J., Yu, B., Du, L., Li, C. G., and Hu, D. Q. 2008. Research on the selectable disassembly strategy of mechanical parts based on the generalized CAD model. *International Journal of Advanced Manufacturing Technology*, 37:5–6, 599–604.

Yokota, K. and Brough, D. R. 1992. Assembly/disassembly sequence planning. *Assembly Automation*, 12:3, 31–38.

Zhang, H. C. and Kuo, T. C. 1996. A graph based approach to disassembly model for end-of-life product recycling. Paper presented at *19th International Electronics Manufacturing Technology Symposium*, October 1996, Austin, TX, pp. 247–254.

Zhang, H. C. and Kuo, T. C. 1997. A graph-based disassembly sequence planning for EOL product recycling. Paper presented at *21st IEEE/CPMT International Electronics Manufacturing Technology Symposium*, October 1997, Austin, TX, pp. 140–151.

Zussman, E., Kriwet, G., and Seliger, G. 1994. Disassembly-oriented assessment methodology to support design for recycling. *Ann CIRP*, 43:1, 9–14.

Zussman, E. and Zhou, M. C. 1999. A methodology for modeling and adaptive planning of disassembly processes. *IEEE Transactions on Robotics and Automation*, 15:1, 190–194.

Zussman, E. and Zhou, M. C. 2000. Design and implementation of an adaptive process planner for disassembly processes. *IEEE Transactions on Robotics and Automation*, 16:2, 171–179.

Zussman, E., Zhou, M. C., and Caudill, R. 1998. Disassembly Petri net approach to modeling and planning disassembly processes of electronic products. *Proceedings of 1998 IEEE International Symposium on Electronics and Environment*, Oak Brook, IL, pp. 331–336.

7 Integrated Inventory Models for Retail Pricing and Return Reimbursements in a JIT Environment for Remanufacturing a Product

Xiangrong Liu, Avijit Banerjee,
Seung-Lae Kim, and Ruo Du

CONTENTS

7.1   INTRODUCTION

With incentive of lowering production costs, along with concerns in environmental issues, possibilities of product recovery and reuse are focuses of most manufacturers today. "Green" manufacturing includes all the practices ranging from waste paper and scrap metal recycling, reuse of containers, to, more recently, recovery of electronic components, etc. The Xerox Green World Alliance reports that more than 90%, 25%, and 100%, respectively, of remanufactured print cartridges, new toners, and plastic parts have been made from remanufacturing and recycling. This program has lead to significant environmental and financial benefits for Xerox (Xerox 2005).

The efficient incorporation of used products and/or materials into manufacturing processes can contribute substantially toward achieving "waste free" goals in a sustainable manner. One issue a manufacturer often faces in this context involves the process of collecting the returns, named as acquisition management. Our attention is confined only to the case of customer returns at the retail level, which is cited to be the most effective method for used products collection (Savaskan et al. 2004). Generally, retailers are responsible for providing incentives to end customers and are subsequently reimbursed by the manufacturer to ensure that end-of-life products are returned for remanufacture in a timely manner. This is a common practice for disposable cameras, mobile phones, and printer ink cartridges, etc. While retailers give certain level of incentives to take the used product back, new products will get promoted at the same time. For an instance, SEARS provide incentives to end customers by proposing a trade-in program that gives 5% of the price of the old units to the customers' gift card (Recycling Today Online). Consequently, the return rate of used product will be determined not only by the reimbursement price of the used product but also by the selling price of the new product. Some retailers involve into this type of program. Sony TV loyalty program and Xerox's trade-in rebate program show successful examples in industry.

Another concern for the manufacturer as well as the retailer pertains to inventory issues. Economies of scale may dictate that manufacturers collect and take back returns at periodic intervals, requiring retailers to have storage space for holding the products returned by customers. By the same token, manufacturers also need to allocate storage space for such items. Needless to say that a product's retail price, customer incentive for returns (both determined by the retailer), as well as the transfer price paid by the producer to the retailer for collecting returns are likely to shape the inventory policies for returned items at both the retailer's and the manufacturer's ends. The efficient incorporation of used products and/or materials into manufacturing processes, regarding the decision of order quantity and the appropriate incentives for collecting used products can contribute substantially toward achieving "waste free" goals in a sustainable manner. This study is an attempt to examine these issues from an integrated supply chain perspective.

7.2   LITERATURE REVIEW

Fleischmann et al. (1997), Gungor and Gupta (1999), Guide et al. (2000, 2003), Rubio et al. (2008), Porharel and Mutha (2009), Guide and Van Wassenhove (2009), Ilgin and Gupta (2010), Lage and Filho (2012), and Ilgin and Gupta (2012) provide thorough surveys of existing research involving remanufacturing. A significant portion of the work on product recovery addresses lot sizing problem in inventory control. In one of the earliest works, Schrady (1967) presents a deterministic model for repairable items and derives EOQ-type fixed lot sizes for recovery and reorder. Based on Schrady's work, Mabini et al. (1992) study shortage issue with a single item under limited capacity and further extend to multiple items. Richter (1996a,b) develops a different control approach considering multiple repair setups. More recently, Koh et al. (2002) proposed a deterministic recovery model with a fixed return rate deriving optimal quantity for new material ordering and inventory level for recoverable products. Investigating the secondary market to sell refurbished product, Konstantaras et al. (2010) extend Koh's work by including the consideration of inspection and sorting process. Although the aforementioned works apply the traditional economic ordering/production model, all of them use a constant demand and/or a constant return rate. Recently, El Saadany and Jaber (2010) depicted an inventory model with return rate associated with collection price of returns. However, the relationship between the return rate and the price factors, including both the price of new product and the collection price of returns, is not clear. Our chapter is the first work to specify this relationship, which is close to the practices where loyalty program have been established.

The major thrust of the extant remanufacturing literature is on the timing and sizing decisions for manufacturing and remanufacturing activities, with primary attention on inventory related matters. Efforts toward integrating the decisions of inventory replenishment, product pricing, and customer incentive for returning used items (in the form of a cash refund or a discount coupon) in a remanufacturing environment have been relatively rare. As a notable exception, in a recent study Savaskan et al. (2004) have dealt with the questions of pricing and return incentives from a game theoretic perspective, in examining alternative reverse logistics structures for the collection of recoverable products. Ray et al. (2005) focused on the decision making in optimal pricing decisions in trade-in program, however, they did not consider the influence of order quantity and inventory control. Although Bhattacharya et al. (2006) conducted the integration of optimal order quantities for unsold new products and reused products, which can fully substitute the new components, in four different decision-making structures, reflecting the various relationships among retailer, OEM, and remanufacturer. Also, Vorasayan and Ryan (2006) outline procedures for deriving the pricing for refurbished products and the percentage of returns for refurbish. Inventory control decisions, which are intertwined with such questions, however, have been only superficially treated in the pricing related research.

Our chapter is an attempt to address this deficiency in the current body of work involving remanufacturing. We address some of the major issues concerning inventories, pricing, used product collection, materials procurement, product delivery, and planning for manufacturing and remanufacturing in an integrated manner. Specifically, we develop procedures for developing such integrated policies toward achieving a well-coordinated supply chain, incorporating a lean production process. Our emphasis is on the mathematical modeling of product remanufacturing under a scenario involving a single retailer and a single manufacturer, dealing with a single recoverable product. Furthermore, as mentioned earlier, the models developed in this chapter attempt to establish an integrated policy, simultaneously specifying decisions concerning inventory replenishment at various stages of the supply chain, retail pricing, as well as the appropriate incentive level for inducing customer returns of used items, from the perspectives of, maximizing the profits of, respectively, the retailer, the manufacturer, and the entire supply chain. For simplicity of analysis and implementation, we assume a deterministic environment. Furthermore, the product's demand and return rates are modeled as simple linear functions.

7.3 NOTATION AND ASSUMPTIONS

7.3.1 Notation

We use the following notational scheme throughout the chapter:

For the retailer:

d demand rate of the product in units/time unit
h_r inventory holding cost of the product ($/unit/time unit)
h_{rr} inventory holding cost of the used (returned) product ($/unit/time unit)
p_s unit selling price of the product (new or remanufactured) in $/unit
r_c unit reimbursement to customers for returns in $/unit
S_r fixed ordering cost ($/order) for retail stock replenishment
x rate at which customers return the used item to the retailer (units/time unit)
X total quantity of returns in a replenishment cycle (units)

For the manufacturer:

c_m variable cost of manufacturing new product ($/unit)
c_r variable cost of transporting, cleaning, preparation, etc. for returned items ($/unit)
c_{rm} variable cost of remanufacturing a returned used product into a new one ($/unit)
c_s variable transportation cost of shipping new product to the retailer ($/unit)
h_i inventory holding cost of input materials necessary for the production of a unit of the new product in $/unit/time unit
h_{ir} inventory holding cost of input materials necessary for remanufacturing a unit of the used product ($/unit/time unit)
h_m inventory holding cost of finished product (new or remanufactured) in $/unit/time unit
m manufacturing or remanufacturing rate of the product (unit/time unit)
p_w wholesale price charged to retailer for the new product ($/unit)
r_m transfer price paid to retailer by manufacturer for collecting used products ($/unit)
S_i fixed ordering cost of input materials ($/lot) for manufacturing and/or remanufacturing

S_m fixed manufacturing/remanufacturing setup cost per replenishment lot ($/setup)

S_{rm} total fixed cost of shipping a replenishment lot of new products to the retailer and transporting the returned items collected back to the manufacturing facility ($/cycle)

Common to both

Q total replenishment quantity (units) consisting of new and/or remanufactured items

T inventory replenishment cycle time (time units), common to retailer and manufacturer

7.3.2 Assumptions

1. The supply chain under study consists of a single retailer and a single manufacturer involved in the production and sale of a single recoverable product. Customers are refunded a part of the purchase price by the retailer as an incentive to return used products, which can be restored to "as new" condition for resale through a remanufacturing process deployed by the manufacturer. The manufacturing/remanufacturing environment is a batch production system where each batch of the new product may consist of a mix of remanufactured and new manufactured items. The used items, after cleaning, restoration, etc., are completely reincorporated in the existing production process, so that remanufacturing and new product manufacturing rates are the same, although their variable costs may differ.
2. For coordination purposes, the lot-for-lot policy is in effect for input materials ordering, manufacturing and remanufacturing, product delivery and retail inventory replenishment, with a common cycle time of T. In other words, the necessary input materials procurement, production (including remanufacture), delivery and retail stock replenishment cycles are one and the same. This lot-for-lot feature is commonly found in JIT based lean manufacturing systems, where minimal levels of material and product inventories are desired.
3. All input materials for manufacturing or remanufacturing are treated as a composite bundle. In each case, the total bundle of inputs necessary for producing (or remanufacturing) a unit of the end product is defined as a "unit." All of the input materials (for manufacturing and remanufacturing) are ordered on a lot-for-lot basis with a single procurement order prior to the setup of a batch.
4. The retailer is responsible for collecting returned items and holding them in inventory until picked up by the producer. In our decentralized models, the manufacturer pays the retailer a unit transfer price for the returned items, in order to induce the latter to engage in the collection activity. Without loss of generality, it is assumed that the retailer's cost of this collection effort is negligibly small, although the cost of holding the returned products in

inventory at the retail level is taken into account. Under the centralized scenario, the used product transfer price and the producer's wholesale price become irrelevant for avoiding double marginalization. In the decentralized models, the retailer sets the item's selling price and the unit reimbursement to customers for returns. The wholesale price, where applicable, is the same for new or remanufactured items.

5. We assume that the market demand, the customer return rate, and all lead times are deterministic. Thus, a production batch of Q units consists of $Q - X$ items manufactured from raw material and X units remanufactured from used product as shown in Figure 7.1, which depicts the process flow schema of the supply chain under consideration. Figure 7.2 shows the various inventory-time plots at the retail and manufacturing facilities. Without loss of generality, these plots are constructed with the assumption that the setup and transit times, as well as the cleaning and refurbishment times for the recovered items are zero. Before setting up a production batch, the X units of returned items collected during the cycle are transported back to the plant for remanufacturing. Therefore, the value of the quantity $Q - X$ is known prior to each setup. After completion of the manufacturing and remanufacturing process, the replenishment lot of Q is delivered to the retailer for sale. All transportation costs are paid by the producer.

6. Consistent with classical microeconomic theory, we model the retail demand rate, d, as a decreasing function of its selling price, p_s, i.e., $d = A - Bp_s$. Furthermore, the product's return rate, x, and the total units returned, X, during a cycle are expressed, respectively, as $x = ar_c - bp_s$ and $X = Tx = Qx/d$. The parameters A, B, a, and b are known or can be estimated empirically. It is reasonable to assume that the average rate of used product returns is likely to increase as the return incentive, r_c, as well as the overall demand level, d, increase (or, alternately, as the retail price decreases). Furthermore, as mentioned earlier, we adopt linear structures for both d and x for simplicity of analysis and implementation.

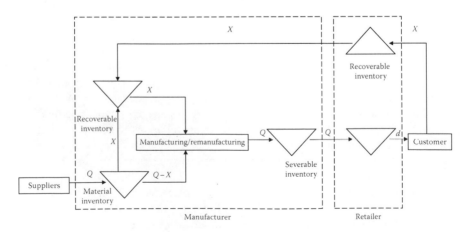

FIGURE 7.1   Recovery and remanufacturing process.

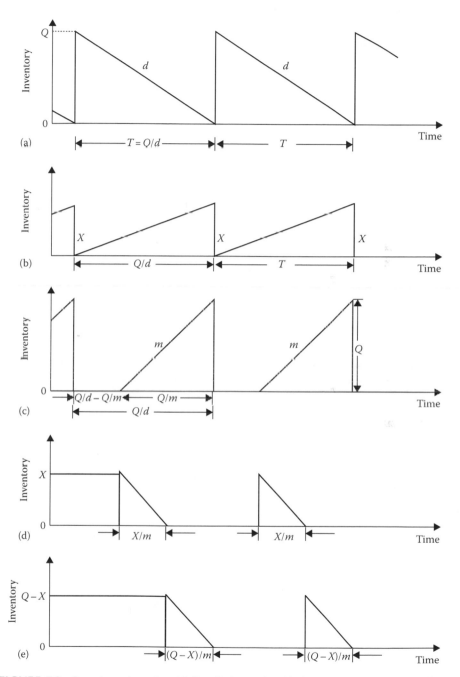

FIGURE 7.2   Inventory-time plot. (a) Retailer's serviceable inventory, (b) returns product inventory, (c) manufacturing inventory, (d) remanufacturing input material inventory, and (e) manufacturing input material inventory.

7.4   DEVELOPMENT OF MODELS AND ANALYSES

7.4.1   DECENTRALIZED MODELS WITH WHOLESALE PRICE SET BY MARKET

In some industries, due to intense competition, the wholesale price for the manufacturer is determined by the existing market conditions and is, consequently, treated as a constant parameter. The exposition in the following subsections pertains to such cases.

7.4.1.1   Decentralized Model for Retailer's Optimal Policy with Given p_w

Suppose that the retailer can set the integrated optimal order quantity and pricing policy for sales and returns, independent of the manufacturer, with the assumption that it wields sufficient power as a dominant member of the supply chain. In such a case, its profit per time unit can be expressed as

$$\Pi_r = (p_s - p_w)d + (r_m - r_c)x - S_r\frac{d}{Q} - h_r\frac{Q}{2} - h_{rr}\frac{Qx}{2d} \tag{7.1}$$

The retailer has two sources of revenue, captured by the first two terms in the previous profit function. The first of these represents the revenue from the sales of new products and the second term expresses the net revenue, through reimbursements from the manufacturer, for collecting the used items. The next term represents the average ordering cost and the remaining two terms show, respectively, the costs of holding new product and returned item inventories per time unit at the retailer's end (see Figure 7.2a and b). Substituting $x = ar_c - bp_s$ and $d = A - Bp_s$ into (7.1), the retailer's average profit per time unit can be rewritten as

$$\Pi_r = (p_s - p_w)(A - Bp_s) + (r_m - r_c)(ar_c - bp_s) - S_r\left[\frac{A - Bp_s}{Q}\right]$$
$$- \frac{Q}{2}\left[h_r + h_{rr}\left(\frac{ar_c - bp_s}{A - Bp_s}\right)\right] \tag{7.2}$$

The first-order optimality conditions are shown in the following, obtained by setting $\partial\Pi_r/\partial Q$, $\partial\Pi_r/\partial r_c$, and $\partial\Pi_r/\partial p_s$, respectively, equal to 0, i.e.,

$$Q = \sqrt{\frac{2S_r(A - Bp_s)}{h_r + h_{rr}(ar_c - bp_s)/(A - Bp_s)}} \tag{7.3}$$

$$r_c = \frac{1}{2}\left[r_m + \frac{b}{a}p_s - \frac{Qh_{rr}}{2(A - Bp_s)}\right] \tag{7.4}$$

$$(A - Bp_s)^2\left[B(p_w - 2p_s) + A - b(r_m - r_c) + S_r\frac{B}{Q}\right]$$
$$- \frac{Q}{2}h_{rr}\left[B(ar_c - bp_s) - b(A - Bp_s)\right] = 0 \tag{7.5}$$

Since the return rate cannot exceed the demand rate of the item, i.e., $d \geq x$, the roots of p_s that are negative or violate this feasibility condition are disregarded in this and subsequent models for computational purposes. Or we can add the following constraints directly:

$$0 \leq ar_c - bp_s \leq A - Bp_s \tag{7.6}$$

The aforementioned conditions (7.3 through 7.6) can be solved simultaneously by any standard equation solving software such as MATLAB®, in order to obtain Q, r_c, and p_s.

Proposition 7.1:

Q, r_c, and p_s obtained from (7.3 through 7.5) represent the local optimal solution if the following conditions are satisfied:

(a) $\dfrac{d}{Q} \geq \left(\dfrac{ah_{rr}^2}{16 S_r} \right)^{1/3}$

(b) $\dfrac{ah_{rr}^2(Ab - aBr_c)(Ab - aBr_c - bd)}{2d^4} + \dfrac{aBh_{rr}[2S_r(Ab - aBr_c) + Q^2 a/2 - dbS_r]}{Q^2 d^2}$

$\quad + \dfrac{2aS_r(B^2 S_r - ABdQ + b^2 dQ)}{Q^4} \leq 0$

A proof of the previous proposition is provided in the Appendix.

7.4.1.2 Decentralized Model for Manufacturer's Optimal Policy with Given p_w

If the manufacturer, instead of the retailer, is in a position of dictating supply policy, it would prefer to implement a production and delivery policy (assuming the lot-for-lot operating framework) that is optimal from its own perspective. In this case, the supplier's wholesale price is treated as a given parameter. The retailer, nevertheless, is likely to be free to set its own selling price and the level of incentive to induce customers to return the used products, given the manufacturer's preferred replenishment lot size. Thus, in order to develop the manufacturer's profit function, we need to determine the average inventories at the manufacturing facility. From Figure 7.1c, the average inventory of the finished product at the manufacturer's end

$$= \dfrac{(Q/2)(Q/m)}{(Q/d)} = \dfrac{Qd}{2m}$$

Also, from Figure 7.1d it can be shown that the average inventories of the input materials necessary for remanufacturing and manufacturing purposes, respectively, are $x^2 Q/2md$ and $Q/2m[d - 2x + x^2/d]$.

Incorporating these results, the profit per time unit for the manufacturer can be expressed as

$$\Pi_m = (p_w - c_s)d - \frac{d}{Q}(S_m + S_{rm} + S_i) - (r_m + c_r)x - \frac{h_m Qd}{2m} - \frac{h_{ir} Qx^2}{2md}$$

$$- \frac{h_i Q}{2m}\left(d - 2x + \frac{x^2}{d}\right) - c_m(d - x) - c_{rm}x \tag{7.7}$$

The first term in (7.7) shows the manufacturer's revenue based on the wholesale price, less the variable shipping cost to the retailer. The second term includes the fixed costs involving production set up, transportation of new products to and used items from the retailer and ordering of input raw materials. The third term expresses the reimbursement cost to retailer, as well as the variable transportation, cleaning, and preparation costs for the returned items. The next three terms represent the holding costs, respectively, for the finished product and input materials inventories necessary for remanufacturing and manufacturing. The final two terms in (7.7) are the variable costs per time unit for manufacturing and remanufacturing, separately. Substituting for d and x into (7.7), and collecting terms, the manufacturer's profit per time unit is rewritten as follows:

$$\Pi_m = (p_w - c_m - c_s)(A - Bp_s) - \frac{A - Bp_s}{Q}(S_m + S_{rm} + S_i)$$

$$- (r_m + c_r + c_{rm} - c_m)(ar_c - bp_s) - \frac{Q}{2m}\left\{ h_m(A - Bp_s) + \frac{h_{ir}(ar_c - bp_s)^2}{(A - Bp_s)} \right.$$

$$\left. + h_i\left[A - Bp_s - 2(ar_c - bp_s) + \frac{(ar_c - bp_s)^2}{A - Bp_s} \right] \right\} \tag{7.8}$$

Note that the item's selling price, p_s, and the customer return reimbursement, r_c are set by the retailer. Then the manufacturer's optimal batch size and the consequent retailer's policy variable values are established by the scheme of a sequential game, which can be solved by combining the optimality conditions (7.4) and (7.5) resulting from the retailer's problem as constraints with the optimality condition with respect to order quantity in the upper level problem. The set of Equations 7.9 shown in the following

$$Q = \sqrt{\frac{2m(A - Bp_s)(S_m + S_{rm} + S_i)}{h_m(A - Bp_s) + \left((h_{ir}(ar_c - bp_s)^2)/(A - Bp_s)\right)}}$$

$$\overline{+ h_i\left[A - (B - 2b)p_s - 2ar_c + \left((ar_c - bp_s)^2/(A - Bp_s)\right) \right]}$$

$$r_c = \frac{1}{2}\left[r_m + \frac{b}{a}p_s - \frac{Qh_{rr}}{2(A - Bp_s)} \right] \tag{7.9}$$

$$(A - Bp_s)^2 \left[B(p_w - 2p_s) + A - b(r_m - r_c) + S_r \frac{B}{Q} \right]$$

$$- \frac{Q}{2} h_{rr} \left[B(ar_c - bp_s) - b(A - Bp_s) \right] = 0$$

$$0 \le ar_c - bp_s \le A - Bp_s$$

can be solved simultaneously. Neither the retailer nor the manufacturer would benefit from any deviation from the aforementioned optimal solution, which represents the equilibrium state of the whole system.

Proposition 7.2:

The manufacturer would adopt a remanufacturing strategy only when the condition

$$r_m + c_r + c_{rm} \le c_m - \frac{Q}{2m} \left[\frac{h_{ir} x}{d} + h_i \left(\frac{x}{d} - 2 \right) \right] \tag{7.10}$$

is satisfied (see the Appendix for proof).

7.4.2   DECENTRALIZED MODEL WITH WHOLESALE PRICE SET BY THE MANUFACTURER

Under monopolistic market conditions, manufacturers may lower the wholesale price in order to encourage retailers to increase their order quantities. As discussed before, under a decentralized policy, the retailer determines its order quantity as one of the decision variables along with the selling price and the customer return incentive. It will make these decisions after the observation of a wholesale price set by the manufacturer. Initially, the manufacturer would anticipate the optimal response from the retailer when it decides on the wholesale price, resulting in the following model:

$$\max_{p_w} \Pi_m = (p_w - c_m - c_s)(A - Bp_s) - \frac{A - Bp_s}{Q}(S_m + S_{rm} + S_i)$$

$$- (r_m + c_r + c_{rm} - c_m)(ar_c - bp_s) - \frac{Q}{2m} \left\{ h_m(A - Bp_s) + \frac{h_{ir}(ar_c - bp_s)^2}{(A - Bp_s)} \right.$$

$$+ h_i \left[A - Bp_s - 2(ar_c - bp_s) + \frac{(ar_c - bp_s)^2}{A - Bp_s} \right] \right\} \quad \text{s.t.} \tag{7.11}$$

$$Q = \sqrt{\frac{2S_r(A - Bp_s)}{h_r + h_{rr}(ar_c - bp_s)/(A - Bp_s)}}$$

$$r_c = \frac{1}{2}\left[r_m + \frac{b}{a}p_s - \frac{Qh_{rr}}{2(A - Bp_s)}\right]$$

$$(A - Bp_s)^2\left[B(p_w - 2p_s) + A - b(r_m - r_c) + S_r\frac{B}{Q}\right]$$

$$- \frac{Q}{2}h_{rr}\left[B(ar_c - bp_s) - b(A - Bp_s)\right] = 0$$

$$0 \le ar_c - bp_s \le A - Bp_s$$

All the decision variables here are nonnegative. Equalities constraints are the optimality conditions of the retailer's problem, which are treated as the constraints on the manufacturer's problem. If the manufacturer, instead of the retailer, has control of the order quantity, the aforementioned model may be written as a bilevel problem, as shown in the following:

$$\max_{p_w, Q} \Pi_m = (p_w - c_m - c_s)(A - Bp_s) - \frac{A - Bp_s}{Q}(S_m + S_{rm} + S_i)$$

$$- (r_m + c_r + c_{rm} - c_m)(ar_c - bp_s) - \frac{Q}{2m}\left\{h_m(A - Bp_s) + \frac{h_{ir}(ar_c - bp_s)^2}{(A - Bp_s)}\right.$$

$$+ h_i\left[A - Bp_s - 2(ar_c - bp_s) + \frac{(ar_c - bp_s)^2}{A - Bp_s}\right]\right\} \quad \text{s.t.} \qquad (7.12)$$

$$r_c = \frac{1}{2}\left[r_m + \frac{b}{a}p_s - \frac{Qh_{rr}}{2(A - Bp_s)}\right]$$

$$(A - Bp_s)^2\left[B(p_w - 2p_s) + A - b(r_m - r_c) + S_r\frac{B}{Q}\right]$$

$$- \frac{Q}{2}h_{rr}\left[B(ar_c - bp_s) - b(A - Bp_s)\right] = 0$$

$$0 \le ar_c - bp_s \le A - Bp_s$$

This constrained nonlinear problem could be solved by one of several widely available optimization software packages, such as MATLAB.

7.4.3 Centralized Model for Supply Chain Optimality

Suppose that the retailer and the manufacturer agree to cooperate toward formulating a jointly optimal integrated policy, involving inventory replenishment, retail pricing, and customer return reimbursement decisions, for the supply chain as a whole. The focus of such a centralized policy, where both parties are willing to freely share their cost and other relevant information, is to maximize the profitability of the entire system, rather than that of either party. We illustrate in the next section that this centralized joint optimization approach can be economically attractive from the standpoint of both the parties through an equitable profit sharing methodology. In this centralized approach, we propose that in order to avoid double marginalization, the parameters wholesale price p_w and manufacturer's rebate for returned items r_m need not be considered and are omitted. Thus, combining (7.2) and (7.8), without an explicit wholesale price and a direct manufacturer's reimbursement to the retailer for product returns, the total supply chain profit is

$$
\begin{aligned}
\Pi_s = {} & (p_s - c_m - c_s)(A - Bp_s) - (r_c + c_r + c_{rm} - c_m)(ar_c - bp_s) \\
& - \frac{(A - Bp_s)}{Q}(S_r + S_m + S_{rm} + S_i) \\
& - \frac{Q}{2m}\left[m\left\{ h_r + h_{rr}\frac{(ar_c - bp_s)}{(A - Bp_s)} \right\} + (h_{ir} + h_i)\left\{ \frac{(ar_c - bp_s)^2}{(A - Bp_s)} \right\} \right. \\
& \left. + (h_m + h_i)(A - Bp_s) - 2h_i(ar_c - bp_s) \right]
\end{aligned}
\tag{7.13}
$$

The first-order optimality conditions yield the optimal values of the replenishment lot size, Q, unit customer reimbursement for returns, r_c, and the unit selling price, p_s, which maximize the total supply chain profit under the proposed centralized policy, as shown in the following:

$$
Q = \sqrt{\frac{2m(A - Bp_s)(S_r + S_m + S_{rm} + S_i)}{\begin{array}{l} m\left\{ h_r + h_{rr}\left((ar_c - bp_s)/(A - Bp_s)\right) \right\} + (h_{ir} + h_i)\left\{ \left((ar_c - bp_s)^2/(A - Bp_s)\right) \right\} \\ + (h_m + h_i)(A - Bp_s) - 2h_i(ar_c - bp_s) \end{array}}}
\tag{7.14}
$$

$$
r_c = \frac{\left(Q/4m(A - Bp_s)\right)\left[(h_{ir} + h_i)bp_s + 2h_i d - h_{rr}m\right] - 1/2\left(c_r + c_{rm} - c_m - (b/a)p_s\right)}{1 + aQ\left[(h_{ir} + h_i)/2m(A - Bp_s)\right]}
\tag{7.15}
$$

$$
\begin{aligned}
& (A - Bp_s)^2\left[A - B(2p_s - c_m - c_s) + b(r_c + c_r + c_{rm} - c_m) + \frac{B}{Q}(S_r + S_m + S_{rm} + S_i) \right] \\
& - \frac{Q}{2m}\left[(h_{ir} + h_i)\left\{ B(ar_c - bp_s)^2 - 2b(ar_c - bp_s)(A - Bp_s) \right\} + h_{rr}m(aBr_c - bA) \right. \\
& \left. - Bh_m(A - Bp_s)^2 + h_i(2b - B)(A - Bp_s)^2 \right] = 0
\end{aligned}
\tag{7.16}
$$

Once again, conditions (7.14 through 7.16) can be solved via any appropriate equation solving software, such as MATLAB, for determining the centrally controlled inventory replenishment, retail pricing, and return reimbursement decisions.

Proposition 7.3:

Q, r_c, and p_s obtained from solving (7.14 through 7.16) are local optimum if the following conditions, in addition to conditions (a) and (b) under Proposition 7.1, are satisfied:

$$(c) \quad \frac{4(S_m + S_{rm} + S_i)(db - xB)^2(h_{ir} - h_i)}{mQ^2 d^2} - \left[-\frac{B(S_m + S_{rm} + S_i)}{Q^2} + \frac{Bh_m - h_i(2b - B)}{2m} \right.$$

$$\left. + \frac{(2db - Bx)x(h_{ir} - h_i)}{2md^2} \right]^2 \geq 0$$

$$(d) \quad \frac{a^2 B^2 Q(h_{ir} - h_i)(h_i + h_m)^2}{4m^3 d} - \frac{a^2 B^2 Q x h_i(h_{ir} - h_i)(h_i - h_m)}{m^3 d^2}$$

$$+ \frac{a^2 B^2 (h_{ir} - h_i)(S_m + S_{rm} + S_i)^2}{Q^3 md}$$

$$- \frac{a^2 B Q x^2 (h_{ir} - h_i)(Bh_{ir}h_i - 6bh_{ir}h_i + Bh_{ir}h_m - 5Bh_i^2 + 6bh_i^2 - Bh_ih_m)}{4m^3 d^3}$$

$$- \frac{a^2 B^2 Q x^3 h_i(h_{ir} - h_i)^2}{m^3 d^4} + \frac{a^2 B^2 Q x^4 (h_{ir} - h_i)^3}{4m^3 d^5} - \frac{a^2 B^2 (h_{ir} - h_i)(h_i - h_m)(S_m + S_{rm} + S_i)}{Qm^2 d}$$

$$+ \frac{2a^2 B^2 x h_i(h_{ir} - h_i)(S_m + S_{rm} + S_i)}{Qm^2 d^2} - \frac{a^2 B^2 x^2 (h_{ir} - h_i)^2(S_m + S_{rm} + S_i)}{Qm^2 d^3} \leq 0$$

7.5   NUMERICAL ILLUSTRATION AND DISCUSSIONS

To illustrate our models outlined earlier, a numerical example is provided. The following information pertaining to the two parties in the supply chain are available:

Retailer:
S_r = \$50/order, h_r = \$0.015/unit/day, h_{rr} = \$0.002/unit/day, A = 120, B = 3.0, a = 15, b = 0.1
    That is, the daily demand rate is d = $120 - 3p_s$ and the daily return rate is $x = 15r_c - 0.1p_s$.

Manufacturer:
m = 100 units/day, S_m = \$300/batch, S_{rm} = \$200/batch, S_i = \$30/batch
h_m = \$0.01/unit/day, h_i = \$0.009/unit/day, h_{ir} = \$0.007/unit/day, p_w = \$20/unit
r_m = \$2.80/unit, c_s = \$2/unit, c_m = \$4/unit, c_{rm} = \$2/unit, c_r = \$1.20/unit.

    All the results obtained from the various perspectives are summarized in Table 7.1. It can be easily verified that the chosen parameters satisfy the joint concavity conditions (a), (b), (c), and (d). Hence, all the solutions shown in Table 7.1 are local optimum.

product in a two-echelon supply chain consisting of a single retailer and a single lean manufacturer. Items returned by customers at the retail level are refurbished and totally reintegrated into the manufacturer's existing production system for remanufacturing and are sold eventually as new products. As in many lean manufacturing (a JIT) environments, we assume a lot-for-lot operating mode for production, procurement, and distribution, as an effective mechanism for supply chain coordination.

Decentralized models are developed and solved for determining profit maximizing optimal policies from the perspectives of both members of the supply chain. A centralized, jointly optimal procedure for maximizing total supply chain profitability is also presented. A numerical example illustrates that the centralized approach is substantively superior to individual optimization, due to the elimination of double marginalization. The example also outlines a fair and equitable proportional profit sharing scheme, which is economically desirable from the standpoint of either member of the supply chain, for the purpose of implementing the proposed centrally controlled model.

Of necessity, the simplifying assumptions made here (e.g., deterministic parameters and the lot-for-lot modality), are the major limitations of this study. Embellishments by future researchers, such as relaxation of the lot-for-lot assumption, incorporation of uncertainty, more realistic and complex demand and product return functions, multiple products, manufacturers, etc., will, undoubtedly, lead to more refined remanufacturing and related models. Furthermore, future efforts in this area should consider the development of integrated decision models under stochastic conditions, which are likely to be more realistic from an implementation standpoint. Nevertheless, the results obtained in this study are likely to be of some value to practitioners as broad guidelines for integrated pricing, recoverable product collection, production planning and inventory control decisions, as well as for designing more streamlined, well-coordinated supply chains toward gaining competitive advantage. We also hope that our efforts will prove to be useful for researchers in shedding light on some of the intricate and interrelated aspects of product remanufacturing toward developing more effective decision-making models for supply chain and reverse logistics management.

APPENDIX

Note: All the proofs outlined are based on the following concept:

f is concave on D if and only if $D^2 f(x)$ is a negative semidefinite matrix for all $x \in D$. An $n \times n$ symmetric matrix A is negative semidefinite if and only if $(-1)^k |A_k| \geq 0$ for all $k \in \{1, \ldots, n\}$ where A_k is the upper left k-by-k corner of A.

Proof of Proposition 7.1:

If Q^*, r_c^* and p_s^* obtained from (7.3 through 7.5) are local optimum, the sufficient condition is that the objective function (7.2) should be jointly concave in these three variables. The Hessian matrix for (7.2) is

TABLE 7.1
Summary of Results

		Q (units)	p_s ($/unit)	r_c ($/unit)	p_w ($/unit)	d (units/day)	X (units/day)	Π_r ($/day)	Π_m ($/day)	Π_s ($/day)
Retailer's optimal policy	Given p_w	428.456	30.032	1.486	23.000	29.904	19.283	318.360	342.675	661.035
	Variable p_w	388.392	31.680	1.498	23.286	24.960	19.299	228.215	358.616	586.830
Manufacturer's optimal policy	Given p_w	6083.126	29.992	1.399	23.000	30.227	17.994	275.698	377.596	653.293
	Variable p_w	12145.911	31.893	1.257	24.138	24.138	15.660	113.753	399.039	512.792
Centralized optimal policy		1562.425	23.251	0.519	—	50.246	5.455	400.271[a]	430.842[a]	831.112

[a] Allocated on the basis of proportional shares of total supply chain profit under retailer's optimal policy.

From this table it is clear that if the retailer has sufficient policy implementation power in the supply chain, it attempts to keep the replenishment lot size comparatively small (i.e., 428.456 units), in view of its relatively low fixed ordering cost. Furthermore, through its retail pricing ($p_s = \$30.032/$unit), in conjunction with a customer return reimbursement price of $1.486/unit, it prefers to achieve daily market demand and customer return rates of 29.904 and 19.283 units, respectively, that attempt to balance the gains from sales and returns against the ordering and inventory carrying (for both new and used items) costs. The maximum attainable daily profit for the retailer is, thus, $318.360, resulting in a profit of $342.675/day for the manufacturer. Note that as every unit of the returned product represents a net gain of $1.314 (i.e., the difference between the amount, r_m, compensated by the manufacturer and the customer reimbursement, r_c) for the retailer, it attempts to achieve a relatively high used item return rate about 64.483%.

If, on the other hand, the manufacturer is in a position to exert a greater level of negotiating power in the supply chain, its individual optimal policy would dictate a significantly larger replenishment batch of 6083.126 units, due to the relatively high fixed setup and transportation costs. In spite of a more than sixfold increase in the lot size, however, the selling price and return reimbursement, set by the retailer in response, are both only slightly lower than their values under its own optimal policy, i.e., $29.992 and $1.399 per unit, respectively. It is interesting to note that, consequently, the retail demand rate increases slightly to 30.227 units/day and the average product returns decline slightly to 17.994 units/day. The returns, however, now decline slightly to 59.530% of sales. Not unexpectedly, implementing the manufacturer's optimal replenishment policy reduces the retailer's profit to $275.698/day, whereas the manufacturer's profit increases to $377.596/day. Nevertheless, in terms of total supply chain profitability, the difference between adopting any one party's optimal policy over the other's amount to only about 1.171%.

Table 7.1 shows that if the retailer and the manufacturer decide to cooperate through the sharing of necessary information and adopt a jointly optimal policy that maximizes the total supply chain profit, instead of optimizing either party's position, both parties stand to gain considerably from such an approach. As mentioned earlier, the centralized model attempts to avoid double marginalization, i.e., the manufacturer does not explicitly charge the retailer a wholesale price, nor does it explicitly offer the latter a reimbursement for collecting the returns (implying that $p_w = r_m = 0$). Without these cost factors, the centrally controlled approach results in a maximum supply chain profit of $831.112/day, representing more than 25.728% improvement in total system profitability, compared to the retailer's optimal policy or over 27.219% improvement *vis-à-vis* the manufacturer's optimal policy. As expected, the jointly optimal replenishment quantity now is 1562.425 units, which is less than the manufacturer's optimal batch size, but larger than the retailer's optimal order quantity. More interestingly, the retail price is reduced to $23.251/unit and the return reimbursement is decreased to $0.519/unit, respectively, resulting in a considerably larger demand rate of 50.246 units/day, as well as a smaller average product return rate of 5.455 units/day (i.e., about 10.857% of items sold are returned by customers). The implication of our centralized model

is that under a jointly optimal policy, relatively fewer products sold are remanufactured items. Under the given set of problem parameters, it appears desirable to increase the overall market demand through a lower retail price. Also, there is a lesser emphasis on collecting customer returns for remanufacturing. The centralized model reduces the incentive for customer returns, which maximizes the total supply chain profitability.

The absence of a wholesale price and an explicit incentive for the retailer to collect returned items raises some interesting questions concerning a fair and equitable sharing of the total gain resulting from the centralized cooperative policy shown in Table 7.1. Although this can be achieved in several possible ways, we propose a profit sharing plan under a scenario where the retailer is the more powerful member of the supply chain and can dictate the implementation of its own optimal policy. The task of the manufacturer is then to offer sufficient incentive to the retailer in order for the latter to adopt the results of this procedure. Note that under its own individual optimal policy, the retailer's share is 48.161% of the total profit for both the parties. Therefore, it would be reasonable if the retailer is allocated the same percentage of the total supply chain profit of $831.112/day yielded by the centralized model. In other words, the retailer's share of the total profit is $400.271/day and that of the manufacturer is $430.842/day. With this profit sharing arrangement, each party's daily profit is more than 8% larger than that achieved under the retailer's optimal policy. Thus, it is economically attractive for both parties to adopt the jointly optimal policy yielded by our centralized model. If the manufacturer is more powerful of the two parties, the terms of a corresponding profit sharing arrangement, can also be derived easily along similar lines.

Finally, Table 7.1 also shows the results for the decentralized models where under monopolistic competition, the manufacturer can set its wholesale price, which is now treated as a decision variable. Compared with the results for a given wholesale price, the retailer's individual optimal policy dictates increasing both the selling price from $30.032/unit to $31.680/unit and the customer return reimbursement from $1.485/unit to $1.498/unit. Consequently, the order quantity is reduced from 428.46 units to 388.392 units. These changes indicate that the retailer would expend less effort to increase market demand and would tend to compensate by attempting to increase its revenue from returns. This appears to be a rational response to a higher wholesale price. Also, as expected, the manufacturer's share of the total supply chain profit now increases from 51.839% to 61.111%, while, the total profits for the supply chain declines to $586.830/day. These effects, not unexpectedly, tend to be magnified when the supplier is in a position to dictate the adoption of its own optimal policy by the retailer. Now the total supply chain profit shrinks further to 512.792$/day, although the manufacturer's relative share of this, as well as its own daily profit go up substantially, albeit at the expense of the retailer.

7.6   SUMMARY AND CONCLUSIONS

In this study, we have developed mathematical models under deterministic condition, for simultaneously determining the production/delivery lot size, the retail price, and the customer return reimbursement level for a single recoverable

$$D^2\Pi_r(Q, r_c, p_s) = \begin{bmatrix} \dfrac{\partial^2 \Pi_r}{\partial Q^2} & \dfrac{\partial^2 \Pi_r}{\partial Q \partial r_c} & \dfrac{\partial^2 \Pi_r}{\partial Q \partial p_s} \\[3mm] \dfrac{\partial^2 \Pi_r}{\partial Q \partial r_c} & \dfrac{\partial \Pi_r^2}{\partial r_c^2} & \dfrac{\partial^2 \Pi_r}{\partial p_s \partial r_c} \\[3mm] \dfrac{\partial^2 \Pi_r}{\partial Q \partial p_s} & \dfrac{\partial^2 \Pi_r}{\partial p_s \partial r_c} & \dfrac{\partial \Pi_r^2}{\partial p_s^2} \end{bmatrix}$$

$$= \begin{bmatrix} -2S_r \dfrac{d}{Q^3} & -\dfrac{ah_{rr}}{2d} & -\dfrac{BS_r}{Q^2} + \dfrac{bh_{rr}}{2d} - \dfrac{Bh_{rr}x}{2d^2} \\[3mm] -\dfrac{ah_{rr}}{2d} & -2a & b - \dfrac{aBQh_{rr}}{2d^2} \\[3mm] -\dfrac{BS_r}{Q^2} + \dfrac{bh_{rr}}{2d} - \dfrac{Bh_{rr}x}{2d^2} & b - \dfrac{aBQh_{rr}}{2d^2} & -2B + \dfrac{bBh_{rr}Q}{d^2} - \dfrac{h_{rr}B^2Q}{d^3} \end{bmatrix}$$

Thus,

$$| A_1 | = \frac{\partial^2 \Pi_r}{\partial Q^2} = -2S_r \frac{d}{Q^3} \le 0$$

Since all the parameters are strictly positive, (A-1) always holds. Also,

$$| A_2 | = \begin{vmatrix} \dfrac{\partial^2 \Pi_r}{\partial Q^2} & \dfrac{\partial^2 \Pi_r}{\partial Q \partial r_c} \\[3mm] \dfrac{\partial^2 \Pi_r}{\partial Q \partial r_c} & \dfrac{\partial \Pi_r^2}{\partial r_c^2} \end{vmatrix} = \begin{vmatrix} -2S_r \dfrac{d}{Q^3} & -\dfrac{ah_{rr}}{2d} \\[3mm] -\dfrac{ah_{rr}}{2d} & -2a \end{vmatrix}$$

$$= \left(\sqrt{\frac{4aS_r d}{Q^3}} + \frac{ah_{rr}}{2d} \right) \left(\sqrt{\frac{4aS_r d}{Q^3}} - \frac{ah_{rr}}{2d} \right) \ge 0$$

Simplification of this leads to condition (a) under Proposition 7.1:

$$| A_3 | = \begin{vmatrix} \dfrac{\partial^2 \Pi_r}{\partial Q^2} & \dfrac{\partial^2 \Pi_r}{\partial Q \partial r_c} & \dfrac{\partial^2 \Pi_r}{\partial Q \partial p_s} \\[3mm] \dfrac{\partial^2 \Pi_r}{\partial Q \partial r_c} & \dfrac{\partial \Pi_r^2}{\partial r_c^2} & \dfrac{\partial^2 \Pi_r}{\partial p_s \partial r_c} \\[3mm] \dfrac{\partial^2 \Pi_r}{\partial Q \partial p_s} & \dfrac{\partial^2 \Pi_r}{\partial p_s \partial r_c} & \dfrac{\partial \Pi_r^2}{\partial p_s^2} \end{vmatrix}$$

$$= -2S_r \frac{d}{Q^3} \left[(-2a) \left(-2B + \frac{bBh_{rr}Q}{d^2} - \frac{h_{rr}B^2Q}{d^3} \right) - \left(b - \frac{aBQh_{rr}}{2d^2} \right)^2 \right]$$

$$- \left(-\frac{ah_{rr}}{2d} \right) \left[\left(-\frac{ah_{rr}}{2d} \right) \left(-2B + \frac{bBh_{rr}Q}{d^2} - \frac{h_{rr}B^2Q}{d^3} \right) \right.$$

$$\left. - \left(-\frac{BS_r}{Q^2} + \frac{bh_{rr}}{2d} - \frac{Bh_{rr}x}{2d^2} \right) \left(b - \frac{aBQh_{rr}}{2d^2} \right) \right]$$

$$+ \left(-\frac{BS_r}{Q^2} + \frac{bh_{rr}}{2d} - \frac{Bh_{rr}x}{2d^2} \right) \left[\left(-\frac{ah_{rr}}{2d} \right) \left(b - \frac{aBQh_{rr}}{2d^2} \right) \right.$$

$$\left. - (-2a) \left(-\frac{BS_r}{Q^2} + \frac{bh_{rr}}{2d} - \frac{Bh_{rr}x}{2d^2} \right) \right]$$

$$= \frac{2aBh_{rr}S_r(Ab - aBr_c)}{2Q^2d^3} + \frac{2DS_rb^2}{Q^3} + \frac{a^2Bh_{rr}^2}{2d^2} - \frac{8aBDS_r}{Q^3}$$

$$+ \frac{2bah_{rr}S_r}{Q^2d} - \frac{abh_{rr}^2(Ab - aBr_c)}{2d^3} + \frac{2DB^2S_r^2}{Q^4} + \frac{ah_{rr}^2(Ab - aBr_c)^2}{2d^4} \le 0$$

This simplifies to condition (b) under Proposition 7.1.

Proof of Proposition 7.2:

Only when all the remanufacturing cost parameters are smaller than the corresponding cost parameters associated with producing the product afresh, the producer would benefit from a remanufacturing strategy. In addition, we check whether it is profitable for the producer to resort to such a strategy. The profit per time unit for the manufacturer without remanufacturing is

$$\Pi'_m = (p_w - c_s)d - \frac{d}{Q}(S_m + S_{rm} + S_i) - \frac{h_m Qd}{2m} - h_m \frac{Qd}{2m} - h_i \frac{Qd}{2m} - c_m d$$

Comparing this with Equation 7.7 it is clear that the manufacturer can benefit from remanufacturing if $\Pi_m - \Pi'_m \ge 0$, i.e.,

$$-(r_m + c_r)x - h_{ir} \frac{Qx^2}{2md} - \frac{h_i Q}{2m} \left(\frac{x^2}{d} - 2x \right) + c_m x - c_{rm}x \ge 0$$

This directly yields condition (7.11) under Proposition 7.2.

Proof of Proposition 7.3:

If Q^{**}, r_c^{**} and p_s^{**} obtained from (7.13)–(7.16) are to be local optimum, the sufficient condition is that the objective function (7.13) should be jointly concave in these three variables. The Hessian matrix for the manufacturer's profit function (7.8) is

$$D^2\Pi_m(Q,r_c,p_s) = \begin{vmatrix} \dfrac{\partial^2\Pi_m}{\partial Q^2} & \dfrac{\partial^2\Pi_m}{\partial Q\partial p_s} & \dfrac{\partial^2\Pi_m}{\partial Q\partial r_c} \\[2mm] \dfrac{\partial^2\Pi_m}{\partial Q\partial p_s} & \dfrac{\partial^2\Pi_m}{\partial p_s^2} & \dfrac{\partial^2\Pi_m}{\partial p_s\partial r_c} \\[2mm] \dfrac{\partial^2\Pi_m}{\partial Q\partial r_c} & \dfrac{\partial^2\Pi_m}{\partial p_s\partial r_c} & \dfrac{\partial\Pi_m^2}{\partial r_c^2} \end{vmatrix} = \begin{bmatrix} F11 & F12 & F13 \\ F21 & F22 & F23 \\ F31 & F32 & F33 \end{bmatrix}$$

$$\begin{bmatrix} -\dfrac{2d(S_m+S_{rm}+S_i)}{Q^3} & -\dfrac{B(S_m+S_{rm}+S_i)}{Q^2} \\ & +\dfrac{Bh_m-h_i(2b-B)}{2m} \\ & +\dfrac{(2db-Bx)x(h_{ir}-h_i)}{2md^2} & \dfrac{ah_i}{2m}-\dfrac{ax(h_{ir}-h_i)}{md} \\[4mm] -\dfrac{B(S_m+S_{rm}+S_i)}{Q^2} \\ +\dfrac{Bh_m-h_i(2b-B)}{2m} & \dfrac{-(h_{ir}-h_i)Q((db-xB)^2}{md^3} & \dfrac{aQ(h_{ir}-h_i)(db-xB)}{md^2} \\ +\dfrac{(2db-Bx)x(h_{ir}-h_i)}{2md^2} \\[4mm] \dfrac{ah_i}{2m}-\dfrac{ax(h_{ir}-h_i)}{md} & \dfrac{aQ(h_{ir}-h_i)(db-xB)}{md^2} & \dfrac{-a^2Q(h_{ir}-h_i)}{md} \end{bmatrix}$$

Set $H = D^2\Pi_m(Q, r_c, p_s)$, then if H is negative semidefinite, the following must hold:

$$|H_1| = \frac{\partial^2\Pi_m}{\partial Q^2} = -\frac{2d(S_m+S_{rm}+S_i)}{Q^2} \le 0$$

Since all variables and parameters are nonnegative, this requirement is always satisfied.

$$|H_{12}| = \begin{vmatrix} \dfrac{\partial^2 \Pi_m}{\partial Q^2} & \dfrac{\partial^2 \Pi_m}{\partial Q \partial p_s} \\[3mm] \dfrac{\partial^2 \Pi_m}{\partial Q \partial p_s} & \dfrac{\partial^2 \Pi_m}{\partial p_s^2} \end{vmatrix}$$

$$\times \begin{vmatrix} -\dfrac{2d(S_m + S_{rm} + S_i)}{Q^2} & \begin{matrix} -\dfrac{B(S_m + S_{rm} + S_i)}{Q^2} \\[3mm] +\dfrac{Bh_m - h_i(2b - B)}{2m} \\[3mm] +\dfrac{(2db - Bx)x(h_{ir} - h_i)}{2md^2} \end{matrix} \\[12mm] \begin{matrix} -\dfrac{B(S_m + S_{rm} + S_i)}{Q^2} \\[3mm] +\dfrac{Bh_m - h_i(2b - B)}{2m} \\[3mm] +\dfrac{(2db - Bx)x(h_{ir} - h_i)}{2md^2} \end{matrix} & -\dfrac{(h_{ir} - h_i)Q((db - xB)^2}{md^3} \end{vmatrix} \geq 0$$

It can be easily verified that the aforementioned inequality results in condition (c) in Proposition 7.3.

$$|H_3| = \begin{vmatrix} F11 & F12 & F13 \\ F21 & F22 & F23 \\ F31 & F32 & F33 \end{vmatrix} = \begin{vmatrix} \dfrac{\partial^2 \Pi_m}{\partial Q^2} & \dfrac{\partial^2 \Pi_m}{\partial Q \partial p_s} & \dfrac{\partial^2 \Pi_m}{\partial Q \partial r_c} \\[3mm] \dfrac{\partial^2 \Pi_m}{\partial Q \partial p_s} & \dfrac{\partial^2 \Pi_m}{\partial p_s^2} & \dfrac{\partial^2 \Pi_m}{\partial p_s \partial r_c} \\[3mm] \dfrac{\partial^2 \Pi_m}{\partial Q \partial r_c} & \dfrac{\partial^2 \Pi_m}{\partial p_s \partial r_c} & \dfrac{\partial \Pi_m^2}{\partial r_c^2} \end{vmatrix}$$

$$= \frac{a^2 B^2 Q(h_{ir} - h_i)(h_i + h_m)^2}{4m^3 d} - \frac{a^2 B^2 Qxh_i(h_{ir} - h_i)(h_i - h_m)}{m^3 d^2}$$

$$+ \frac{a^2 B^2 (h_{ir} - h_i)(S_m + S_{rm} + S_i)^2}{Q^3 md}$$

$$- \frac{a^2 BQx^2(h_{ir} - h_i)(Bh_{ir}h_i - 6bh_{ir}h_i + Bh_{ir}h_m - 5Bh_i^2 + 6bh_i^2 - Bh_ih_m)}{4m^3 d^3}$$

$$- \frac{a^2 B^2 Qx^3 h_i(h_{ir} - h_i)^2}{m^3 d^4} + \frac{a^2 B^2 Qx^4(h_{ir} - h_i)^3}{4m^3 d^5}$$

$$- \frac{a^2 B^2 (h_{ir} - h_i)(h_i - h_m)(S_m + S_{rm} + S_i)}{Qm^2 d}$$

$$+ \frac{2a^2 B^2 xh_i(h_{ir} - h_i)(S_m + S_{rm} + S_i)}{Qm^2 d^2} - \frac{a^2 B^2 x^2(h_{ir} - h_i)^2(S_m + S_{rm} + S_i)}{Qm^2 d^3} \leq 0$$

Again, this inequality reduces to condition (d) in Proposition 7.3.

Conditions (c) and (d) are necessary to ensure that the profit function (7.8) of the manufacturer is jointly concave with respect to Q^*, r_c^*, and p_s^*. Also, as shown earlier, conditions (a) and (b) are sufficient for joint concavity of the retailer's profit function (7.2). Since the total supply chain profit (7.13) is the sum of (7.2) and (7.8), all four of these conditions are necessary for it to be jointly concave in Q^{**}, r_c^{**}, and p_s^{**}.

REFERENCES

Bhattacharya, S., V. D. R. Guide Jr., and L. N. Van Wassenhove. 2006. Optimal order quantities with remanufacturing across new product generations. *Production and Operations Management* 15: 421–431.

El Saadany, A. M. A. and M. Y. Jaber. 2010. A production/remanufacturing inventory model with price and quality dependant return rate. *Computers & Industrial Engineering* 58: 352–362.

Fleischmann, M., J. M. Bloemhof-Ruwaard, R. Dekker et al. 1997. Quantitative models for reverse logistics: A review. *European Journal of Operational Research* 103: 1–17.

Guide, Jr., V. D., V. Jayaraman, R. Srivastava, and W. C. Benton. 2000. Supply chain management for recoverable manufacturing systems. *Interfaces* 30: 125–142.

Guide, Jr., V. D. R., R. H. Teunter, and L. N. Van Wassenhove. 2003. Matching demand and supply to maximum profits from remanufacturing. *Manufacturing and Service Operations Management* 5: 303–316.

Guide, Jr., V. D. R. and L. N. Van Wassenhove. 2009. The evolution of closed-loop supply chain research. *Operations Research* 57: 10–18.

Gungor, A. and S. M. Gupta. 1999. Issues in environmentally conscious manufacturing and product recovery: A survey. *Computers & Industrial Engineering* 36: 811–853.

Ilgin, M. A. and S. M. Gupta. 2010. Environmentally conscious manufacturing and product recovery (ECMPRO): A review of the state of the art. *Journal of Environmental Management* 91: 563–591.

Ilgin, M. A. and S. M. Gupta. 2012. *Remanufacturing Modeling and Analysis*. Boca Raton, FL: CRC Press.

Koh, S. G., H. Hwang, S. Kwon-IK, and C. S. Ko. 2002. An optimal ordering and recovery policy for reusable items. *Computers & Industrial Engineering* 43: 59–73.

Konstantaras, I., K. Skouri, and M. Y. Jaber. 2010. Lot sizing for a recoverable product with inspection and sorting. *Computers & Industrial Engineering* 58: 452–462.

Lage, Jr., M. and M. G. Filho. 2012. Production planning and control for remanufacturing: Literature review and analysis. *Production Planning and Control* 23: 419–435.

Mabini, M. C., L. M. Pintelon, and L. F. Gelders. 1992. EOQ type formulations for controlling repairable inventories. *International Journal of Production Economics* 28: 21–33.

Porharel, S. and A. Mutha. 2009. Perspectives in reverse logistics: A review. *Resources, Conservation and Recycling* 53: 175–182.

Ray, S., T. Boyaci, and N. Aras. 2005. Optimal prices and trade-in rebates for durable, remanufacturable products. *Manufacturing and Service Operations Management* 7: 208–228.

Richter, K. 1996a. The EOQ repair and waste disposal model with variable setup numbers. *European Journal of Operational Research* 95: 313–324.

Richter, K. 1996b. The extended EOQ repair and waste disposal model. *International Journal of Production Economics* 45: 443–447.

Rubio, S., A. Chamorro, and F. J. Miranda. 2008. Characteristics of the research on reverse logistics. *International Journal of Production Research* 46: 1099–1120.

Savaskan, R. C., S. Bhattacharya, and L. N. Van Wassenhove. 2004. Closed-loop supply chain models with product remanufacturing. *Management Science* 50: 239–252.

Schrady, D. A. 1967. A deterministic inventory model for repairable items. *Naval Research Logistics Quarterly* 14: 391–398.

Vorasayan, J. and S. M. Ryan. 2006. Optimal price and quantity of refurbished products. *Production and Operations Management* 15: 369–383.

Xerox. 2005. http://www.xerox.com/downloads/usa/en/e/ehs_2005_progress_report.pdf

8 Advanced Remanufacturing-to-Order and Disassembly-to-Order System under Demand/Decision Uncertainty

Onder Ondemir and Surendra M. Gupta

CONTENTS

8.1 INTRODUCTION

For the last decade, environmentally conscious manufacturing and product recovery has been focused on by many researchers and practitioners because of the disturbing depletion of natural resources and emerging environmental problems such as ozone layer depletion, global warming, air pollution, and toxic wastes. One of the major factors in this downfall is the shortened economic life of products. This is mostly due to the awe-inspiring advancements in technology making higher quality products more affordable than ever, hence opening a new consumption era. In this new era, consumers discard their products in the interest of obtaining newer ones even before the current products reach their technological end-of-lives (EOLs). Rapid turnover of products diminishes landfills, exhausts virgin resources, and potentially poisons the earth. Product recovery is the most effective way of reversing this environmental breakdown owing to its simple philosophy: "reuse the value trapped in discarded products." This simple philosophy has been widely adopted by lawmakers and environmental activists. This has led to several legislations that hold manufacturers responsible for their products even after the point of sale, thus imposing optimal management of end-of-life products (EOLPs).

Product recovery may be of several forms such as remanufacturing, refurbishing, repairing, and recycling. All recovery options involve disassembly operations up to a certain level. Disassembly is a labor-intensive operation carried out to extract parts from EOLPs for several purposes including elimination of hazardous parts, reusable component recovery, component testing, and content inspection. Of all recovery operations, remanufacturing and disassembly are considered to be the most complex ones. This is mostly due to the lack of information about the quality and quantity of EOLPs and their components. When there is no information available on the components' quality status, comprehensive testing is needed to collect that. After testing, if an EOLP is found not suitable for remanufacturing, the time and resources spent on determining that are wasted; otherwise, necessary recovery operations and spare parts are listed based on the testing results. EOLPs, however, do not show typical qualities since they originate from various sources where they had been subject to different working conditions. As a result, it is highly likely that each EOLP has its own quality condition exhibiting unique remanufacturing needs. Hence, finding the EOLPs with minimal recovery costs requires testing the whole EOLP inventory, which can be very expensive. However, emerging information technology devices, such as sensors and radio-frequency identification (RFID) tags, can be used to mitigate EOL recovery decisions.

Sensors and RFID tags, once incorporated with the products, can monitor the critical components throughout products' economic lives and deliver the collected life cycle information when the products reach recovery facilities. In the beginning-of-life (BOL) phase, bill of materials, model and serial numbers, manufacturing date, location, warranty terms, maintenance instructions, and EOL processing guidelines (static data) are saved in the tags. In the middle-of-life (MOL) phase, sale date and customer number (static data), run cycles, working temperatures, failures, environmental sensory inputs such as dust, vibration, humidity levels (dynamic data), and maintenance information (e.g., dates, operations, center IDs, and technician IDs) are logged. By means of the central information sharing (Figure 8.1), collected information is utilized to improve product design, ensure on-time maintenance, and establish an early warning system in the BOL and MOL phases. EOL operations benefit the most from this information. Complete knowledge on the condition and quantity of EOLPs and remaining life determination eliminate costly preliminary disassembly and inspection operations, and enable optimal remanufacturing planning.

Although uncertainty about the quality and quantity of EOLPs is removed by means of the sensors and RFID tags, there are two other sources of uncertainties that need to be addressed: system goals and demands.

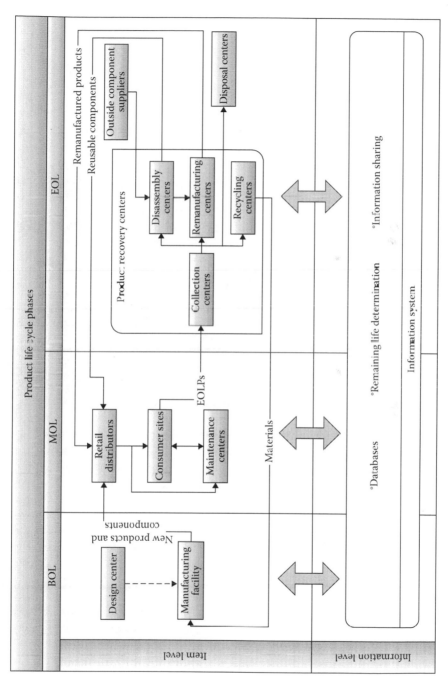

FIGURE 8.1   Material and information flow among product life cycle phases.

In this chapter, we propose an ARTODTO model for EOL processing of SEPs under demand and decision uncertainty. The proposed model is formulated as a fuzzy goal programming (FGP) model to achieve a variety of financial, environmental, and physical goals. A sensor-embedded water-heater recovery case is considered to illustrate the methodology.

8.2 LITERATURE REVIEW

Relevant publications in product recovery, sensor and RFID tag embedded devices, and fuzzy optimization areas are explored and a brief review is presented in the following sections.

8.2.1 PRODUCT RECOVERY

The state of the art on Environmentally Conscious Manufacturing and Product Recovery (ECMPRO) through 1999 is presented by Gungor and Gupta (1999). Ilgin and Gupta (2010b) complemented this work exploring the state of the art through 2009. A recent book by Wang and Gupta (2011) addressed the issues regarding environmental thinking in every phase of products' life cycle. Lage Jr. and Godinho Filho (2012) reviewed 76 journal articles published on *Production Planning and Control* (PPC) in remanufacturing between 2000 and 2009. The authors also provided a comparative analysis between this literature review and the review conducted by Guide Jr.(2000). Ilgin and Gupta (2012) covered the design, planning, and processing issues in remanufacturing area.

Uncertainties in recovery operations are taken into account by many authors. Galbreth and Blackburn (2006) sought for optimal sorting and acquisition policies by considering the variability in the condition of the used products. Zikopoulos and Tagaras (2007) analyzed the effect of uncertainty in quality levels of returned products on the profitability of a single period refurbishing operation. Huang et al. (2009) developed dynamic closed-loop supply chain models by considering the uncertainty due to time-delay in remanufacturing and returns, system cost parameters, and customer demand's disturbances. Several methods to deal with remanufacturing operations were proposed in the literature. Li et al. (2009) integrated a hybrid cell evaluated GA with a DES model to optimize the PPC policies for dedicated remanufacturing. DePuy et al. (2007) presented a production planning method which estimates the expected number of remanufactured units to be completed in each future period together with the number of components needed to be purchased to avoid any projected shortages. Xanthopoulos and Iakovou (2009) proposed an MILP-based aggregate production planning model which can determine how many EOL products and components should be collected, nondestructively or destructively disassembled, recycled, remanufactured, stored, backordered, and disposed in each period. Remanufacturing planning models, in general, use estimated or assumed quality levels.

Disassembly is another frequently researched topic in ECMPRO owing to its importance in all recovery operations. Although disassembly problems have been studied under several subtopics, disassembly-to-order (DTO) research exhibits the most relevance to this work. DTO systems exhibit multiple conflicting objectives where multi-criteria decision making (MCDM) can be applied. Kongar and Gupta (2002) presented a preemptive goal programming (PGP) model of an electronic products DTO system considering a variety of physical, financial, and environmental constraints and goals. Kongar and Gupta (2009b) proposed a linear physical programming (LPP) model to solve the DTO problem with multiple physical targets. Heuristics, meta-heuristics, and expert models have been used in the DTO problem because of the increased complexity especially when multiple criteria and/or periods are considered. Kongar and Gupta (2009a) presented a multi-objective TS algorithm and Gupta et al. (2009) proposed an artificial neural network model in order to solve the DTO problem.

For additional reading on disassembly operations and problems, the reader is referred to the papers by Taleb and Gupta (1997), Taleb et al. (1997), Tang et al. (2002), Veerakamolmal and Gupta (1999), Lambert and Gupta (2005), and Gupta and Taleb (1994).

8.2.2 Sensor and RFID Technologies

Several publications (Chang and Hung 2012, Cheng et al. 2004, Ilgin and Gupta 2010c, 2011a, Karlsson 1997, Petriu et al. 2000, Scheidt and Shuqiang 1994) have been made on embedded sensors for after-sale monitoring and end-of-life decision making. Klausner et al. (1998) proposed an information system for product recovery (ISPR) where a sensor is integrated in a product to record and store data strongly correlated with the degradation of components during the use stage of a product. The data recorded and processed during the use stage are retrieved and analyzed by the ISPR when the product is returned. Klausner et al. (2000) extended the previous study by including the economic efficiency of the sensor in a reuse scenario. Yang et al. (2009) performed field trials to show that sensor-embedded products can enable EOL treatment and other suitable product-related services. Vadde et al. (2008) conducted a simulation study showing the use of sensor-based life cycle data in EOL decisions. Ilgin and Gupta (2010a) investigated the impact of SEPs on the various performance measures of a multi-product (i.e., refrigerators and washing machines) disassembly line using simulation analysis. Via a cost–benefit analysis, authors showed that SEPs not only reduce the total system cost, but also increase the revenue and profit. Ilgin and Gupta (2011b) simulated the recovery of sensor-embedded washing machines using a multi-kanban controlled disassembly line. Ilgin et al. (in press) explored the financial benefits of embedding sensors into products as an extension to the multi-kanban controlled just-in time remanufacturing system.

Europe took the lead in imposing environmental regulations. It has become mandatory to tag the disassembly instructions with products (Vadde et al. 2008) as defined in waste electrical and electronic equipment (WEEE) directive. Since the memory offered by ultra-high frequency RFID tags is large enough to include disassembly and recycling information (Luttropp and Johansson 2010), RFID tags become prominent in environmentally conscious product recovery. Also, in ZeroWIN (Towards Zero Waste in Industrial Networks) project, RFID technology will be investigated, designed, and tested (Curran and Williams 2012). Parlikad and McFarlane (2007) discussed how RFID-based product identification technologies can be employed to provide the necessary information and showed the positive impacts on product recovery decisions. RFID technology also permits product life cycle monitoring when active or semi-active tags are used (Dolgui and Proth 2008). Gonnuru (2010) proposed an RFID integrated fuzzy based disassembly planning and sequencing model and showed the use of life cycle information for optimal disassembly decisions. Kulkarni et al. (2005) examined the benefits of information provided by RFID tagged monitoring systems in decision making during product recovery. Zhou et al. (2007) proposed an RFID-based remote monitoring system for enterprise internal production management.

Product life cycle data, once captured, allow remaining life time estimation (Engel et al. 2000, Lee et al. 1999, Middendorf et al. 2003, 2005, Rugrungruang 2008, Wang and Zhang 2008). Herzog et al. (2009) proved the advantage of using condition-based data in remaining life prediction. Mazhar et al. (2007) proposed a comprehensive two-step approach combining Weibull analysis and artificial neural networks (ANNs) for remaining life estimation of used components in consumer products. Byington et al. (2004) proposed a data-driven neural network methodology to remaining life predictions for aircraft actuator components. Ondemir et al. (2012) proposed a mathematical DTO model utilizing life cycle data in order to fulfill remaining life time–based product and component demands as well as material demands assuming the availability of sensor-based information.

8.2.3 FUZZY OPTIMIZATION IN PRODUCT RECOVERY

Product recovery operations are known to have high degrees of uncertainty. Inderfurth (2005) showed the negative effects of return, quality, and demand uncertainties on the recovery fraction via a numerical analysis. The author stated that purely deterministic approaches as employed by Ferrer and Whybark (2001), in many situations, will not be sufficient to completely understand the economics of remanufacturing and respective consequences for recovery behavior. Therefore, deterministic models gave way to fuzzy, stochastic, heuristic, meta-heuristic, and expert models. Barba-Gutierrez and Adenso-Diaz (2009) extended Gupta and Taleb's (1994) method using a fuzzy logic approach, incorporating imprecision and subjectivity into the model formulation and solution process. Langella (2007) developed a multi-period heuristic considering holding costs and external procurement of items. Inderfurth and Langella (2006) developed two heuristic procedures to investigate the effect of stochastic yields on the DTO system. Kongar and Gupta (2006) presented a FGP model for DTO systems under uncertainty.

8.3 FUZZY GOAL PROGRAMMING

The notion of fuzzy sets was first introduced by Zadeh (1965) in order to accommodate vague (uncertain) information in mathematical operations. "A fuzzy set is a class of objects with a continuum of grades of membership. Such a set is characterized by a membership (characteristic) function which assigns to each object a grade of membership ranging between zero and one." After the emergence of fuzzy mathematics, researchers discovered its use in many different areas. Zimmermann (1983, 2001) investigated the application of fuzzy sets in various areas including optimization and decision making. Ross (1995) presented the application of fuzzy logic in engineering problems such as simulation, decision making, classification, and pattern recognition with imperfect information.

Goal programming (GP) is a MCDM approach introduced by Charnes and Cooper (1961). A broad literature reviews of GP through 2000 were presented by Schniederjans (1995) and Jones and Tamiz (2002). In its generic form, GP is not capable of dealing with imprecise target values. To this end, researchers investigated the use of fuzzy sets in operation research (Zimmermann 1983) and developed several FGP (Chen and Tsai 2001, Hannan 1981, Narasimhan 1980, Tiwari et al. 1986, 1987) approaches for the situations where the decision makers (DM) are uncertain about the goals or the collected data are not reliable. A recent book by Jones and Tamiz (2010) also provided in-depth information on GP and FGP.

In a GP model, these membership functions can be used to incorporate uncertain goals and constraints. In this context, each goal is defined as a fuzzy set $(\tilde{G}_s)$ and each feasible solution as a member of these sets at some degree. Fuzzy goals and constraints are expressed using "$\tilde{>}$," "$\tilde{<}$," and "$\tilde{=}$" symbols. These symbols correspond to right-sided, left-sided, and triangular membership functions, respectively. Although fuzzy membership functions may have many shapes, right-sided, left-sided, triangular, and trapezoidal linear membership functions (Figure 8.2) are the most common ones. The degree of membership represents the achievement level of a goal and are calculated using the membership functions whose mathematical formulations are given as follows.

The symbols used in the mathematical formulation are given in Table 8.1.

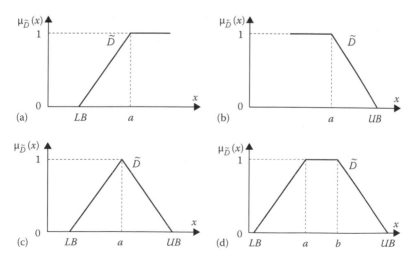

FIGURE 8.2 (a) Right-sided, (b) left-sided, (c) triangular, and (d) trapezoidal membership functions.

TABLE 8.1

Variables and Parameters Used in the Chapter

Variables	Definition	Parameter	Definition
CSL	Customers' satisfaction level	I	Set of EOLPs on hand
MSR	Material sales revenue	B	Set of remaining-life-bins
TC	Total cost		(for components)
TDC	Total disassembly cost	J	Set of components dealt with
$TDIC$	Total disposal cost	M	Alias for B (for products)
THC	Total holding cost	K	Set of material types dealt with
TRC	Total recycling cost	P_j	Set of predecessor components
$TRMC$	Total remanufacturing cost		of component j
$WDIS$	Total weight of components	b, i, j, k, m	Running numbers
	disposed of	α	Destructive disassembly cost factor
$cdis_{jb}$	Number of component js in	a_{ij}	1 if component j of EOLP i is
	remaining-life-bin b that are		functional
	disposed of	β_{ij}	The highest life-bin that
def_{ijb}	1 if component j in EOLP i is		component j of EOLP i can be
	disassembled because of remaining		placed in
	life deficiency and placed in	c_{jb}	Outside procurement cost of a
	remaining-life-bin b during		component j for life-bin b
	remanufacturing, zero otherwise	ca_j	Assembly cost of a component j
fd_j	Number of nonfunctional component	cd_j	Disassembly cost
	js that are disposed of		of a component j
fr_j	Number of nonfunctional component	cds_j	Disposal cost of a component j
	js that are recycled	ch_j	Holding cost of a component j
l_{jb}	Number of component js purchased	cin_{ijb}	1 when remaining life of
	for remaining-life-bin		component j is adequate for
μ_t	Achievement level of the tth goal,		remaining-life-bin b, zero
	$t = \{1, 2, 3, 4, 5\}$		otherwise
r_{jb}	Number of component js in	crc_j	Recycling cost of a component j
	remaining-life-bin b that are	dc_{jb}	Demand for component j in
	recycled		remaining-life-bin b
rep_{imjb}	1 if a component j from life-bin b	dp_m	Demand for product in remaining-
	needs to be used to remanufacture		life-bin m
	EOLP i in order to make a product	dm_k	Demand for material k
	for life-bin m, zero otherwise	dfc_{imj}	1 if component j of EOLP i is
rp_{ij}	1 if component j in EOLP i is		remaining-life-deficient for
	disassembled during		life-bin m, zero otherwise
	remanufacturing, zero otherwise	f_{ij}	1 if component j of EOLP i is
s_i	1 if EOLP i is stored, zero otherwise		nonfunctional, zero otherwise
sc_{jb}	Number of component js in	γ_{jk}	Material k yield of a component j,
	remaining-life-bin b that are stored	h	Unit EOLP holding cost

TABLE 8.1 (continued)
Variables and Parameters Used in the Chapter

Variables	Definition	Parameter	Definition
sm_k	Amount of material k stored	LB_t	Lower bound of the tth goal
u	The model objective to be minimized	mh_k	Unit holding cost for material k
		mis_{ij}	Binary parameter taking 1 if component j is missing in
w_i	1 if EOLP i is recycled, zero otherwise		EOLP i, zero otherwise
$\bar{x}_i$	1 if EOLP i is disassembled for components, zero otherwise	prc_k	Unit sales price of material k
		τ	Percent tolerance of demand
x_{ijb}	1 if component j in EOLP i is disassembled and placed in remaining-life-bin b, zero otherwise		forecasts
		UB_t	Upper bound of the tth goal
		ω_j	Weight of a component j
$\bar{y}_i$	1 if EOLP i is remanufactured, zero otherwise		
y_{im}	1 if EOLP i is remanufactured to make a product for remaining-life-bin m, zero otherwise		
z_t	1 if EOLP i is disposed of, zero otherwise		

Membership function $\mu_{\tilde{D}}(x)$ represents the membership of element x in fuzzy set $\tilde{D}$. Mathematical expression of $\mu_{\tilde{D}}(x)$ for common membership shapes can be written as follows:

Right-sided linear membership function

$$\mu_{\tilde{D}}(x) = \begin{cases} 0, & x \le LB \\ \dfrac{x - LB}{a - LB}, & LB \le x \le a \\ 1, & x \ge a \end{cases} \tag{8.1}$$

Left-sided linear membership function

$$\mu_{\tilde{D}}(x) = \begin{cases} 1, & x \le a \\ \dfrac{UB - x}{UB - a}, & a \le x \le UB \\ 0, & x \ge UB \end{cases} \tag{8.2}$$

Triangular linear membership function

$$\mu_{\tilde{D}}(x) = \begin{cases} 0, & x \le LB \\ \dfrac{x - LB}{a - LB}, & LB \le x \le a \\ \dfrac{UB - x}{UB - a}, & a \le x \le UB \\ 0, & x \ge UB \end{cases} \tag{8.3}$$

Trapezoidal linear membership function

$$\mu_{\tilde{D}}(x) = \begin{cases} 0, & x \le LB \\ \dfrac{x - LB}{a - LB}, & LB \le x \le a \\ 1, & a \le x \le b \\ \dfrac{UB - x}{UB - b}, & b \le x \le UB \\ 0, & x \ge UB \end{cases} \tag{8.4}$$

There are several FGP approaches in the literature. Tiwari et al. (1987) proposed an additive FGP model that maximizes the sum of achievement levels. Zimmermann (2001) presented another MCDM approach that maximizes the minimum achievement level. Chen and Tsai (2001) presented the limitations of both approaches in accommodating different priority levels and proposed an additive FGP model that considers preemptive priority levels with a single objective function. General form of this model is given next:

$$\text{Max. } u = \sum_{t \in \{R \cup L \cup T\}} \mu_t$$

$$\frac{g_t - LB_t}{a - LB_t} \ge \mu_t, \qquad \forall t \in R \text{ (for right-sided goals)}$$

$$\frac{UB_t - g_t}{UB_t - a} \ge \mu_t, \qquad \forall t \in L \text{ (for left-sided goals)}$$

$$1 - \left| \frac{2(g_t - a)}{UB_t - LB_t} \right| \ge \mu_t, \quad \forall t \in T \text{ (for isosceles triangular goals)} \tag{8.5}$$

$$AX \le B$$

$$\mu_t \ge \mu_{t'}, \qquad \forall t, t' \in \{R \cup L \cup T\} \wedge \Pr(t) \ge \Pr(t')$$

$$\mu_t \le 1, \qquad \forall t \in \{R \cup L \cup T\}$$

$$x, \mu_t \ge 0, \qquad \forall t \in \{R \cup L \cup T\}$$

where

μ_t ($\mu_{\tilde{G}_t}$) and g_t are the achievement degree and value of the tth goal
UB, LB, and a are the upper bounds, lower bounds, and aspiration values of the goals
$AX \le B$ is the crisp system constraints
R, L, and T are the sets of right-sided, left-sided, and triangular-shaped goals
$\Pr(t)$ is the priority level of the tth goal

It should be noted that the model is nonlinear when triangular goals are present. However, each nonlinear constraint can be re-formulated as two linear constraints when the goals have isosceles triangular shapes.

8.4   ADVANCED REMANUFACTURING-TO-ORDER AND DISASSEMBLY-TO-ORDER SYSTEM

The ARTODTO system is a product recovery system where EOLPs are embedded with sensors and RFID tags. The ARTODTO system involves remanufacturing option to meet product demands. Necessary spare parts for remanufacturing operations can be taken from disassembled part inventory and/or purchased from part vendors. Customers' component demands are also answered using these two sources.

Since life cycle data collected by sensors enable remaining useful life determination, the ARTODTO system can respond to remaining-life based demands. As a result, customers are provided with the products that meet/exceed the minimum remaining-life requirements of their orders. Therefore, the problem is to find an optimal set of the EOLPs to remanufacture, disassemble, recycle, store, and dispose of so that the remaining-life based demands are satisfied while the uncertain goals of the system are achieved as much as possible. The ARTODTO system described is portrayed in Figure 8.3.

All the information stored on embedded tags is retrieved wirelessly as the EOLPs arrive at the communication perimeter of RFID tag readers at the recovery facility.

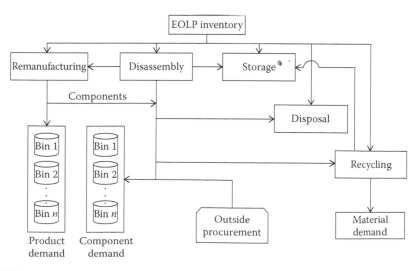

FIGURE 8.3   Material and information flows of the ARTODTO system.

Information retrieved from each EOLP is fed to central the information system. The information system interprets the data and supports the optimal management of EOLPs. This information can also be shared among the parties of a manufacturing alliance (Huang et al. 2011).

Although uncertainty about the quality and quantity of EOLPs is removed by means of sensors and RFID tags, there are two other sources of uncertainties that need to be addressed. One of them is the uncertainty about the system goals, i.e., total cost, weight of disposed items, material sales revenue, and customers' satisfaction level. It is easier for DMs to set target ranges (fuzzy sets) than precise values. The relative importance levels of the goals are imposed by preemptive modeling. The second type of uncertainty stems from the imprecision in demand forecasting. Therefore, forecasted demands are taken into account as triangular fuzzy numbers.

8.5 PROPOSED MATHEMATICAL MODEL

8.5.1 Goals

Proposed ARTODTO model has four fuzzy goals. The first goal is related to the total cost (TC) of ARTODTO activities. It is desired to have a total cost value of at most TC^*. However, values above the aspiration level are allowed up to an upper bound with decreasing achievement degrees. The second goal imposes the environmentally responsible aspect of the ARTODTO system. Total weight of disposed items ($WDIS$) is desired to be less than or equal to $WDIS^*$. The third goal is to attain or exceed the target material sales revenue (MSR^*) value. The final objective of the system is related with the customers' satisfaction level (CSL) on remaining life of ordered items. Customers' satisfaction level is the difference between the actual remaining lives and the minimum life requirements of orders. The goal is to reach the target customers' satisfaction level (CSL^*) that is defined by the DMs.

The aforementioned four goals of the system can be formulated as follows:

$$g_1: \ TC \ \tilde{<} \ TC^* \tag{8.6}$$

$$g_2: \ WDIS \ \tilde{<} \ WDIS^* \tag{8.7}$$

$$g_3: \ MSR \ \tilde{>} \ MSR^* \tag{8.8}$$

$$g_4: \ CSL \ \tilde{>} \ CSL^* \tag{8.9}$$

8.5.2 Constraints

Constraints of the model include achievement levels (membership values) of goals and other system constraints. The first goal is a left-sided fuzzy set, hence the achievement level of the first goal can be written as follows:

$$\mu_1 \leq \frac{UB_1 - TC}{UB_1 - TC^*} \tag{8.10}$$

where totalcost (TC) is the sum of total disassembly cost (TDC), total remanufacturing cost ($TRMC$), total outside procurement cost ($TOPC$), total disposal cost ($TDIC$), total recycling cost (TRC), and total holding cost (THC). Therefore, the total cost function can be given as follows:

$$TC = TDC + TRMC + TOPC + TDIC + TRC + THC \qquad (8.11)$$

Each term is described next.

Total disassembly cost, TDC, is incurred by completely disassembled EOLPs. Cost of disassembly activities performed in remanufacturing processes is accounted in $TRMC$. TDC can be formulated as follows:

$$TDC = \sum_{i \in I, j \in J} \left(\overline{x}_i (a_{ij} \, cd_j + f_{ij} \, \alpha cd_j) + (z_i + w_i)(a_{ij} + f_{ij}) \alpha cd_j \right) \qquad (8.12)$$

Remanufacturing activity comprises the disassembly of broken and remaining-lifetime deficient components, and assembly of required ones. Hence, total remanufacturing cost, $TRMC$, is defined as the sum of the costs of these two activities. One should also note that the disassembly of components is subject to a set of precedence relationships. Therefore, the related cost should reflect the cost of disassembling and reassembling all preceding components. This situation is taken into consideration by introducing the binary variable rp_{ij}. Therefore

$$TRMC = \sum_{i \in I, j \in I} \left[rp_{ij}(a_{ij}(cd_j + ca_j) + f_{ij}(\alpha cd_j + ca_j) + mis_{ij} ca_j) \right] \qquad (8.13)$$

and,

$$y_{im}(f_{ij} + mis_{ij} + dfc_{imj}) \le rp_{ik}, \quad \forall i, m, \{j, k \,|\, k \in P_j\} \qquad (8.14)$$

where P_j is the set of components preceding component j.

Total outside procurement cost, $TOPC$, is a function of the number of procured component js whose remaining life is within the range of component life-bin b. Mathematical expression for $TOPC$ can be written as follows:

$$TOPC = \sum_{j \in J, b \in B} c_{jb} l_{jb} \qquad (8.15)$$

Total disposal cost, $TDIC$, is defined as the cost of product, component, and material disposal. Products chosen to be disposed of are broken into their components by destructive disassembly and obtained parts are discarded along with other disassembled functional and nonfunctional components that are to be disposed of. Hence,

$$TDIC = \sum_{j \in J} cds_j \left(\sum_{b \in B} cdis_{jb} + \sum_{i \in I} z_i(a_{ij} + f_{ij}) + fd_j \right) \qquad (8.16)$$

Total recycling cost, *TRC*, is calculated by summing the costs of product and component recycling. Products chosen to be recycled are broken into their components by destructive disassembly. Obtained components are recycled together with the other disassembled functional and nonfunctional components that are separated for recycling. Therefore,

$$TRC = \sum_{j \in J} crc_j \left(\sum_{b \in B} r_{jb} + \sum_{i \in I} w_i(a_{ij} + f_{ij}) + fr_j \right) \qquad (8.17)$$

Finally, total holding cost, *THC*, is a function of stored EOLPs, components, and materials. So, *THC* can be expressed mathematically as follows:

$$THC = h \sum_{i \in I} s_i + \sum_{j \in J} ch_j \sum_{b \in B} sc_{jb} + \sum_{k \in K} mh_k sm_k \qquad (8.18)$$

The achievement level of the second goal can be written as follows:

$$\mu_2 \leq \frac{UB_2 - WDIS}{UB_2 - WDIS^*} \qquad (8.19)$$

Disposal weight (*WDIS*) can be obtained by multiplying all components to be disposed of with their corresponding weights (ω_j):

$$WDIS = \sum_{j \in J} \left(\sum_{b \in B} cdis_{jb} + \sum_{i \in I} z_i(a_{ij} + f_{ij}) + fd_j \right) \omega_j \qquad (8.20)$$

The achievement level of the third goal can be written as follows:

$$\mu_3 \leq \frac{MSR - LB_3}{MSR^* - LB_3} \qquad (8.21)$$

Material sales revenue (*MSR*) can be calculated by multiplying the sum of material demand and the amount of stored materials with the unit material sale price factor, and finally summing over all material types. Note that all stored recycled material is assumed to be sold later on. Therefore,

$$MSR = \sum_{k \in K} prc_k(dm_k + sm_k) \qquad (8.22)$$

Finally, the achievement level of the fourth goal can be written as follows:

$$\mu_4 \leq \frac{CSL - LB_4}{CSL^* - LB_4} \qquad (8.23)$$

Customers' satisfaction level (CSL) is a conceptual notion and calculated as the sum of two terms, namely, CSL^1 and CSL^2. CSL^1 is the sum of the differences between the highest life-bins that components could be placed and the life-bins they are actually placed in. For remanufacturing case (CSL^2), the same calculation is performed for all components in remanufactured products using the products' target life-bin. Hence,

$$CSL = CSL^1 + CSL^2 \qquad (8.24)$$

$$CSL^1 = \sum_{i \in I, j \in J, b \in B} x_{ijb}(\beta_{ij} - b) \qquad (8.25)$$

$$CSL^2 = \sum_{i \in I, m \in M, j \in J, b \in B} rep_{imjb}(b - m) + \sum_{i \in I, m \in M, j \in J} \left(a_{ij} y_{im} - \sum_{b \in B} rep_{imjb} \right)(\beta_{ij} - m) \qquad (8.26)$$

Assuming the remaining lives of all components are known and three life-bins are defined for the system, the parameter β_{ij} can be calculated as follows:

$$\beta_{ij} = \begin{cases} 1, & 0 < \tau_{ij} \le n_1 \\ 2, & n_1 < \tau_{ij} \le n_2, \quad \forall i, j \text{ and } 0 < n_1 < n_2 \\ 3, & \tau_{ij} > n_2 \end{cases} \qquad (8.27)$$

The equation set given next ensures that an EOLP in the inventory is disassembled, remanufactured, disposed of, recycled or left untouched (stored). Thus,

$$\bar{x}_i + \bar{y}_i + z_i + w_i + s_i = 1, \quad \forall i \qquad (8.28)$$

In this chapter, complete disassembly is considered. Complete disassembly implies that all components in the product structure will be extracted, if an EOLP is chosen to be disassembled. Since a functional component can be placed in only one life-bin after disassembly, related constraints can be expressed as follows:

$$\sum_{b \in B} x_{ijb} = \bar{x}_i a_{ij}, \quad \forall i, j \qquad (8.29)$$

The equation set given next assures that an EOLP is remanufactured to produce only one product and that product is evaluated in only one product life-bin:

$$\sum_{m \in M} y_{im} = \bar{y}_i, \quad \forall i \qquad (8.30)$$

Sophisticated product demand is satisfied by remanufactured EOLPs. Thus, the number products in product life-bin m (that are produced by remanufacturing the EOLPs) must be—at least—equal to the corresponding product demand. Hence,

$$\sum_{i \in I} y_{im} = dp_m, \quad \forall m \tag{8.31}$$

Component demand is satisfied by recovered and procured operable components that meet certain remaining life criteria. Recovered components are obtained from the disassembled and remanufactured EOLPs. During the remanufacturing operation, remaining-life-time deficient components are taken out and placed in component bins based on their remaining useful lives. Hence, a remanufacturing activity adds components to the bins and consumes some components from the bins at the same time. For each life-bin b and component j, the number of recovered and procured components must be at least equal to the number of demanded components after components used in remanufacturing operations, recycled, stored, and disposed of are taken out. It should be noted that component demands are assumed to be imprecise. In other words, demand forecasts are used as the aspiration values and a percent tolerance (τ) is introduced to calculate upper and lower bounds. This isosceles triangular fuzzy constraint can be introduced to the model as the fifth goal using the following statements. The achievement level of this constraint, in its original form, is in fact nonlinear (see Equation 8.5). However, it is possible to reformulate this constraint linearly as seen in the following statements.

$$\mu_5 \leq 1 + \frac{(DA_{jb} - cdis_{jb})}{\tau cdis_{jb}}, \quad \forall b, j \tag{8.32}$$

$$\mu_5 \leq 1 - \frac{(DA_{jb} - cdis_{jb})}{\tau cdis_{jb}}, \quad \forall b, j \tag{8.33}$$

$$DA_{jb} = \sum_{\{i \in I | cin_{ijb} = 1\}} (x_{ijb} + def_{ijb}) - \sum_{i \in I, m \in M} (rep_{imjb}) + l_{jb} - r_{jb} - sc_{jb} - cdis_{jb}, \quad \forall b, j \tag{8.34}$$

where

$$\sum_{\{b \in B | cin_{ijb} = 1\}} def_{ijb} = \sum_{m \in M} (dfc_{imj} y_{im}), \quad \forall i, j \tag{8.35}$$

and

$$\sum_{b \in B} def_{ijb} \leq 1, \quad \forall i, j \tag{8.36}$$

$$\sum_{\{b \in B | cin_{ijb} \neq 1\}} def_{ijb} = 0, \quad \forall i, j \tag{8.37}$$

Nonfunctional, missing, and remaining-life-time deficient components must be filled in with components having a remaining life time that is sufficient for producing a product for product life-bin m. Therefore,

$$\sum_{\{b\in B, m\in M|b\geq m\}} rep_{imjb} = y_{im}(f_{ij} + mis_{ij} + dfc_{imj}), \quad \forall i,j,m \tag{8.38}$$

Replacement of a component can be taken from only one bin.

$$\sum_{b\in B, m\in M} rep_{imjb} \leq 1, \quad \forall i,j \tag{8.39}$$

$$\sum_{\{b\in B, m\in M|b<m\}} rep_{imjb} = 0, \quad \forall i,j \tag{8.40}$$

Material demand is satisfied by recycled components and products. Hence,

$$\sum_{j\in J} \gamma_{jk} \left(\sum_{b\in B} r_{jb} + \sum_{i\in I} w_i(a_{ij} + f_{ij}) + fr_j \right) + sm_k = dm_k, \quad \forall k \tag{8.41}$$

All nonfunctional components have to be either disposed of or recycled. This is assured by the following equation:

$$fr_j + fd_j = \sum_{i\in I} (\bar{x}_i + y_i) f_{ij}, \quad \forall j \tag{8.42}$$

Therefore, the FGP model can be written as follows:

$$\text{Max. } u = \sum_{t=\{1,2,3,4,5\}} \mu_t \tag{8.43}$$

$$\text{Subject to Equations } 8.10 - 8.42$$

8.6 NUMERICAL EXAMPLE

In this section, proposed approach is illustrated considering a device-embedded (i.e., sensors and RFID tags) water heater ARTODTO system. Each heater contains eight components that are subject to precedence relationships. The heater and its components are shown in Figure 8.4. Components and products are assumed to fall

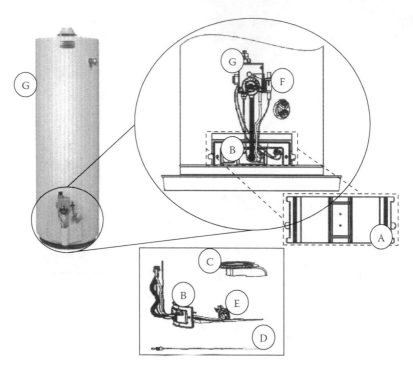

FIGURE 8.4 Components of the water heater.

into three remaining life categories. These categories are called remaining-life-bins. The first bin holds those items that have a remaining-life time of at least 1 year, the second bin holds those items whose remaining-life is at least 4 years. The last bin holds the items having 6 years or more remaining life.

Component weights, disassembly, assembly costs, recycling, holding, and disposal costs are given in Table 8.2. Remaining life-based forecasted component demands and outside procurement prices are given in Table 8.3. Component disassembly relationships are depicted in Figure 8.5.

Remaining life-based remanufactured product demands are assumed to be 12, 4, and 6 for the bins 1, 2, and 3, respectively. Water heaters are recycled to obtain steel. Steel yields, demands, holding costs, and sale prices are given in Table 8.4.

Additional data include $\alpha = 0.80$, $h = 5.00$ (\$/unit), $\tau = 0.15$, $B = \{1, 2, 3\}$, $I = \{1, 2, \ldots, 200\}$, $J = \{1, 2, \ldots, 8\}$, $K = \{1, 2\}$, $M = \{1, 2, 3\}$. Item level sensor data including a_{ij}, f_{ij}, mis_{ij}, and def_{imj} are populated from a database table. A small portion of this database is given in Table 8.5. "*" and "0.00" notations indicate missing and nonfunctional (broken) components, respectively.

Aspiration value and upper bound for total cost are given as \$1500 ($TC^* = 1500$) and \$1800 ($UB_1 = 1800$). Due to environmental responsibility, disposal goal is set at 0 lbs. ($WDIS^* = 0$), however, regulations allow disposal up to 50 lbs. ($UB_2 = 50$). DMs believe that revenue from the material sales should be at least \$200 ($LB_3 = 200$), while the aspiration value is \$500 ($MSR^* = 500$). Finally, the aspiration value for customers' satisfaction level is set at 650 ($CSL^* = 650$) because DMs aim to meet

TABLE 8.2
Component Related Data

Component	Weight (lb)	Disassembly Cost ($)	Assembly Cost ($)	Recycling Cost ($)	Holding Cost ($)	Disposal Cost ($)
Outer door (A)	1.00	0.50	0.50	0.30	0.12	0.12
Door + manifold (B)	1.00	2.00	2.00	0.26	0.12	0.12
Burner (C)	7.00	0.15	0.15	—	1.44	1.44
Thermocouple (D)	2.00	0.25	0.25	—	0.66	0.66
Pilot assembly (E)	3.00	1.00	1.00	—	0.25	0.25
Igniter (F)	0.30	0.15	0.15	—	0.10	0.10
Gas valve (G)	2.50	1.00	1.00	—	0.25	0.25
Tank (H)	62.00	0.00	0.00	12.00	10.00	10.00

TABLE 8.3
Component Related Remaining Life Specific Data

Component	Demand			Procurement Price ($)		
	Bin 1	Bin 2	Bin 3	Bin 1	Bin 2	Bin 3
Outer door (A)	24	24	9	13.99	13.99	13.99
Door + manifold (B)	15	14	11	20.99	20.99	20.99
Burner (C)	15	28	14	12.99	27.49	41.99
Thermocouple (D)	2	2	12	7.99	7.99	7.99
Pilot assembly (E)	27	30	16	9.99	17.99	25.99
Igniter (F)	30	22	9	6.99	11.74	16.49
Gas valve (G)	54	22	14	74.99	134.49	193.99
Tank (H)	0	0	0	56.24	100.87	145.49

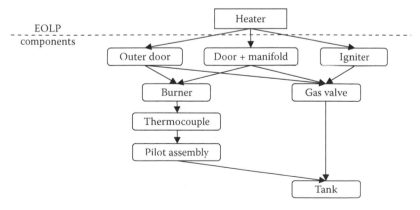

FIGURE 8.5   Disassembly precedence relationships among the heater components.

TABLE 8.4
Material Related Data

	Components	Steel
Yield (lb)	Outer door	1.00
	Door	1.00
	Burner	0.00
	Thermocouple	0.00
	Pilot assembly	0.00
	Igniter clip	0.00
	Gas valve	0.00
	Tank	50.00
Demand (lb)		895.00
Holding cost ($/lb)		0.95
Sale price ($/lb)		0.50

TABLE 8.5
Remaining Life Times (years) of Components

	Components							
EOLP #	A	B	C	D	E	F	G	H
1	6.67	5.12	5.97	*	6.76	6.43	6.69	7.56
2	6.61	5.51	6.84	5.88	4.86	6.45	*	3.95
3	5.88	4.62	6.78	5.48	6.34	5.84	8.06	5.85
4	7.22	6.47	7.86	5.28	6.22	4.04	4.22	0.00
⋮	⋮	⋮	⋮	⋮	⋮	⋮	⋮	⋮
100	6.22	5.73	5.26	6.19	0.00	6.94	5.20	5.87
⋮	⋮	⋮	⋮	⋮	⋮	⋮	⋮	⋮
196	0.00	*	6.81	0.00	6.24	6.67	6.40	5.60
197	6.08	6.45	7.12	6.21	7.10	5.67	6.48	6.79
198	6.33	5.03	*	7.52	6.07	6.04	6.46	5.87
199	6.74	4.57	4.28	6.26	0.00	6.68	4.31	7.17
200	7.14	0.00	0.00	6.29	5.69	4.92	6.83	6.31

the component demands with superior components. DMs also believe that minimum acceptable *CSL* is 400 ($LB_4 = 400$). By imposing the last goal, the model seeks for a solution in which all components provided to the customers are in fact eligible for bins that are higher than demanded bin. More specifically, when all component and product demands for remaining life-bins 1 and 2 are met, the total number of components supplied to customer becomes 437 including the ones in remanufactured products. Since *CSL* is defined as the difference between the best bin that a component can be placed and the bin it is actually placed, the aspiration level for *CSL* imposes this difference to be between 0.9 (400/437) and 1.5 (650/437) on the average for each component.

8.7   RESULTS

The model was solved using CPLEX Interactive Optimizer. Optimal achievement values and levels of the goals are given in Table 8.6. Component demand achievement level (μ_5) is 1 in the optimal solution. Table 8.7 shows the serial numbers of the EOLPs that are subject to various operations according to the optimal solution.

Solution also provides the detailed optimal remanufacturing plan indicating which EOLP is to be remanufactured and in what bin it is to be placed. Complete remanufacturing plan is given in Table 8.8. According to this table, the model picks EOLP #5

TABLE 8.6
Goals, Tolerances, Achievement Values and Levels

	Description	Lower Bounds	Aspiration Values	Upper Bounds	Achievement Value	Achievement Level
Goals	Total cost	—	$1500	$1800	$1507.42	0.9753
	Disposal quantity	—	0 lb	50 lb	1.2 lb	0.9753
	Material sales revenue	$200	$500	—	$477	0.9233
	Customers' satisfaction level	400	650	—	632	0.9233
Other measures	Total revenue	—	—	—	$20,610.06	—
	Total disassembly cost	—	—	—	$423.18	—
	Total remanufacturing cost	—	—	—	$12.10	—
	Total recycling cost	—	—	—	$230.44	—
	Total holding cost	—	—	—	$841.30	—
	Total disposing cost	—	—	—	$0.40	—

TABLE 8.7
Allocation of EOLPs to Recovery Operations

Activity	EOLP Serial Numbers
Disassembled	1, 4, 8, 9, 10, 13, 16, 17, 21, 22, 24, 28, 29, 32, 33, 35, 36, 42, 45, 50, 52, 59, 60, 64, 65, 66, 69, 77, 84, 89, 90, 92, 93, 94, 95, 96, 98, 99, 104, 105, 106, 111, 112, 113, 115, 116, 118, 119, 122, 123, 124, 125, 126, 127, 129, 130, 131, 133, 135, 136, 138, 140, 141, 143, 146, 148, 150, 151, 152, 153, 154, 157, 158, 159, 162, 165, 166, 167, 171, 175, 176, 179, 180, 184, 186, 187, 190, 192, 195, 198
Remanufactured	5, 7, 12, 25, 30, 37, 39, 41, 47, 53, 55, 56, 67, 76, 79, 87, 103, 117, 132, 173, 185, 197
Disposed of	No EOLPs are allocated for disposal
Recycled	No EOLPs are allocated for recycle
Stored	2, 3, 6, 11, 14, 15, 18, 19, 20, 23, 26, 27, 31, 34, 38, 40, 43, 44, 46, 48, 49, 51, 54, 57, 58, 61, 62, 63, 68, 70, 71, 72, 73, 74, 75, 78, 80, 81, 82, 83, 85, 86, 88, 91, 97, 100, 101, 102, 107, 108, 109, 110, 114, 120, 121, 128, 134, 137, 139, 142, 144, 145, 147, 149, 155, 156, 160, 161, 163, 164, 168, 169, 170, 172, 174, 177, 178, 181, 182, 183, 188, 189, 191, 193, 194, 196, 199, 200

TABLE 8.8

Complete Remanufacturing Plan

EOLP #	Bin	EOLP #	Bin
5	2	56	1
7	1	67	1
12	3	76	1
25	3	79	2
30	2	87	1
37	2	103	3
39	1	117	3
41	1	132	1
47	1	173	3
53	1	185	1
55	1	197	3

TABLE 8.9

Portion of the Disassembly Plan

EOLP #	Components							
	A	B	C	D	E	F	G	H
1	3	1	2	—	3	3	1	1
4	3	3	3	1	1	2	2	—
8	1	1	1	3	1	3	3	1
192	2	3	—	2	2	3	2	—
195	2	3	2	3	2	3	3	2
198	3	2	—	3	1	3	3	2

to fulfill the demand for a product having at least four years of remaining life (bin 2). Optimal disassembly plan showing the bin numbers in which disassembled components are placed is also obtained. A sample from the optimal disassembly plan is given in Table 8.9.

For instance, when EOLP #1 is disassembled, outer door, pilot assembly and igniter are placed in bin 3, burner is placed in bin 2, and the others are placed in bin 1.

8.8 CONCLUSIONS

Environmentally conscious manufacturing and product recovery (ECMPRO) has recently gained a lot of attention from both researchers and practitioners as a response to the rising public awareness on distressing downfall of natural resources. Majority of the complexity encountered in recovery operations stems from the high degrees of uncertainty. Sensor and RFID embedded EOLPs mitigate end-of-life (EOL) management and product recovery by providing valuable static and dynamic information about their history, thus eliminating product-related uncertainties. However, there are other sources of uncertainty, viz. goals and parameters that are set by the DMs.

In this chapter, a fuzzy multi-criteria advanced remanufacturing-to-order and disassembly-to-order (ARTODTO) system was developed. The proposed system utilized the life cycle data that are collected and delivered by sensors and RFID tags, in order to obtain the optimum disassembly, remanufacturing, disposal, recycling, and storage plans under uncertain conditions. Optimal solution of the model provided the serial numbers of the products to be subjected to various recovery operations. The solution also provided the detailed item-based remanufacturing plan. The goals were related to total system cost (TC), disposal weight ($WDIS$), material sales revenue (MSR), and customers' satisfaction level (CSL). The objective function was to maximize the sum of the achievement levels of all goals while accommodating their relative importance. A case example was considered to illustrate the application of the proposed approach and solved using CPLEX Interactive Optimizer. Achievement levels were within the 10% distance from the maximum possible values, thus showing how realistic the goals were set.

As a conclusion, proposed ARTODTO model produced a comprehensive recovery plan optimizing the overall achievement of not only financial goals (i.e., TC and MSR), but also environmental and qualitative goals (i.e., $WDIS$ and CSL) under uncertainty.

REFERENCES

Barba-Gutierrez, Y. and Adenso-Diaz, B. 2009. Reverse MRP under uncertain and imprecise demand. *International Journal of Advanced Manufacturing Technology*, 40: 413–424.

Byington, C. S., Watson, M., and Edwards, D. 2004. Data-driven neural network methodology to remaining life predictions for aircraft actuator components. *Proceedings of IEEE Aerospace Conference, 2004*, Big Sky, MT, Vol. 6, pp. 3581–3589.

Chang, C.-Y. and Hung, S.-S. 2012. Implementing RFIC and sensor technology to measure temperature and humidity inside concrete structures. *Construction and Building Materials*, 26: 628–637.

Charnes, A. and Cooper, W. W. 1961. *Management Models and Industrial Applications of Linear Programming*. New York: Wiley.

Chen, L.-H. and Tsai, F.-C. 2001. Fuzzy goal programming with different importance and priorities. *European Journal of Operational Research*, 133: 548–556.

Cheng, F.-T., Huang, G.-W., Chen, C.-H., and Hung, M.-H. 2004. A generic embedded device for retrieving and transmitting information of various customized applications. *IEEE International Conference on Robotics and Automation*, April 26–May 1, 2004, New Orleans, LA, pp. 978–983.

Curran, T. and Williams, I. D. 2012. A zero waste vision for industrial networks in Europe. *Journal of Hazardous Materials*, 207–208: 3–7.

Depuy, G. W., Usher, J. S., Walker, R. L., and Taylor, G. D. 2007. Production planning for remanufactured products. *Production Planning and Control*, 18: 573–583.

Dolgui, A. and Proth, J.-M. 2008. RFID technology in supply chain management: State of the art and perspectives. *The 17th International Federation of Automatic Control World Congress, 2008*, Seoul, South Korea. Elsevier, Amsterdam, the Netherlands, pp. 4465–4475.

Engel, S. J., Gilmartin, B. J., Bongort, K., and Hess, A. 2000. Prognostics, the real issues involved with predicting life remaining. *IEEE Aerospace Conference, 2000*, Big Sky, MT, Vol. 6, pp. 457–469.

Ferrer, G. and Whybark, D. C. 2001. Material planning for a remanufacturing facility. *Production & Operations Management*, 10: 112–124.

Galbreth, M. R. and Blackburn, J. D. 2006. Optimal acquisition and sorting policies for remanufacturing. *Production & Operations Management*, 15: 384–392.

Gonnuru, V. K. 2010. Radio-frequency identification (RFID) integrated fuzzy based disassembly planning and sequencing for end-of-life products. Masters thesis, The University of Texas, San Antonio, TX.

Guide Jr., V. D. R. 2000. Production planning and control for remanufacturing: Industry practice and research needs. *Journal of Operations Management*, 18: 467–483.

Gungor, A. and Gupta, S. M. 1999. Issues in environmentally conscious manufacturing and product recovery: A survey. *Computers & Industrial Engineering*, 36: 811–853.

Gupta, S. M., Imtanavanich, P., and Nakashima, K. 2009. Using neural networks to solve a disassembly-to-order problem. *International Journal of Biomedical Soft Computing and Human Sciences (Special Issue on Total Operations Management)*, 15: 67–71.

Gupta, S. M. and Taleb, K. N. 1994. Scheduling disassembly. *International Journal of Production Research*, 32: 1857–1866.

Hannan, E. L. 1981. On fuzzy goal programming. *Decision Sciences*, 12: 522.

Herzog, M. A., Marwala, T., and Heyns, P. S. 2009. Machine and component residual life estimation through the application of neural networks. *Reliability Engineering & System Safety*, 94: 479–489.

Huang, G. Q., Qu, T., Fang, M. J., and Bramley, A. N. 2011. RFID-enabled gateway product service system for collaborative manufacturing alliances. *CIRP Annals—Manufacturing Technology*, 60: 465–468.

Huang, X. Y., Yan, N. N., and Qiu, R. Z. 2009. Dynamic models of closed-loop supply chain and robust H control strategies. *International Journal of Production Research*, 47: 2279–2300.

Ilgin, M. A. and Gupta, S. M. 2010a. Comparison of economic benefits of sensor embedded products and conventional products in a multi-product disassembly line. *Computers & Industrial Engineering*, 59: 748–763.

Ilgin, M. A. and Gupta, S. M. 2010b. Environmentally conscious manufacturing and product recovery (ECMPRO): A review of the state of the art. *Journal of Environmental Management*, 91: 563–591.

Ilgin, M. A. and Gupta, S. M. 2010c. Evaluating the impact of sensor-embedded products on the performance of an air conditioner disassembly line. *The International Journal of Advanced Manufacturing Technology*, 53: 1199–1216.

Ilgin, M. A. and Gupta, S. M. 2011a. Performance improvement potential of sensor embedded products in environmental supply chains. *Resources, Conservation and Recycling*, 55: 580–592.

Ilgin, M. A. and Gupta, S. M. 2011b. Recovery of sensor embedded washing machines using a multi-kanban controlled disassembly line. *Robotics and Computer-Integrated Manufacturing*, 27: 318–334.

Ilgin, M. A. and Gupta, S. M. 2012. *Remanufacturing Modeling and Analysis*. Boca Raton, FL: CRC Press.

Ilgin, M. A., Ondemir, O., and Gupta, S. M. in press. An approach to quantify the financial benefit of embedding sensors into products for end-of-life management: A case study. *Production Planning and Control*, 1–18.

Inderfurth, K. 2005. Impact of uncertainties on recovery behavior in a remanufacturing environment: A numerical analysis. *International Journal of Physical Distribution and Logistics Management*, 35: 318–336.

Inderfurth, K. and Langella, I. 2006. Heuristics for solving disassemble-to-order problems with stochastic yields. *OR Spectrum*, 28: 73–99.

Jones, D. F. and Tamiz, M. 2002. Goal programming in the period 1990–2000. In *Multiple Criteria Optimization: State of the Art Annotated Bibliographic Surveys*, eds. Ehrgott, M. and Gandibleux, X., pp. 129–170. Boston, MA: Kluwer Academic Publishers.

Jones, D. and Tamiz, M. 2010. *Practical Goal Programming*. New York: Springer.

Karlsson, B. 1997. A distributed data processing system for industrial recycling. *IEEE Instrumentation and Measurement Technology Conference (IMTC) 'Sensing, Processing, Networking'*, May 19–21, 1997, Ottawa, Ontario, Canada, pp. 197–200.

Klausner, M., Grimm, W. M., Hendrickson, C., and Horvath, A. 1998. Sensor-based data recording of use conditions for product takeback. *IEEE International Symposium on Electronics and the Environment*, 1998, Oak Brook, IL, pp. 138–143.

Klausner, M., Grimm, W. M., Horvath, A., Lee, H. G., and Wendy, M. 2000. Sensor-based data recording for recycling: A low-cost technology for embedded product self-dentification and status reporting. In *Green Electronics/Green Bottom Line*, L. H. Goldberg and W. Middleton, eds., pp. 91–101. Woburn, MA: Butterworth-Heinemann.

Kongar, E. and Gupta, S. M. 2002. A multi-criteria decision making approach for disassembly-to-order systems. *Journal of Electronics Manufacturing*, 11: 171–183.

Kongar, E. and Gupta, S. M. 2006. Disassembly to order system under uncertainty. *Omega*, 34: 550–561.

Kongar, E. and Gupta, S. M. 2009a. A Multiple objective tabu search approach for end-of-life product disassembly. *International Journal of Advanced Operations Management*, 1: 177–202.

Kongar, E. and Gupta, S. M. 2009b. Solving the disassembly-to-order problem using linear physical programming. *International Journal of Mathematics in Operational Research*, 1: 504–531.

Kulkarni, A. G., Parlikad, A. K. N., Mcfarlane, D. C., and Harrison, M. 2005. Networked RFID systems in product recovery management. *IEEE International Symposium on Electronics and the Environment & the IAER Electronics Recycling Summit*, May 16–19, 2005, New Orleans, LA, pp. 66–71.

Lage Jr., M. and Godinho Filho, M. 2012. Production planning and control for remanufacturing: Literature review and analysis. *Production Planning and Control*, 23: 419–435.

Lambert, A. J. D. and Gupta, S. M. 2005. *Disassembly Modelling for Assembly Maintenance, Reuse, and Recycling*. Boca Raton, FL: CRC Press.

Langella, I. M. 2007. Heuristics for demand-driven disassembly planning. *Computers & Operations Research*, 34: 552–577.

Lee, B. S., Chung, H. S., Kim, K.-T., Ford, F. P., and Andersen, P. L. 1999. Remaining life prediction methods using operating data and knowledge on mechanisms. *Nuclear Engineering and Design*, 191: 157–165.

Li, J., González, M., and Zhu, Y. 2009. A hybrid simulation optimization method for production planning of dedicated remanufacturing. *International Journal of Production Economics*, 117. 286–301.

Luttropp, C. and Johansson, J. 2010. Improved recycling with life cycle information tagged to the product. *Journal of Cleaner Production*, 18: 346–354.

Mazhar, M. I., Kara, S., and Kaebernick, H. 2007. Remaining life estimation of used components in consumer products: Life cycle data analysis by Weibull and artificial neural networks. *Journal of Operations Management*, 25: 1184–1193.

Middendorf, A., Griese, H., Grimm, W. M., and Reichl, H. 2003. Embedded life cycle information module for monitoring and identification of product use conditions. *3rd International Symposium on Environmentally Conscious Design and Inverse Manufacturing (EcoDesign), 2003*, Tokyo, Japan, pp. 733–740.

Middendorf, A., Reichl, H., and Griese, H. 2005. Lifetime estimation for wire bond interconnections using life cycle-information modules with implemented models. *4th International Symposium on Environmentally Conscious Design and Inverse Manufacturing (EcoDesign)*, December 12–14, 2005, Tokyo, Japan, pp. 614–619.

Narasimhan, R. 1980. Goal Programming in a fuzzy environment. *Decision Sciences*, 11: 325.

Ondemir, O., Ilgin, M. A., and Gupta, S. M. 2012. Optimal end-of-life management in closed loop supply chains using RFID & sensors. *IEEE Transactions on Industrial Informatics*, 8(3): 719–728.

Parlikad, A. K. and Mcfarlane, D. 2007. RFID-based product information in end-of-life decision making. *Control Engineering Practice*, 15: 1348–1363.

Petriu, E. M., Georganas, N. D., Petriu, D. C., Makrakis, D., and Groza, V. Z. 2000. Sensor-based information appliances. *IEEE Instrumentation & Measurement Magazine*, 3: 31–35.

Ross, T. J. 1995. *Fuzzy Logic with Engineering Applications*. West Sussex, U.K.: John Wiley.

Rugrungruang, F. 2008. An integrated methodology for assessing physical & technological life of products for reuse. PhD thesis, The University of New South Wales, Kensington, New South Wales, Australia.

Scheidt, L. and Shuqiang, Z. 1994. An approach to achieve reusability of electronic modules. *IEEE International Symposium on Electronics and the Environment*, May 2–4, 1994, San Francisco, CA, pp. 331–336.

Schniederjans, M. J. 1995. *Goal Programming: Methodology and Applications*. Norwell, MA: Kluwer Academic Publishers.

Taleb, K. N. and Gupta, S. M. 1997. Disassembly of Multiple Product Structures. *Computers & Industrial Engineering*, 32: 949–961.

Taleb, K. N., Gupta, S. M., and Brennan, L. 1997. Disassembly of complex products with parts and materials commonality. *Production Planning and Control*, 8: 255–269.

Tang, Y., Zhou, M., Zussman, E., and Caudill, R. 2002. Disassembly modeling, planning, and application. *Journal of Manufacturing Systems*, 21: 200–217.

Tiwari, R. N., Dharmar, S., and Rao, J. R. 1986. Priority structure in fuzzy goal programming. *Fuzzy Sets and Systems*, 19: 251–259.

Tiwari, R. N., Dharmar, S., and Rao, J. R. 1987. Fuzzy goal programming—An additive model. *Fuzzy Sets and Systems*, 24: 27–34.

Vadde, S., Kamarthi, S. V., Gupta, S. M., and Zeid, I. 2008. Product life cycle monitoring via embedded sensors. In *Environment Conscious Manufacturing*, eds. Gupta, S. M. and Lambert, A. J. D., pp. 91–104. Boca Raton, FL: CRC Press.

Veerakamolmal, P. and Gupta, S. M. 1999. Analysis of design efficiency for the disassembly of modular electronic products. *Journal of Electronics Manufacturing*, 9: 79–95.

Wang, H.-F. and Gupta, S. M. 2011. *Green Supply Chain Management: Product Life Cycle Approach*. New York: McGraw Hill.

Wang, W. and Zhang, W. 2008. An asset residual life prediction model based on expert judgments. *European Journal of Operational Research*, 188: 496–505.

Xanthopoulos, A. and Iakovou, E. 2009. On the optimal design of the disassembly and recovery processes. *Waste Management*, 29: 1702–1711.

Yang, X., Moore, P., and Chong, S. K. 2009. Intelligent products: From life cycle data acquisition to enabling product-related services. *Computers in Industry*, 60: 184–194.

Zadeh, L. A. 1965. Fuzzy sets. *Information and Control*, 8: 338–353.

Zhou, S., Ling, W., and Peng, Z. 2007. An RFID-based remote monitoring system for enterprise internal production management. *International Journal of Advanced Manufacturing Technology*, 33: 837–844.

Zikopoulos, C. and Tagaras, G. 2007. Impact of uncertainty in the quality of returns on the profitability of a single-period refurbishing operation. *European Journal of Operational Research*, 182: 205–225.

Zimmermann, H. J. 1983. Using fuzy sets in operational research. *European Journal of Operational Research*, 13: 201–216.

Zimmermann, H. J. 2001. *Fuzzy Set Theory—And Its Applications*. Norwell, MA: Kluwer Academic Publishers.

9 Importance of Green and Resilient SCM Practices for the Competitiveness of the Automotive Industry
A Multinational Perspective

Susana G. Azevedo, V. Cruz-Machado,
Joerg S. Hofstetter, Elizabeth A. Cudney,
and Tian Yihui

CONTENTS

9.1   INTRODUCTION

Among the various supply chain management (SCM) paradigms, the green and the resilient paradigms are considered critical for the competitiveness and success of supply chains (SCs) (Azevedo et al. 2010; Carvalho et al. 2010, 2011). In the current business environment, the competitiveness of a company and a SC depends not only on the cost, quality, lead time, and service level but also on their ability to avoid and overcome the numerous disturbances that jeopardize their performance. In addition, companies and SCs are forced to adopt ecologically responsive practices to meet legislative requirements; this ecological responsiveness can also lead to a sustained competitive advantage, thus improving their long-term profitability. Recently, the resilience and green topics have been studied in the SCM context. Carter and Rogers (2008) proposed that risk management, including contingency, planning, and supply disruptions, are critical issues that should be considered simultaneous to the environmental performance in order to achieve a sustainable SC.

Briefly, the green paradigm is concerned with environmental risk and impact reduction (Zhu et al. 2008), and the resilient paradigm focuses on the ability of the SC to recover to a desired state after a disruption (Christopher and Peck 2004).

The automotive SC provides a rich context to explore this issue. There is evidence that the tendency of many automotive companies to seek out low-cost solutions may have led to leaner but also more vulnerable SC (Azevedo et al. 2008; Svensson 2000). The automotive SC is also under pressure to become more sustainable and, therefore, more environmentally friendly while also incurring the expected economic benefits from greener practices (Koplin et al. 2007; Thun and Muller 2010).

Consequently, this chapter aims to explore the importance of green and resilient SCM practices for the competitiveness of the automotive SC.

The chapter is organized as follows. Following the introduction, the two paradigms of green and resilient are described from a SCM perspective as a set of management practices. Then, the research questions and hypotheses are formulated. Next, the research methodology is defined with the description of the survey instrument and also the sample selection process. After that, the data are analyzed and the main findings are discussed by research question. Finally, several conclusions are drawn.

9.2   BACKGROUND

Green supply chain management (GSCM) has emerged as an organizational philosophy by which organizations and their partners achieve corporate profit and market-share objectives by reducing environmental risks and impacts while improving ecological efficiency (de Figueiredo and Mayerle 2008; Rao and Holt 2005; Zhu et al. 2008). The main influences that have led to the increased use of this philosophy by SCs worldwide are central government environment regulations, supplier's advances in developing environmentally friendly goods, environmental partnerships with suppliers, competitors' green strategies, cost for the disposal of hazardous materials (Zhu et al. 2005), scarcity of resources, degradation of the environment, and increased pressure from consumers (Christmann and Taylor 2001; Wang and Gupta 2011).

According to Lee (2008), the main factor facilitating the participation of small and medium-size suppliers in green SC initiatives is inter-organizational initiatives, which attempt to improve environmental performance throughout the SC. However, these initiatives have primary origins stemming from external pressures such as regulations on take-back and the use of certain hazardous substances. Some authors defend such initiatives, asserting that this is mainly due to the following reasons: (1) disruption risks engendered by environmental issues can be passed on through suppliers; and (2) there is a broad range of practices from green purchasing to integrated SCs flowing from suppliers to customers and to the reverse SC, effectively "closing the loop" (Zhu and Sarkis 2004, 2006).

A growing number of research studies on GSCM have dealt not only with the drivers and practices of GSCM but also with the relationship between GSCM and operational and/or economic performance (e.g., Lamming and Hampson 1996; Vachon and Klassen 2006; Zhu and Sarkis 2006). Although some organizations have adopted ecologically responsive practices to meet legislative requirements, ecological responsiveness can also lead to a sustained competitive advantage thus improving their long-term profitability (Paulraj 2009). Accordingly, Srivastava (2007) defined GSCM as "integrating environmental thinking into SCM, including product design, material sourcing and selection, manufacturing processes, final product delivery, and end-of-life management of the product after its useful life."

According to Rao and Holt (2005), GSCM practices should cover all SC activities, from green purchasing to integrated life cycle management, through the manufacturer and customer and closing the loop with reverse logistics. Reverse logistics has become a new frontier of management that is composed of the reverse distribution of materials, recycling, reusing, as well as reducing the amount of materials in forward systems (Carter and Elram 1998). It includes activities such as reclaim, recycle, remanufacture, reuse, take-back, and disposal (Meade and Sarkis 2002). More recently, closed-loop SCs, which involve the simultaneous consideration of forward and reverse flows, have become an alternative for cost-effective management of reverse logistics operations (Ilgin and Gupta 2010a). Research involving reverse logistics in a SC context is also increasing (Lee 2008; Murphy and Poist 2003; Sahay et al. 2006; Zhu et al. 2007; Zhu and Sarkis 2006). Table 9.1 presents several of the main green practices available in the literature on the automotive industry.

Considering the literature review and the supporting evidence from the automotive context, the green practices focused on in this study are as follows:

Environmental collaboration with suppliers—The interaction between organizations in the SC pertaining to joint environmental planning and shared environmental know-how or knowledge (Vachon and Klassen 2008). This may represent environmental programs that include technological and organizational development projects with suppliers (Sarkis 2003)

Environmental monitoring of suppliers—The continuous auditing and monitoring of supplier performance to determine, over time, whether the green supplier development programs contribute to performance (Bai and Sarkis 2010)

ISO 14001 certification—It defines the criteria for an environmental management system and requires commitment to comply with applicable legislation,

TABLE 9.1
Green Practices in the Automotive Supply Chain

Reference	Methodology	Experts/Professionals	Key Green Practices Included
Azevedo et al. (2011a)	Case study on automotive industry	Perspective of professionals from the automotive industry	• Reverse logistics • To minimize waste • ISO 14001
Nunes and Bennett (2010)	Benchmarking of green operational initiatives based on an analysis of secondary data from the automotive industry	Perspective of professionals from the automotive industry	• To move shipments from truck to rail and reduced the number of miles trucks run empty between shipments • To improve packaging and reusable metal shipping containers • To reduce initiatives related to water and energy, materials, and toxic substances • ISO 14001 certification • Reverse logistics • To recycle end-of-life parts • To establish partnerships with suppliers in terms of environment • To reduce energy, water, and raw material consumption per unit of output • To increase renewable sources of energy • To shift transport from road to rail and sea
Thun and Müller (2010)	Empirical study of companies from the automotive industry	Perspective of professionals from the automotive industry	• To improve packaging and waste reduction • To use a reusable packaging system • To consider environmental impact • Criteria when selecting suppliers • Conjoint development and perception of eco-friendly technology
Zhu et al. (2007)	Empirical study with a focus on a Chinese automobile engine manufacturer	Perspective of professionals from the automotive industry	• Internal environmental management • Green purchasing • Eco-design • To use equipment for emission purification, noise elimination, and wastewater treatment

Azevedo et al. (2011b)	Delphi technique	Academics	• Waste water reuse • To cooperate closely with research institutes and universities on eco-design projects • To implement cleaner production activities • To implement collaborative development efforts with suppliers • To collaborate with the customer to develop improved engines that consume less fuel, while maintaining suitable performance standards • To implement joint research on substitute materials and technologies to improve environmental practices with those partners and collaborate in innovation programs with competitors • To reduce energy consumption • To reuse/recycling materials and packaging • Environmental collaboration with suppliers • Reverse logistics
Seuring and Muller (2008)	Delphi technique	Researchers Non-governmental organizations (NGO) Practitioners	• Closed-loop SCM • Reverse logistics • Development and implementation of international standards (e.g. ISO 14001)
Liu et al. (2011)	Delphi technique	Automobile research institutes Automobile companies Departments from the government Universities	Top new energy automobile technologies which aim to reduce air and noise pollution: • Manufacturing and production technologies of mini pure electric automobiles • Battery technology of electric auto with single energy electromagnetic compatibility/electromagnetic interference (EMC/EMI) • Technology standard for vehicle and related accessories • Electrical control technology of diesel engine clean diesel technology with high quality

regulations, and continuous improvement. In the SC context, it can act indirectly by influencing all partners to adopt more environmentally friendly practices (Nishitani 2010)

Reduction of energy consumption—It consists of improving environmental performance throughout the SC with more efficient processes that reduce energy consumption (Tate et al. 2011)

Reuse/recycle materials and packaging—The use or reuse of packages requires cooperation with suppliers and helps reduce storage and recovery delays, which represent operational cost savings and are also environmentally conscious (Rao and Holt 2005)

Environmental collaboration with the customer—This type of collaboration comprises a set of environmental activities engaged in by companies and their customers in order to develop a mutual understanding of environmental performance responsibilities aimed to reduce the environmental impact of their activities, resolve environmental-related problems, and reduce the environmental impact of their product (Vachon and Klassen 2006)

Reverse logistics—It represents "the management of the flow of products or parts destined for remanufacturing, recycling, or disposal by effectively use of resources" (Dowlatshahi 2000). Reverse logistics involves all of the activities associated with the collection and either recovery or disposal of used products (Ilgin and Gupta 2010b)

The current business environment is characterized by high levels of turbulence and volatility. As a result, SCs are vulnerable to disruption and, consequently, the risk to business continuity has increased (Azevedo et al. 2008). This increased vulnerability of SCs is motivated mainly by globalization, outsourcing, and the reduction of a supplier base (Thun and Hoenig 2011).

Whereas in the past, the principal objective in SC design was cost minimization or service optimization, the emphasis today must rely on resilience (Tang 2006). Resilient SCs may not be the lowest-cost SCs, but they are more capable of coping with an uncertain business environment.

Resilience is referred to as the ability of a SC to cope with unexpected disturbances. The goal of SC resilience analysis and management is to prevent the shifting to undesirable states, i.e., the ones where failure modes could occur. In SC systems, the purpose is to react efficiently to the negative effects of disturbances (which could be more or less severe). The aim of resilience strategies has two key aspects (Haimes 2006): (1) to recover the system that has been disturbed back to the desired state within an acceptable time period and at an acceptable cost and (2) to reduce the disturbance impact by changing the effectiveness level of a potential threat.

The impact of an incident on a SC depends on the type of incident and on the design of the SC. The latter refers to the aspect of vulnerability of a SC. Christopher and Peck (2004) define vulnerability as "an exposure to serious disturbance, arising from risks within the supply chain as well as risks external to the supply chain."

The ability to recover from a disturbance is related to the development of responsiveness capabilities through flexibility and redundancy (Rice and Caniato 2003). Hansson and Helgesson (2003) proposed that robustness can be treated as a special case of resilience, since it implies that the system returns to the original state after a disturbance. In addition, Tang (2006) proposed the use of robust SC strategies to enable an organization to deploy the associated contingency plans efficiently and effectively when facing a disruption, which makes the SC more resilient. The strategies proposed were based on (1) postponement; (2) strategic stock; (3) flexible supply base; (4) make-and-buy trade-off; (5) economic supply incentives; (6) flexible transportation; (7) revenue management; (8) dynamic assortment planning; and (9) silent product rollover.

Christopher and Peck (2004) stated that resilience in SCs should be designed according to the following principles: (1) selecting SC strategies that keep several options open; (2) re-examining the efficiency and redundancy trade off; (3) developing collaborative work across SCs to help mitigate risk; (4) developing visibility to create a clear view of upstream and downstream inventories, demand and supply conditions, and production and purchasing schedules; (5) improving SC velocity through streamlined processes, reduced in-bound lead times, and non-value added time reduction.

A representative sample of key resilience practices in the SC context found in the literature is shown in Table 9.2.

TABLE 9.2
Resilient SCM Practices in the Automotive Industry

Reference	Methodology	Experts/ Professionals	Most Important Resilient Practices Pointed Out
Tsiakouri (2008)	Automotive manufacturing production line case study	Professionals	• Flexible sourcing strategies • SC visibility • Flexible transportation • Postponement strategy • Stockpiling • Economic supply incentives • Standardization
Azevedo (2011b)	Delphi technique and case study	Academics and professionals	• Sourcing strategies to allow switching of suppliers • Flexible supply base/flexible sourcing • Strategic stock • Lead time reduction • Creating total supply chain visibility • Flexible transportation • Developing visibility to a clear view of downstream inventories and demand conditions
Pettit et al. (2010)	Focus group	Functional experts	• Flexibility sourcing • Flexibility in order fulfillment • Adaptability • Collaboration • Financial Strength

Considering some of the practices identified in the literature and the evidence from the automotive context, the resilient practices focused on in this chapter are the following:

Sourcing strategies to allow switching of suppliers—the ability to switch suppliers quickly supports a recovery when compared to a less dense network (Greening and Rutherford 2011)

Flexible supply base/flexible sourcing—consists of the availability of a range of options and the ability of the purchasing process to effectively exploit them in order to respond to changing requirements related to the supply of purchased components (Tachizawa et al. 2007)

Strategic stock—consists of holding some inventories at certain strategic locations (e.g., warehouse, logistics hubs, and distribution centers) to be shared by multiple SC partners (e.g., retailers and repair centers) (Tang 2006)

Lead time reduction—when the lead time is long the SC is more vulnerable to disruption. To reduce the exposure to this risk, the lead time can be reduced by redesigning the SC network (Tang 2006)

Creating total SC visibility—a clear picture of inventories and flows in the SC, the status of vendors, manufacturers, intermediaries and customers, and the logistics network can provide the first step for effective management (Iakovou et al. 2007)

Flexible transportation—this practice includes multi-modal transportation, multi-carrier transportation, and multiple routes to ensure a continuous flow of materials even when transportation disruptions occur (Tang 2006)

Developing visibility of downstream inventories and demand conditions—SC partners who exchange information regularly are able to work as a single entity and, therefore, can respond to disruptions more quickly by rerouting shipments, adjusting capacities, and/or revising the original production plans (Iakovou et al. 2007)

9.3   RESEARCH QUESTIONS AND HYPOTHESES

After completing a review of the literature, the following research questions and hypothesis were formulated:

The first research question (RQ) suggested in this chapter is the following:

RQ$_1$: To what extent is the green paradigm considered more important than the resilient paradigm for the competitiveness of the automotive industry?

Organizations worldwide are constantly trying to develop new and innovative ways to enhance their competitiveness. According to Bacallan (2000), some of these organizations are enhancing their competitiveness through improvements in their environmental performance to comply with increasing environmental regulations, to address the environmental concerns of their customers, and to mitigate the environmental impact of their production activities. Many organizations are adopting GSCM to address such environmental issues. It is generally perceived that GSCM promotes efficiency and synergy among business partners and their lead corporations, and helps to enhance environmental performance, minimize waste, and achieve cost savings. This synergy is expected to enhance the corporate image, competitive advantage, and marketing exposure. Greening the SC contributes to cost reduction and integrating suppliers in a participative decision-making process that promotes environmental innovation (Bowen et al. 2001; Rao 2002).

However, many organizations still look at green initiatives as involving trade-offs between environmental performance and economic performance (Klassen and McLaughlin 1996). The financial performance is affected by environmental performance in a variety of ways. When waste, both hazardous and non-hazardous, is minimized as part of environmental management, it results in increased utilization of natural resources, improved efficiency, higher productivity, and reduced operating costs. Again, when the environmental performance of the organization improves, it leads to a tremendous marketing advantage, improved revenue, increased market share, and new market opportunities. Organizations that minimize the negative environmental impacts of their products and processes, recycle post-consumer waste, and establish environmental management systems are poised to expand their markets or displace competitors that fail to promote strong environmental performance (Klassen and Mclaughlin 1996).

With respect to the relationship between the resilient paradigm and the organizations' competitiveness, according to Sheffi and Rice (2005), a company's resilience is a function of its competitive position and the responsiveness of its SC. Fast-responding companies can gain market share and slow responders risk losing market share. Companies with market power that respond quickly to disruptions have the opportunity to solidify their leadership positions. The investment in resilience for such companies is typically justified due to the high margins associated with such strong market position and because market leaders that are slow to respond may invite regulatory intervention. Therefore, considering the resilient paradigm, this can only contribute to the improvement of an organization's competitiveness if they are already competitive.

The second proposed research question is as follows:

RQ₂: Does the importance given to green and resilient SCM practices for the competitiveness of the automotive SC vary with academics and professionals?

This is not the case as the importance given by experts to green practices is the same as that given by professionals. Azevedo et al. (2011a) concluded that professionals from the automotive industry considered the most important green practices to the greenness of the automotive industry are the following: reverse logistics, minimizing waste, and ISO 14001 certification. In addition, the academic community highlighted the following green practices as the most important: reverse logistics, reusing/recycling materials and packaging, reducing energy consumption, and using environmentally friendly raw materials. Therefore, in comparing the two perspectives, only reverse logistics is considered by both groups as important to the greenness of the automotive industry.

Moreover, Azevedo et al. (2011b) analyzed the green and resilient paradigms in the automotive industry and also concluded that academics and professionals attribute different levels of importance to the green and resilient practices. In considering green practices, the attributes considered most important by academics for the greenness of the automotive industry are the following: reduce energy consumption, reuse/recycling materials and packaging, and environmental collaboration with suppliers. On the other hand, industry professionals highlighted the following green practices: energy consumption reduction, ISO 14001 certification, and reverse logistics. Only one practice was considered important for the greenness of the automotive industry by both groups. In the same research, the importance of the resilient practices based on country of the automotive industry was also analyzed between academics and professionals. The results are the following: the resilient practices considered most important by academics are flexible supply base/flexible sourcing, sourcing strategies to allow switching of suppliers, and creating total SC visibility; while the resilient practices considered most important by professionals are strategic stock, flexible base of suppliers, and flexible transportation modes and routes. Again, only one resilient practice was considered important to the resilience of the automotive industry by both groups.

Based on the literature review the following hypothesis is formulated:

H_1. There are no differences in the importance given to green and resilient SCM practices between academics and professionals.

The third suggested research question is as follows:

RQ_3: To what extent does the importance given to green and resilient SCM practices for the competitiveness of the automotive SC vary by country?

The pressure and motivation to implement green and resilient programs and the improvement of environmental performance arise from a wide range of sources including customers, regulators, suppliers, and competitors within an organization's SC (Zhu et al. 2008). According to Fineman and Clarke (1996), green practices are adopted by organizations pressured either by regulators or by non-governmental organizations (NGO), rather than embracing them voluntarily and proactively. For example, U.K. organizations tend to react to specific regulatory acts or the need to obtain some form of industry standard to maintain commercial contracts, rather than adopt a holistic view of green practices. Considerable differences may exist between the Chinese and the U.K. automotive industry, as vehicle production and use have dramatically increased in China (Wells 2005). Chinese automobile organizations have to compete with their international counterparts. As a result, Chinese automobile SCs must work to improve their economic and environmental performance simultaneously.

In diverse countries, however, these pressures are exerted differently. For example, many leading organizations from environmentally aware and developed markets evaluate not only their direct suppliers but also their second-tier suppliers (Walton et al. 1998). Pressures deriving from both regulators and tax policies are likely to elicit substantially different responses from individual organizations. Thus, overall pressures arising from regulators, SC partners, competitors, and the market are likely to be interpreted differently in different countries.

When industrial production is spread across various countries and all segments of the SC are critical, a disturbance affecting one segment of the chain will reverberate throughout the chain. The effects of a disturbance move forward from the supplier to the customer rather than backward as in the demand-driven model. During 2000–2006, Japan's organizations were the largest potential exporters of SC disturbances, because they were large suppliers of intermediate goods to other economies. On the contrary, Malaysia and Thailand were the largest importers of such disturbances because of the high degree of integration of their manufacturing sectors in international SCs and their reliance on imported inputs rather than domestic. Also, between 2000 and 2006, organizations from China notably increased their forward international linkages and their domestic backward linkages. Chinese organizations became large exporters of disturbances in 2006, on par with Japan, but their vulnerability to an imported disturbance remained relatively stable because Chinese manufacturers are increasingly relying on domestic suppliers (Escaith 2010). In summary, the different roles performed by organizations in the global economy lead to the deployment of varying kinds of resilient practices to overcome these disturbances. Based on the literature review, the following hypothesis is formulated:

H_2. *There are no differences in the importance given to green and resilient SCM practices among countries.*

9.4   RESEARCH METHODOLOGY

9.4.1   SURVEY INSTRUMENT

The instrument used to gather data and test the hypotheses was a survey. This survey instrument (questionnaire) was pre-tested by four academics who are experts in green and resilient SCM in the automotive industry and two professionals from the same industry. The modified questionnaire was piloted to check its applicability and appropriateness for the target population prior to distribution. After final modifications, the survey was sent to academics and professionals during November and December of 2011.

The questionnaire consists of three sections. In the first section, data on the respondents profile are collected. In the second section, the perception of respondents on the importance of green and resilient SCM paradigms for the competitiveness of the automotive industry is obtained. Finally, in the third section, two sets of green and resilient practices are suggested in order to gather information on the importance of each one for the greenness and resilience of the automotive SC, respectively (see the Appendix). The answers of academics and professionals were obtained from the completed questionnaires using a 5 point Likert scale ranging from (1) not important to (5) extremely important.

The questionnaire was tested to confirm its reliability. Cronbach's α was used to determine the reliability of the constructs (Chow et al. 2008; Nunnally 1978). According to Nunally (1978), the closer the value of the reliability to 1.00 the more reliable the result. The values of reliability that are less than 0.7 are assumed to be weak while reliability values of 0.70 or higher are acceptable. Table 9.3 provides the reliability values for the green and resilient practices. The results indicate that the values for all the variables involved in the research are above 0.7. Therefore, they are accepted as reliable.

TABLE 9.3

Values of Reliability

Constructs	Items	Cronbach's α	Number of Items
Green practices	Environmental collaboration with suppliers Environmental monitoring of suppliers ISO 14001 certification To reduce energy consumption To reuse/recycling materials and packaging Environmental collaboration with the customer Reverse logistics	0.712	7

Resilient practices	Sourcing strategies to allow switching
	  of suppliers
	Flexible supply base/flexible sourcing
	Strategic stock
	Lead time reduction
	Creating total supply chain visibility
	Flexible transportation
	Developing visibility to a clear view of
	  downstream inventories and demand conditions
		0.783	7

9.4.2 Sample Selection

As the data necessary to develop this study require in-depth knowledge and sound experience of the automotive industry and green and resilient SCM paradigms, a purposeful sampling was adopted to select a group of academics and professionals (Chan et al. 2001). That is, a non-probability sampling technique was used (Polit and Hunglar 1999). Okoli and Pawlowski (2004) provide a detailed discussion of the process of how experts should be selected for a rigorous approach. According to this process, it is crucial to identify the kind of knowledge required for an expert entering the study. In this research, two different sets of experts were selected: academic researchers and professionals. This is justified, as it was intended to achieve a wide range of interest groups and their respective opinions in the research. In order to identify eligible academics for this part of the study, the following two criteria were followed: (1) a sound knowledge and understanding of green and resilient SCM paradigms and (2) current/recent involvement in the automotive industry research topics. In order to obtain the most valuable opinions, only academics who met the two selection criteria were considered. The criterion used to select professionals was that the professionals should work in automotive companies in some capacity.

The study focuses on the automotive industry because this industry seems to be the most developed in terms of environmental and sustainability issues and is also vulnerable to SC disruptions. The automotive sector experiences considerable expectations from customers and society concerning environmental performance, as its products are resource-burning products by nature (Thun and Müller 2010). The automotive SC is also under pressure to become more sustainable and, therefore, more environmentally friendly while achieving the expected economic benefits from greener behavior (Koplin et al. 2007; Thun and Muller 2010). Also, there is evidence that the tendency of many automotive companies to seek out low-cost solutions may have led to leaner but more vulnerable SCs (Azevedo et al. 2008; Svensson 2000). The automotive SC is a typical example of high vulnerability to disturbances (Svensson 2000). According to Thun and Hoenig (2011), the trends in globalization and the necessity to offer many product variants are the key drivers to increasing the vulnerability of this industry.

9.5 DATA ANALYSIS AND FINDINGS

The data were analyzed by examining the distribution of responses based on frequencies and percentages and multivariate analysis. Using the Statistical Package for Social Sciences (SPSS) version 19, frequency, Mann–Whitney test, and Kruskal–Wallis test tables were generated and analyzed as discussed next.

From a total of 42 completed questionnaires received, the academic and professional respondents were equally represented with a total of 21 academics and 21 professionals participating in this study. Also, most of the respondents are from China (31%), followed by Portugal (26.2%), Germany (16.7%), and the United States (16.7%). The respondents from Switzerland represent 7.1% of the sample and Belgium represents 2.4% (Table 9.4). However, there are some unbalanced distributions of respondents based on the two variables of country and expertise. For example, the respondents from Germany are all experts from industry and the Portuguese respondents are all academics. The case of the United States is also unbiased since only one respondent is academic and the other six are professionals.

A further review of the demographics of the professionals that participated in the research revealed five work in production and operations management, four are directors of companies, three work in logistics, three are in project management, two work in quality management, and the remaining represent other areas.

TABLE 9.4
Respondents by Country and Expertise

Country	Academics N (%)	Professionals N (%)	Total N (%)
Belgium	—	1 (4.8)	1 (2.4)
China	8 (38.1)	5 (23.8)	13 (31)
Germany	—	7 (33.3)	7 (16.7)
Portugal	11 (52.4)	—	11 (26.2)
Switzerland	1 (4.8)	2 (9.5)	3 (7.1)
United States	1 (4.8)	6 (28.6)	7 (16.7)
Total	21 (100)	21 (100)	42 (100)

9.5.1 Perception of the Importance of Green and Resilient Paradigms for the Competitiveness of the Automotive Industry

An aggregate statistical analysis combining the responses of academics and professionals was performed to identify the level of importance of each individual SCM paradigm for the competitiveness of the automotive industry.

In order to respond to the first research question (RQ_1): *To what extent is the green paradigm considered more important than the resilient paradigm for the competitiveness of the automotive industry?*), the answers to the first question of the questionnaire were aggregated and descriptive statistics were performed. Table 9.5 shows that the two SCM paradigms received different degrees of importance from the respondents. As shown, the resilient paradigm is considered more important than the green paradigm for the competitiveness of the automotive industry since it has a higher mean value (4.48).

These results highlight the resilient paradigm as the one that most contributes to the competitiveness of the automotive industry. With a mean value of 4.48 it is considered extremely important. With regard to the green paradigm, its mean value is 3.74, which represents moderate importance.

In order to further analyze these results, Table 9.6 provides the perspective of academics and professionals toward these paradigms. According to this

TABLE 9.5

Importance of SCM Paradigms for the Competitiveness of the Automotive Industry

	N	Minimum	Maximum	Mean	Std. Deviation
Importance of green paradigm	42	2	5	3.74	0.83
Importance of resilient paradigm	42	3	5	4.48	0.74
Valid N (listwise)	42				

TABLE 9.6

Importance of Green and Resilient Paradigms among Academics and Professionals

		Academics N (%)	Professionals N (%)	Total N (%)
Importance of green	2	2 (9.5)	1 (4.8)	3 (7.1)
	3	2 (9.5)	10 (47.6)	12 (28.6)
	4	14 (66.7)	6 (28.6)	20 (47.6)
	5	3 (14.3)	4 (19)	7 (16.7)
Total		21 (100)	21 (100)	42 (100)
Importance of resilient	2	—	—	—
	3	3 (14.3)	3 (14.3)	6 (14.3)
	4	5 (23.8)	5 (23.8)	10 (23.8)
	5	13 (61.9)	13 (61.9)	26 (61.9)
Total		21 (100)	21 (100)	42 (100)

table, it is possible to state that the academics consider the green paradigm more important than the resilient paradigm for the competitiveness of the automotive industry. The second column provides the academics' responses, which show that 66.7% consider the green paradigm important and 14.3% as extremely important. With respect to the resilient paradigm, the same level of importance was indicated by academics and professionals. Also, the resilient paradigm is considered extremely important for the competitiveness of the automotive industry by 85.7% of both.

However, these results are not completely supported by the literature. Some works defend the green paradigm as contributing more to enhancing the competitiveness of organizations and SC. For example, Bacallan (2000) highlights the contribution of a green behavior to the competitiveness of organizations; and Bowen et al. (2001) defend greening the SC as a way to reduce costs, enhance corporate image, and increase the competitive advantage of an organization and their corresponding SCs. Others defend that the resilient paradigm can only contribute to improving the competitiveness of organizations if they are already competitive (Sheffi and Rice 2005). Therefore, the results of this research are not supported by these authors. However, one research study (Azevedo et al. 2011b) reached the same results but used only the perception of academics and only in one country.

In summary, according to the statistical results the green paradigm is not considered more important than the resilient paradigm for the competitiveness of the automotive industry. The resilient paradigm is considered the most important.

9.5.2 Importance of Green and Resilient SCM Practices for Academics and Professionals

In order to test hypothesis H_1 (*There are no differences in the importance given to green and resilient SCM practices between academics and professionals*), the non-parametric Mann–Whitney U test was used.

The Mann–Whitney test is the non-parametric equivalent of the independent samples t-test. It is used when the sample data are not normally distributed, as this is the case (Hair et al. 1998). This test has the advantage of being useful for small samples and is also one of the most powerful non-parametric tests (Landers 1981), where the statistical power corresponds to the probability of rejecting a false null hypothesis. The Mann–Whitney test evaluates whether the mean ranks for the two groups (academics and professionals) differ significantly from each other. The Mann–Whitney test was conducted to evaluate the hypothesis that there are differences in the importance given to green and resilient SCM practices between academics and professionals (Tables 9.7 and 9.8).

The results in Table 9.7 indicate which group gives more importance to each paradigm. The academics group places more importance to the green paradigm since it has a mean rank of 23.67. With regard to the importance given to resilient paradigm, there is no difference between groups. Table 9.8 provides the test statistics including the U value as well as the asymptotic significance (two-tailed) p-value.

TABLE 9.7
Rank between Groups

		N	Mean Rank	Sum of Ranks
Importance of green practices	Academics	21	23.67	497.00
	Professionals	21	19.33	406.00
	Total	42		
Importance of resilient practices	Academics	21	21.50	451.50
	Professionals	21	21.50	451.50
	Total	42		

TABLE 9.8
Mann–Whitney Test

Test Statistics[a]		
	Importance of Green Practices	Importance of Resilient Practices
Mann–Whitney U	175.000	220.500
Wilcoxon W	406.000	451.500
Z	−1.231	0.000
Asymp. Sig. (two-tailed)	0.218	1.000
Exact Sig. (two-tailed)	0.229	1.000

[a] Grouping variable: Academic_Professional.

In order to gain a little statistical power, the exact significance value was computed. As this value is a two-tailed p-value and we have one-tailed test, this value must be divided by 2 to get the one-tailed p-value. Therefore, the p-value is 0.114 for the importance of green and 0.500 for the importance of resilient. Since both p-values are greater than the specified α level of 0.05, it is concluded that the hypothesis H_1 is not rejected. In other words, the differences in the importance given to green and resilient SCM practices between academics and professionals are not statistically significant.

9.5.3 IMPORTANCE GIVEN TO GREEN AND RESILIENT PRACTICES FOR THE COMPETITIVENESS OF THE AUTOMOTIVE SC VARY BY COUNTRY

In order to test the hypothesis H_2 (*There are no differences in the importance given to green and resilient SCM practices among countries.*), the non-parametric Kruskal–Wallis test was used.

The Kruskal–Wallis test is a non-parametric test equivalent to the one-way ANOVA and an extension of the Mann–Whitney test. This test is used to compare three or more sets of scores that come from different groups when the assumption of normality or equality of variance is not met (Hair et al. 1998). Using the Kruskal–Wallis test, the mean of the rank for population groups were measured and compared (Sheskin 2004). The Kruskal–Wallis test was computed to green and resilient practices considering the group variable of country for the respondents (Table 9.9).

The chi-square value (Kruskal–Wallis), the degrees of freedom, and the significance level are presented in Table 9.9. According to these values, it is concluded that there is not a statistically significant difference between the importance given to green and resilient practices by academics and professionals according to their origin (country). Only in the green practice "Environmental monitoring of suppliers" and in the resilient practices "Sourcing strategies to allow switching of suppliers" and "Lead time reduction," the differences seem to be statistically significant because the p value is greater than 0.05. In conclusion, the H_2 hypothesis is not rejected. Only for the three identified practices the variable "Country" seems to influence the level of importance given by academics and professionals.

TABLE 9.9
Kruskal–Wallis Test for the Importance of Green and Resilient Practices by Country

Test Statistics[a,b]	Chi-Square	df	Asymp. Sig.
Green practices			
Environmental collaboration with suppliers	7.158	5	0.209
Environmental monitoring of suppliers	14.015	5	0.016*
ISO 14001 certification	7.039	5	0.218
To reduce energy consumption	6.004	5	0.306
To reuse/recycling materials and packaging	10.930	5	0.053
Environmental collaboration with the customer	8.037	5	0.154
Reverse logistics	2.838	5	0.725
Resilient practices			
Sourcing strategies to allow switching of suppliers	14.019	5	0.015*
Flexible supply base/flexible sourcing	8.148	5	0.148
Strategic stock	11.390	5	0.044
Lead time reduction	14.716	5	0.012*
Creating total supply chain visibility	5.339	5	0.376
Flexible transportation	8.740	5	0.120
Developing visibility to a clear view of downstream inventories and demand conditions	9.532	5	0.090

[a] Kruskal–Wallis test.
[b] Grouping variable: Country.
*Significant to $p < 0.05$.

9.6 CONCLUSIONS

This chapter follows an exploratory approach in order to clarify the importance recognized by the academic community and automotive industry professionals of the
green and the resilient SCM paradigms for the competitiveness of the automotive SC.
The main objective of the chapter is to provide an in-depth analysis on the importance of green and resilient SCM practices for the competitiveness of the automotive
supply. In order to attain this objective, a panel of academics and professionals from
six different countries was used. This enlarged scope associated with the collected
data, makes a broad perspective of the research topic possible. Also, the professionals that collaborated with this research are working in leader companies in the automotive industry. This contributes to enforce the reliability of the results reached in
this chapter. Also, experts in the green and resilient paradigms were involved which
further enhances the validity of the research. Based on the literature review, three
research questions were developed and statistical analysis was performed to accom-
plish the main research objective.

For the first research question "To what extent is the green paradigm considered
more important than the resilient paradigm for the competitiveness of the automotive industry?" the descriptive statistics illustrate that the answer to this questions is negative. In other words, according to the research sample the resilient
paradigm is considered more important than the green paradigm for the competitiveness of the automotive industry. As stated in an earlier section, this is not consistent with most of the literature. This could be explained because green practices
deployed by companies are typically implemented due to compulsory regulations,
that is, they are imposed externally (Biermann 2002). On the other hand, resilient
practices are planned, coordinated, and performed by companies as a strategic
weapon for enhancing their competitiveness (Sheffi and Rice 2005). The resilient
practices are viewed as a way of minimizing risks (Rice and Caniato 2003; Tang
2006). It is also interesting to note that academics give more importance to the
green paradigm while the resilient paradigm has the same level of importance for
both groups.

The second research question asks, "Does the importance given to green and
resilient SCM practices for the competitiveness of the automotive SC vary with academics and professionals?" This is translated into the following hypothesis, H_1: *There
are no differences in the importance given to green and resilient SCM practices
between academics and professionals*. After the deployment of the Mann–Whitney
test, it was concluded that this hypothesis is not rejected. Both groups consider the
green and the resilient paradigms to have the same level of importance for the greenness and resilience of the automotive SC. The same conclusion was also reached by
Azevedo et al. (2011b). This could be a reflection of a closer relationship that exists
between academics and the automotive industry that makes the transfer of knowledge possible.

Finally, the third research question asks, RQ_3: *To what extent is the importance
given to green and resilient SCM practices for the competitiveness of the automotive SC vary by country?* Based on the statistical analysis, there are no differences
in the importance given to green and resilient SCM practices among countries.

This could be explained because the automotive industry is a global sector with its SCs crossing several countries (Nunes and Bennett 2010). There exists a universal organizational culture shared all over the world that reflects the specificities of this kind of industry.

The content of this chapter is particularly important for managers when developing a set of green and resilient practices. The information presented enables managers to consider the most important aspects for the greenness and resilience of the individual automotive companies and corresponding SCs. This also informs the automotive professionals of the increased importance that the resilient SCM practices are reaching. Also, the green SCM practices have a lower importance among professionals.

In addition to the important contributions of this chapter, the limitations of the study should be noted. The set of green and resilient practices should be enlarged to consider additional practices that are identified in the literature. Also, the sample size is limited (42 respondents) which restricts the types of statistical analyses that can be performed.

In terms of future research, it would be interesting to deploy the structural equation modeling (SEM) to test more complex relationships that exist among the research variables after suggesting a structural model.

APPENDIX

This framework is intended to get information about experts/professionals perception on the importance of green and resilient paradigms to the competitiveness of the automotive industry. (Green paradigm is concerned with environmental risks and impacts reduction, and the resilient paradigm focuses on the SC ability to recover to a desired state after a disruption occurrence.) Try to answer to the questions, please.

Experts/professionals identification

Country: _____

Affiliation: _____

Job title:_____

Experience: _____

1—For the following supply chain management paradigms, please describe your perception about their importance to the competitiveness of the automotive industry

	1 Not Important	2	3	4	5 Extremely Important
Green					
Resilient					

2—For the following green practices, please describe your perception about their importance to the greenness of the automotive supply chain

	1 Not Important	2	3	4	5 Extremely Important
Environmental collaboration with suppliers					
Environmental monitoring upon suppliers					
ISO 14001 certification					
To reduce energy consumption					
To reuse/recycling materials and packaging					
Environmental collaboration with the customer					
Reverse logistics					

3—For the following resilient practices, please describe your perception about their importance to the resilience of the automotive supply chain

	1 Not Important	2	3	4	5 Extremely Important
Sourcing strategies to allow switching of suppliers					
Flexible supply base/flexible sourcing					
Strategic stock					
Lead time reduction					
Creating total supply chain visibility					
Flexible transportation					
Developing visibility to a clear view of downstream inventories and demand conditions					

Thank you for your collaboration.

REFERENCES

Azevedo, S.G., Carvalho, H., and Cruz-Machado, V. 2011a. The influence of green practices on supply chain performance: A case study approach. *Transportation Research Part E: Logistics and Transportation Review* 47(6): 850–871.

Azevedo, S., Carvalho, H., Cruz-Machado, V., and Grilo, F. 2010. The Influence of agile and resilient practices on supply chain performance: An innovative conceptual model proposal. In *Innovative Process Optimization Methods in Logistics: Emerging Trends, Concepts and Technologies*, eds. Blecker, T., Kersten, W., and Luthje, C., pp. 265–286. Erich Schmidt Verlag GmbH & Co. KG, Berlin, Germany.

Azevedo, S.G., Kannan, G., Carvalho, H., and Cruz-Machado, V. 2011b. GResilient index to assess the greenness and resilience of the automotive supply chain. Working paper 7/2011, Department of Business and Economics, University of Southern Denmark, Odense, Denmark.

Azevedo, S., Machado, V., Barroso, A., and Cruz-Machado, V. 2008. Supply chain vulnerability: Environment changes and dependencies. *International Journal of Logistics and Transport* 1(2): 41–55.

Bacallan, J.J. 2000. Greening the supply chain. *Business and Environment* 6(5): 11–12.

Bai, C. and Sarkis, J. 2010. Integrating sustainability into supplier selection with grey system and rough set methodologies. *International Journal of Production Economics* 124(1): 252–264.

Biermann, F. 2002. Strengthening green global governance in a disparate world society would a world environment organisation benefit the south? *International Environmental Agreements: Politics, Law and Economics* 2(4): 297–315.

Bowen, F., Cousins, P., Lamming, R., and Faruk, A. 2001. Horses for courses: Explaining the gap between the theory and practice of green supply. *Greener Management International* 35: 41–59.

Carter, C. and Ellram, L. 1998. Reverse logistics: A review of the literature and framework for future investigation. *Journal of Business Logistics* 19(1): 85–102.

Carter, C. and Rogers, D.S. 2008. A framework of sustainable supply chain management: Moving toward new theory. *International Journal of Physical Distribution & Logistics Management* 38(5): 360–387.

Carvalho, H., Azevedo, S.G., and Cruz-Machado, V. 2010. Supply chain performance management: Lean and green paradigms. *International Journal of Business Performance and Supply Chain Modelling* 2(3/4): 304–333.

Carvalho, H., Duarte, S., and Cruz-Machado, V. 2011. Lean, agile, resilient and green: Divergencies and synergies. *International Journal of Lean Six Sigma* 2(2): 151–179.

Chan, A., Yung, E., Lam, P., Tam, C., and Cheung, S. 2001. Application of Delphi method in selection of procurement systems for construction. *Construction Management and Economics* 19(7): 699–718.

Chow, W., Madu, C., Kuei, C., Lu, M.H., Lin, C., and Tseng, H. 2008. Supply chain management in the U.S. and Taiwan: An empirical study. *Omega—International Journal of Management Science* 36(5): 665–679.

Christmann, P. and Taylor, G. 2001. Globalization and the environment: Determinants of firm self-regulation in China. *Journal of International Business Studies* 32(3): 439–458.

Christopher, M. and Peck, H. 2004. Building the resilient supply chain. *The International Journal of Logistics Management* 15(2): 1–14.

Dowlatshahi, S. 2000. Developing a theory of reverse logistics. *Interfaces* 30(3): 143–155.

Escaith, H. 2010. Global supply chains and the great trade collapse: Guilty or casualty? *Theoretical and Practical Research in Economic Fields* 1(1): 27–41.

de Figueiredo, J. and Mayerle, S. 2008. Designing minimum—Cost recycling collection networks with required throughput. *Transportation Research Part E: Logistics and Transportation Review* 44: 731–752.

Fineman, S. and Clarke, K. 1996. Green stakeholders: Industry interpretation and response. *Journal of Management Studies* 36(6): 715–730.

Greening, P. and Rutherford, C. 2011. Disruptions and supply networks: A multi-level, multi-theoretical relational perspective. *International Journal of Logistics Management* 22(1): 104–126.

Haimes, Y. 2006. On the definition of vulnerabilities in measuring risks to infrastructures. *Risk Analysis: An Official Publication of the Society for Risk Analysis* 26(2): 293–296.

Hair, J. Jr., Anderson, R., Tatham, R., and Black, W. 1998. *Multivariate Data Analysis*, 5th edn. Prentice Hall, Upper Saddle River, NJ.

Hansson, S. and Helgesson, G. 2003. What is stability? *Synthese* 136(2): 219–235.

Iakovou, E., Vlachos, D., and Xanthopoulos, A. 2007. An analytical methodological framework for the optimal design of resilient supply chains. *International Journal of Logistics Economics and Globalisation* 1(1): 1–20.

Ilgin, M. and Gupta, S. 2010a. Use of sensor embedded products for end of life processing. *Proceedings of the 2010 Winter Simulation Conference*, eds. Johansson, B., Jain, S., Montoya-Torres, J., Hugan, J., and Yücesan, E., pp. 1584–1591. IEEE, Baltimore, MD.

Ilgin, M.A. and Gupta, S.M. 2010b. Environmentally conscious manufacturing and product recovery (ECMPRO): A review of the state of the art. *Journal of Environmental Management* 91(3): 563–591.

Klassen, R. and McLaughlin, C. 1996. The impact of environmental management on firm performance. *Management Science* 42(8): 1199–1213.

Koplin, J., Seuring, S., and Mesterharm, M. 2007. Incorporating sustainability into supply management in the automotive industry—The case of the Volkswagen AG. *Journal of Cleaner Production* 15(11/12): 1053–1062.

Lamming, R. and Hampson, J. 1996. The environment as a supply chain management issue. *British Journal of Management* 7: 45–62.

Landers, J. 1981. *Quantification in History, Topic 4. Hypothesis Testing II-Differing Central Tendency*. All Souls College, Oxford, U.K.

Lee, S.-Y. 2008. Drivers for the participation of small and medium-sized suppliers in green supply chain initiatives. *Supply Chain Management: An International Journal* 13(3): 185–198.

Liu, G.-F., Chen, X.-L., Riedel, R., and Müller, E. 2011. Green technology foresight on automobile technology in China. *Technology Analysis & Strategic Management* 23(6): 683–696.

Meade, L. and Sarkis, J. 2002. A conceptual model for selecting and evaluating third-party reverse logistics providers. *Supply Chain Management: An International Journal* 7(5): 283–295.

Murphy, P. and Poist, R.F. 2003. Green perspectives and practices: A comparative logistics study. *Supply Chain Management: An International Journal* 8(2): 122–131.

Nishitani, K. 2010. Demand for ISO 14001 adoption in the global supply chain: An empirical analysis focusing on environmentally conscious markets. *Resource and Energy Economics* 32(3): 395–407.

Nunes, B. and Bennett, D. 2010. Green operations initiatives in the automotive industry: An environmental reports analysis and benchmarking study. *Benchmarking: An International Journal* 17(3): 396–420.

Nunnally, J.C. 1978. *Psychometric Theory*, 2nd edn. McGraw-Hill, New York.

Okoli, C. and Pawlowski, S. 2004. The Delphi method as a research tool: An example, design considerations and applications. *Information and Management* 42(1): 15–29.

Paulraj, A. 2009. Environmental motivations: A classification scheme and its impact on environmental strategies and practices. *Business Strategy and the Environment* 18(7): 453–468.

Pettit, T., Fiksel, J., and Croxton, K., 2010. Ensuring supply chain resilience: Development of a conceptual framework. *Journal of Business Logistics* 31(1): 1–21.

Polit, D. and Hungler, B. 1999. *Nursing Research: Principles and Methods*, 6th edn. Lippincott Williams & Wilkins, Philadelphia, PA.

Rao, P. 2002. Greening of the supply chain: A new initiative in South East Asia. *International Journal of Operations & Production Management* 22(6): 632–655.

Rao, P. and Holt, D. 2005. Do green supply chains lead to competitiveness and economic performance? *International Journal of Operations & Production Management* 25(9): 898–916.

Rice, J. and Caniato, F. 2003. Building a secure and resilient supply network. *Supply Chain Management Review* 7(5): 22–30.

Sahay, B., Gupta, J., and Mohan, R. 2006. Managing supply chains for competitiveness: The Indian scenario. *Supply Chain Management: An International Journal* 11(1): 15–24.

Sarkis, J. 2003. A strategic decision framework for green supply chain management. *Journal of Cleaner Production* 11(4): 397–409.

Seuring, S. and Muller, M. 2008. Core issues in sustainable supply chain management—A Delphi study. *Business Strategy and the Environment* 17(8): 455–466.

Sheffi, Y. and Rice, J. 2005. A supply chain view of the resilient enterprise. *MIT Sloan Management Review* 47(1): 40–48.

Sheskin, D. 2004. *Handbook of Parametric and Nonparametric Statistical Procedures*, 3rd edn. Chapman & Hall/CRC, Boca Raton, FL.

Srivastava, S. 2007. Green supply-chain management: A state of the-art literature review. *International Journal of Management Reviews* 9(1): 53–80.

Svensson, G. 2000. A conceptual framework for the analysis of vulnerability in supply chains. *International Journal of Physical Distribution & Logistics Management* 30(9): 731–750.

Tachizawa, E., Gime, C., and Thomsen, C. 2007. Drivers and sources of supply flexibility: An exploratory study. *International Journal of Operations & Production Management* 27(10): 1115–1136.

Tang, C.S. 2006. Robust strategies for mitigating supply chain disruptions. *International Journal of Logistics Research and Applications* 9(1): 33–45.

Tate, W.L., Dooley, K.J., and Ellram, L.M. 2011. Transaction cost and institutional drivers of supplier adoption of environmental practices. *Journal of Business Logistics* 32(1): 6–16.

Thun, J. and Hoenig, D. 2011. An empirical analysis of supply chain risk management in the German automotive industry. *International Journal of Production Economics* 131(1): 242–249.

Thun, J.-H. and Müller, A. 2010. An empirical analysis of green supply chain management in the German automotive industry. *Business Strategy and the Environment* 19(2): 119–132.

Tsiakouri, M. 2008. Managing disruptions proactively in the supply chain: The approach in an auto-manufacturing production line. In *Proceedings of the POMS 19 Conference*, May, 9–12, La Jolla, CA, pp. 361–395.

Vachon, S. and Klassen, R. 2006. Extending green practices across the supply chain: The impact of upstream and downstream integration. *International Journal of Operations & Production Management* 26(7): 795–821.

Vachon, S. and Klassen, R. 2008. Environmental management and manufacturing performance: The role of collaboration in the supply chain. *International Journal of Production Economics* 111(2): 299–315.

Walton, S., Handfield, R., and Melnyk, S. 1998. The green supply chain: Integrating suppliers into environmental management process. *International Journal of Purchasing and Materials Management* 34(4): 28–38.

Wang, H.-F. and Gupta, S.M. 2011. *Green Supply Chain Management: Product Life Cycle Approach*. McGraw Hill, New York.

Wells, P. 2005. Exports from China: Will the trickle become a flood? AWKnowledge, April 26.

Zhu, Q., Crotty, J., and Sarkis, J. 2008. A cross-country empirical comparison of environmental supply chain management practices in the automotive industry. *Asian Business & Management* 7(4): 467–488.

Zhu, Q. and Sarkis, J. 2004. Relationships between operational practices and performance among early adopters of green supply chain management practices in Chinese manufacturing enterprises. *Journal of Operations Management* 22(3): 265–289.

Zhu, Q. and Sarkis, J. 2006. An inter-sectoral comparison of green supply chain management in China: Drivers and practices. *Journal of Cleaner Production* 14(17): 472–486.

Zhu, Q., Sarkis, J., and Geng, Y. 2005. Green supply chain management in China: Pressures, practices performance. *International Journal of Operations & Production Management* 25(5/6): 449–469.

Zhu, Q., Sarkis, J., and Lai, K.-H. 2007. Green supply chain management: Pressures, practices and performance within the Chinese automobile industry. *Journal of Cleaner Production* 15(11/12): 1041–1052.

Zhu, Q., Sarkis, J., and Lai, K. 2008. Green supply chain management implications for closing the loop. *Transportation Research Part E: Logistics and Transportation Review* 44(1): 1–18.

10 Balanced Principal Solution for Green Supply Chain under Governmental Regulations

Neelesh Agrawal, Lovelesh Agarwal,
F.T.S. Chan, and M.K. Tiwari

CONTENTS

10.1 INTRODUCTION

The increased public awareness of the environmental issues faced globally has triggered the need for companies to make their supply chains more environmental friendly. Green supply chain management (GSCM) integrates ecological factors with supply chain management principles to address how an organization's supply chain processes affect the environment (www.cognizant.com). With increasing customer awareness and regulatory norms, organizations with greener supply chain management practices will have a competitive edge over companies that are reluctant to embrace GSCM. Thus, a majority of small and large companies understand and embrace the need for protecting the environment and utilizing GSCM practices as means for reducing costs while increasing customer and shareholder value. The environmental regulations, to which the companies must comply with, impose immediate pressure on manufacturing firms along a supply chain to work together toward development of a sustainable supply chain via GSCM.

In green supply chains, the use of cleaner technologies, efficient use, and proper disposal of waste involved in manufacturing a product contribute to greening at the production stage (Agarwal et al. 2011). This investment would lead to increased prices, which in turn would lead to a decrease in demands. But at the same time, it would also promote the marketing of the product by positioning it as an environment friendly entity, which in turn raises the demand. The equilibrium point is that where green benefits of the product balance the prices and hence the supply chain revenue is optimized in a framework with government acting as a regulating body.

The effect of GSCM on industry is significant, in that the green procurement system is built to control suppliers, especially for ODM, so that they satisfy environmental protection standards for their products (Wang and Gupta 2011). GSCM is an emerging field that stands out of traditional supply chain perspective. GSCM has gained popularity with both academics and practitioners to aim in reducing waste and preserving the quality of product life and the natural resources (Fortes 2009). Global market demands and governmental pressures are pushing businesses to become more sustainable (Guide and Srivastava 1998; Gungor and Gupta 1999). Eco efficiency and remanufacturing processes are now important assets to achieve best practice (Ashley 1993; Srivastava 2007).

The remarkable rise of GSCM activity in the practical realm has led to an increased empirical and research work regarding the uptake of GSCM practices and their impact on firm performance (Simpson and Samson 2008). Extensive literature exists on buyer–retailer cooperation, producer–consumer strategies, and government-firm–environment relations. Kelle and Silver (1989) were the first who developed an optimal forecasting system for organizations to use to forecast products that can potentially be reused. Navin-Chandra (1991) was the first to consider the need for a green design to reduce the impact of product waste. Allen and Allenby (1994) and Zhang et al. (1997) expanded the framework of green design. Brass and McIntosh (1999), Sarkis and Cordeiro (2001), and Laan et al. (1996) discuss green production, planning, and manufacturing.

Through this chapter, we aim to target a green supply chain cooperation under the intervention of government via green taxation and subsidization. Game theory provides a mathematical background for defining the system and provides solutions in varying interests. Our proposed model basically incorporates the influences of government regulations via green tax and subsidies on the bargaining framework of green supply chains to yield the equilibrium solutions of negotiations between producers and reverse logistics (RL) suppliers, where the latter is responsible for collection of used products, recycling, and producing end-of-life products. For negotiations between the two tiers over exchange price, we use the balanced principal (BP) bargaining solutions, assuming arbitrary number of competitive firms in each tier.

Selection of feasible environmentally conscious manufacturing or RL projects is a popular multicriteria decision-making problem studied by the researchers

(Ilgin and Gupta 2010). In this work, we consider take-back legislation by the government as one of the important factors in modeling the scenario. The issue of efficient take-back legislation has been previously addressed by Atasu et al. (2009), who considered subsidizing the producers who go for recycling. But differing from this work, we consider a scenario where government model is not limited to interaction between producers and government but mainly targets the negotiations between producers and RL suppliers bargaining under government intervention. The development of bargaining framework considering interaction between producers and RL suppliers has been previously addressed by Sheu (2011), where he proposes an asymmetrical NB product to estimate equilibrium negotiation prices under certain restrictions. Through this chapter, we aim at providing numerical equilibrium solutions and also considering arbitrary number of firms in each tier.

The further sections of the chapter have been organized as follows. Section 10.2 defines negotiations between producers and RL suppliers using a bargaining framework stating the assumptions and the problem scope. Section 10.3 formulates a three-stage game-based model and derives the equilibrium solutions for all these stages. Section 10.4 provides qualitative and quantitative insight into the interdependence of the government, producer, and RL supplier parameters in the proposed bargaining framework. Section 10.5 states the summary and conclusions.

10.2 FRAMEWORK

In this section, we describe a bargaining framework (Figure 10.1) that defines the negotiations between producers and RL suppliers in competitive green supply chains under government's intervention via green taxation and subsidization and develop a bargaining-based solution assuming sole-sourcing within each other. The problem background is that m producers compete in the market to sell their products while competing with the government's taken-back directives and are also subjected to green taxation (denoted by t) for producing a unit product. These producers transfer these EOL product collection, processing, and recycling duties on RL suppliers who also compete away themselves. Thus RL suppliers receive subsidies for recycling unit EOL product. Thus, at equilibrium, a producer and a RL supplier cooperate with each other to form a green supply chain (so that

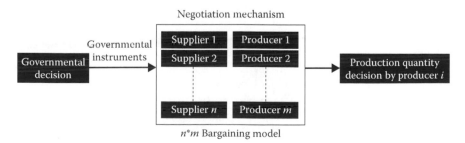

FIGURE 10.1   Bargaining model.

product is ecofriendly) but also have differing opinion (i.e., price the producers will pay to the suppliers in exchange for the recycled components). For the purpose of model formation, we assume the following:

1. We assume single sourcing (only one firm in each tier emerges as an active firm) and complete local information (all firms of a tier are aware of the value of contract to all potential buyers and cost to all potential supplies) (Lovejoy 2010).
2. There exists oligopoly competition where n producers compete among themselves with Cournot oligopoly and each producer transfers its taken-back duty to one of the m RL suppliers who also exists in competition.
3. The final product price ($P(Q)$) is assumed to be a simple Cournot inverse demand (Q) function in which the final product demand market density is normalized to 1. Thus, $P(Q) = 1 - Q$, where $Q = \sum_{i'=1}^{m} q_i$, where q_i is the amount of product sold by its producer (Sheu 2011).
4. To produce a unit product, a producer requires α_x quantities of virgin components (x) and α_y quantities of recycled components (y).
5. The unit manufacturing cost (r_m) is assumed to be equal for all producers and procurement cost of a unit recycled component (r_p) is equal for all RL suppliers.
6. The total green taxes provided by the government are equal to the total green subsidies it provides so that the government is not benefited financially.

To improve interorganizational coordination and product quantity, manufacturing firms often demand that their supply chain partners such as subcontractors or supplier implement common processes, which often require sharing process knowledge (Treleven and Schweikhart 1988). Interorganizational knowledge sharing within a supply chain has thus become a common practice, because it enhances the competitive advantage of the supply chain as a whole.

Single sourcing is broadly defined as fulfillment of all of an organization's needs by a single vendor/supplier (Cheng et al. 2008). Deming (1982), Schonberger (1982), Hall (1983) and, to some extent, Feigenbaum (1983) support the concept of single sourcing, where Deming particularly advocates strongly of this concept.

The existence of equilibrium for Cournot oligopoly has been proved and that the concept in nonambiguous has been shown by Grilo and Mertens (2009).

The model consists of a three-tier structure. The first tier includes the government deciding its financial instruments, that is, the green tax to be taxed on the producers (t) and the subsidies to be provided to the RL suppliers (s). The second tier consists of the main bargaining structure that includes m producers negotiating with n RL suppliers to decide the unit price $p_{y(i)}$ to be paid by the producers in exchange for $y_{(i)}$ quantities of recycled components to be provided by the RL suppliers. The third tier includes the producer's decision of production based on government's intervention and negotiation in the second tier.

10.3 MODEL

This section describes a three-stage bargaining structure to model the bargaining problem mentioned earlier in the chapter. Stage 1 comprises the government's decision and strategies that lead to the equilibrium values for green tax (t) and subsidy (s). Stage 2 aims at finding the bargaining principal solution for negotiations between producers and RL suppliers given the values for t and s. Stage 3 comprises the producer's decision for producing q quantities based on government's influence via t and s and also based on the BP solutions of stage 2. In order to solve this model, we follow the procedure of backward induction (Kreps 1990), where we start from stage 3 with the producer's decision for production and end with stage 1 with government's decision for its financial instruments.

10.3.1 Stage 3

This stage leads to the equilibrium solution for the final production by the producer under the influence of government legislation and his negotiations with RL suppliers.

Let the price for a unit virgin (x) be p_x, which is assumed to be equal for all producers. Let q_i be the equilibrium production quantity decided by the producer, $p_{y(i)}$ be the price negotiated with the RL supplier and c_p be the cost of production. Then the producer i maximizes his profit μ_i as

$$\mu_i = \left[1 - \sum_{i'=1}^{m} q_{i'}\right] \times q_i - p_x \alpha_x q_i - p_{y(i)} \alpha_y q_i - c_p q_i - t q_i, \quad \forall i \quad (10.1)$$

To find the equilibrium solution, we need the first order differential with respect to q_i, i.e.,

$$1 - \sum_{i'=1}^{m} q_{i'} - p_x \alpha_x - p_y \alpha_y - c_p - t = 0, \quad \forall i \quad (10.2)$$

This indeed gives the maxima solution for q_i as the second-order differential yields negative value (i.e., $\partial^2 \mu_i / \partial q_i^2 = -2 < 0, \forall i$).

From our stated assumption regarding the Cournot oligopoly competition adopted by all the producers, the equilibrium solution can be stated as follows:

$$q_i = \frac{(1 - p_x \alpha_x - p_{y(i)} \alpha_y - c_p - t)}{m+1}, \quad \forall i \quad (10.3)$$

10.3.2 Stage 2

This stage leads to the decision of negotiations between m producers and n RL suppliers. Here we consider multiechelon bargaining chains with arbitrary number of competing firms in each tier. We use BP solution for this two tier ($m \times n$) subsystem.

We define r^i as the net revenues or the net amount that the ith most efficient producer is willing to pay. Similarly, we define c^j as the total cost of supply for the jth most efficient RL supplier firm. Then according to Lovejoy (2010), the equilibrium price at which the producer–supplier pair will contract is given by

$$p_{y(i)} = \frac{1}{2}[(r^1 \wedge c^2) + (r^2 \vee c^1)] \tag{10.4}$$

As stated, r^1, r^2 are for the most efficient and the second most efficient firm in producer tier and c^1, c^2 correspond to the most efficient and the second most efficient firm in the RL supplier tier. Since a buyer firm i and selling firm j only contract if $r^i \geq c^j$, we arrive at four possible cases as shown in Table 10.1.

Table 10.1 illustrates the conditions and the BP price under each of these conditions. For example, in third case although the most efficient producer can pair with either of the RL suppliers ($r^1 > c^2 > c^1$), contracting with the second most efficient producer is not feasible ($r^2 < c^1 < c^2$).

The term "r" (net revenues) can be theoretically represented as follows:

net revenue = (selling cost − value adding cost) per unit of product

$$= \left[1 - \sum_{i'=1}^{m} q_{i'}\right] q_i - c_p q_i - t q_i, \quad \forall i \tag{10.5}$$

and the term "c" (total cost of supply) is theoretically given by

total supply cost = (fixed input price + value adding cost) per unit product

$$= 0 + r_{col}(aq_i) + r_{pro}(abq_i) + r_p(\alpha_y - ab)q_i - sabq_i, \quad \forall i \tag{10.6}$$

where
r_{col} represents unit costs for collection of end-of-life products by RL suppliers
r_{pro} represents the unit cost for processing these collected products

---

TABLE 10.1
Various Cases for BP Model

Case	Conditions	Price
1	$r^1 \leq c^2$ $r^2 \leq c^1$	$p_{y(i)} = \frac{1}{2}(r^1 + c^1)$
2	$r^1 \leq c^2$ $r^2 \geq c^1$	$p_{y(i)} = \frac{1}{2}(r^1 + r^2)$
3	$r^1 \geq c^2$ $r^2 \leq c^1$	$p_{y(i)} = \frac{1}{2}(c^1 + c^2)$
4	$r^1 \geq c^2$ $r^2 \geq c^1$	$p_{y(i)} = \frac{1}{2}(r^2 + c^2)$

Thus, profit of a RL supplier can be represented as

$$\text{profit (RL supplier)} = \text{sales cost} - \text{total cost for supply}$$

$$\Rightarrow \mu_j = p_{y(i)} \alpha_y q_i - c \tag{10.7}$$

Replacing the values for net revenues and total cost of supply in the expression for prices obtained from Table 10.1, we obtain the equilibrium solutions as follows:

Case 1:

$$\Rightarrow p_{y(i)} = \frac{(1 - p_x \alpha_x - c_{p1} - t)\left(\left[1 - \sum_{i'=1}^{m} q_{i'} - t\right] - c_{p1} + r_{p1}(\alpha_y - ab) + r_{col1}a + r_{pro1}ab - t\right)}{\alpha_y \left(\left[1 - \sum_{i'=1}^{m} q_{i'} - t\right] - c_{p1} + r_{p1}(\alpha_y - ab) + r_{col1}a + r_{pro1}ab - t\right) + 2(m+1)}, \quad \forall i \tag{10.8}$$

Case 2:

$$\Rightarrow p_{y(i)} - \frac{1}{2} \frac{\left[(1 - p_x \alpha_x - c_{p1} - t)\left(2\left[1 - \sum_{i'=1}^{m} q_{i'} - t\right] - (c_{p1} + c_{p2})\right)\right]}{\left[m + 1 - \alpha_y \left(\left[1 - \sum_{i'=1}^{m} q_{i'} - t\right] - \frac{1}{2}(c_{p1} + c_{p2})\right)\right]}, \quad \forall i \tag{10.9}$$

Case 3:

$$\Rightarrow p_{y(i)} = \frac{\left[(1 - p_x \alpha_x - c_{p1} - t)\left(\begin{array}{c}(r_{p1} + r_{p2})(\alpha_y - ab) + (r_{col1} + r_{col2})a \\ + (r_{pro1} + r_{pro2})ab - 2t\end{array}\right)\right]}{\left[2m + 2 + \alpha_y \left(\begin{array}{c}(r_{p1} + r_{p2})(\alpha_y - ab) + (r_{col1} + r_{col2})a \\ + (r_{pro1} + r_{pro2})ab - 2t\end{array}\right)\right]}, \quad \forall i \tag{10.10}$$

Case 4:

$$\Rightarrow p_{y(i)} = \frac{(1 - p_x \alpha_x - c_{p1} - t)\left(\left[1 - \sum_{i'=1}^{m} q_{i'} - c_{p2} - t\right] + \left(\begin{array}{c}r_{p2}(\alpha_y - ab) \\ + r_{col2}a + r_{pro2}ab - t\end{array}\right)\right)}{\left(2(m+1) + \alpha_y \left(\left[1 - \sum_{i'=1}^{m} q_{i'} - c_{p2} - t\right] + \left(\begin{array}{c}r_{p2}(\alpha_y - ab) \\ + r_{col2}a + r_{pro2}ab - t\end{array}\right)\right)\right)}, \quad \forall i \tag{10.11}$$

where subscripts 1 and 2 represent corresponding parameters for the first and second most efficient firm in each tier, respectively.

10.3.3 Stage 1

This stage leads to the equilibrium solutions of government's financial instruments, i.e., the green taxes levied on the producers and the subsidies offered to the RL suppliers.

According to the assumption, the government doesn't benefit financially and this is achieved by balancing green taxes and subsidies, i.e.,

$$s \times \sum_{i'=1}^{m} q_{i'} ab = t \times \sum_{i'=1}^{m} q_{i'}, \quad \forall i \qquad (10.12)$$

The equilibrium values for t and s have been derived by Sheu (2011) by maximizing the social welfare function. We just state here the equilibrium relation between t and s as obtained from balanced budget condition:

$$s \times ab = t \qquad (10.13)$$

10.4 ANALYSIS

This section analyzes the interdependence of various parameters present in the bargaining chain, i.e., the profits in each tier, the green tax levied by the government, the negotiation price, the production quantities at the equilibrium, and the number of producers. The green tax was varied from 0.2 to 0.7, while conducting the analysis and the preset values for other essential parameters are displayed in Table 10.2.

TABLE 10.2
Values of Parameters

Collection rate, a	0.8
Recycling rate, b	0.8
Number of competitive producers, m	2
Number of competitive RL suppliers, n	2
Unit price of original components, p_x	0.5
Quantities of original components for unit product productions, α_x	0.5
Quantities of recycled components for unit product productions, α_y	0.5
Unit cost of manufacturing, c_p	$c_{p1} = 0.03, \ c_{p2} = 0.04$
Unit cost of recycled component, r_p	$r_{p1} = 0.003, \ r_{p2} = 0.004$
Unit cost for EOL-product collection, r_{col}	$r_{col1} = 0.6, \ r_{col2} = 0.7$
Unit cost for processing EOL product, r_{pro}	$r_{pro1} = 0.5, \ r_{pro2} = 0.6$

1. *Financial instruments v/s negotiation agendas*:

   Based on the negotiation tactics, unit price ($p_{y(i)}$) and quantities ($y_{(i)}$) of recycled components ordered by producer $i(\forall i \in m)$ and supplied by RL suppliers j ($\forall j \in n$) are considered the two primary agendas of negotiation.

   Assuming the value of green tax (t) and green subsidy (s) being the variable parameter and treating values of a, b, α_x, α_y, r_p, c_p and unit costs as constants in all the four cases, the unit price of recycled components ($p_{y(i)}$) decreases with an increase in green tax (Figure 10.2). This is because, keeping other parameters constant, a rise in green tax tempts the producer

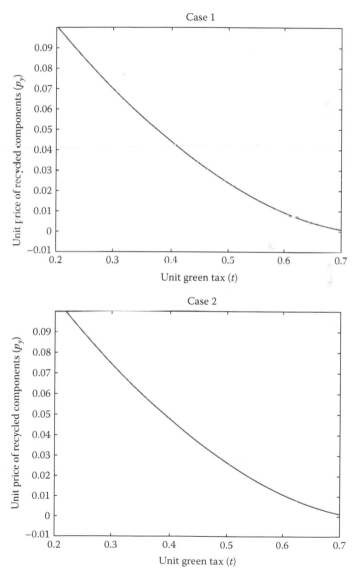

FIGURE 10.2   Unit price of recycled components v/s unit green tax.

to buy the recycled components at a lower price so as to maintain his profit margin. Also, a corresponding increase in subsidy (balanced budget condition) provided by the government causes the RL supplier to sell the recycled components at a lower price.

2. *Negotiation agendas vs. production quantities*:

Because of the governmental financial intervention, a rise in recycled component unit price ($p_{y(i)}$) results in decrease of the quantities of recycled components ($y_{(i)}$), which further results in a decrease of the producer production (q_i) (Figure 10.3). The producer production is an increasing

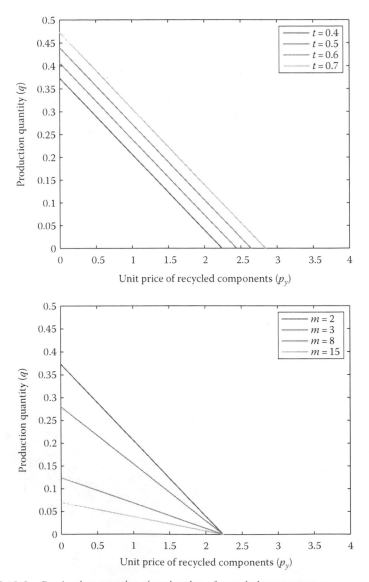

FIGURE 10.3   Production quantity v/s unit price of recycled component.

function of unit price of recycled components ($p_{y(i)}$). This remark is based on the derivative shown in Equations 10.14 and 10.15:

$$\frac{dq_i}{dp_{y(i)}} = \frac{-\alpha_y}{I+1} < 0, \quad \forall i \tag{10.14}$$

$$\frac{dq_i}{dy(i)} = \frac{-\alpha_y(dp_{y(i)}/dy(i))}{I+1} = \frac{1}{\alpha_y} > 0, \quad \forall i \tag{10.15}$$

Thus, with an increase in price of the recycled components ($p_{y(i)}$), the production quantity of the final product decreases. The amount and nature of decrease depends on a number of parameters like no. of producers (m), green tax (t), etc. A change in m leads to a change in slope of the $q_i - p_{y(i)}$ graph and the graph becomes steeper with an increase in number of producer firms. A change in t just shifts the $q_i - p_{y(i)}$ parallelly toward the origin.

3. *Financial instruments v/s profits*:
   The profit of a producer decreases continuously with increase in green tax as expected. On the other hand, the profit of a RL supplier continuously increases due to corresponding increase in government subsidy (balanced budget condition [Figure 10.4]).

4. *Bargaining chain parameters v/s financial instruments*:
   The net revenues (r) and the total cost of supply (c) are termed as bargaining chain parameters as they lead to the equilibrium solutions of the bargaining chains.
   A rise in green tax decreases the net revenue for the producer in spite of a corresponding increase in subsidy (balanced budget condition), which tends to increase the net revenue by decreasing $p_{y(i)}$. Similar is the case for total supply cost, which decreases due to increase in the value of subsidy provided by the government (Figure 10.5).

5. *Financial instruments v/s production quantities*:
   An increase in green tax per unit final product influences the total quantities to be manufactured and leads to the producer's decision of decrease in production quantities (Figure 10.6). This can be shown by the derivative obtained from Equation 10.16:

$$\frac{dq_i}{dt} = \frac{-1}{m+1} < 0, \quad \forall i \tag{10.16}$$

6. *Production quantities v/s number of producers*:
   The equilibrium production quantities decrease hyperbolically with an increase in the number of firms in production tier (Figure 10.7).

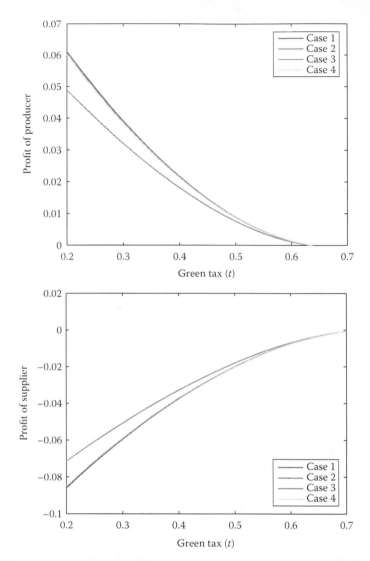

FIGURE 10.4 Profit v/s green tax.

10.5 CONCLUSION

This chapter analyzed a green supply chain consisting of a producer tier and a RL supplier tier both with an arbitrary number of firms. Through a three-stage game, this work elaborates and yields equilibrium solutions for negotiations between the producers and RL suppliers restricted by government's financial instruments—green tax and subsidies.

This work also provides numerical and graphical insight into correlation of the various parameters that decide the negotiation price at equilibrium.

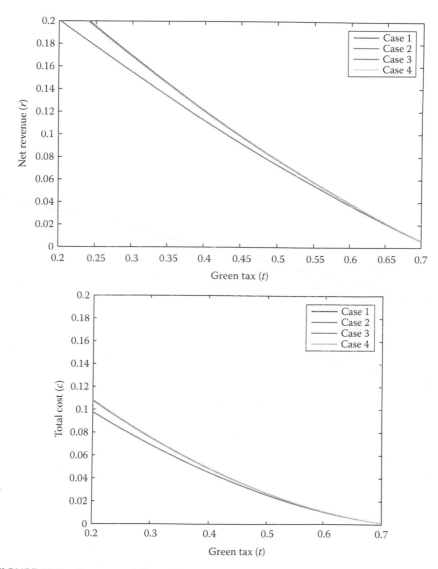

FIGURE 10.5 Green tax v/s bargaining chain parameters.

The financial instruments help the government in regulating the industrial activities responsible for environmental impact and hence are of paramount importance to the government but it should optimize its instruments so as to maximize the supply chain member profits. The green tax levied on the producer plays an important role in his decision of the equilibrium production quantities. The negotiation prices between the producer and RL supplier depend in turn on both of the aforementioned two decisions. This work formulates and provides equilibrium solutions to this problem using a BP bargaining model.

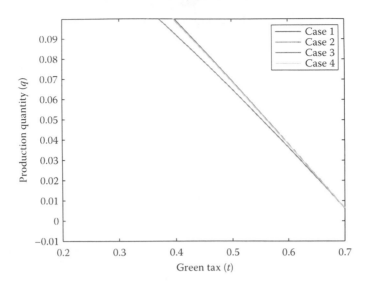

FIGURE 10.6   Production quantity v/s green tax.

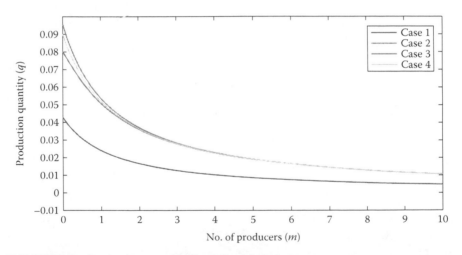

FIGURE 10.7   Production quantity v/s number of producers.

This model can be further extended to one involving the detailed regulations and clauses that are faced by the manufacturing sector and hence practical solutions can be obtained for the environmental problems considering the regional economics.

With more and more attention being drawn toward green sciences, this work attempts to fill certain voids in the field of GSCM. The chapter with the help of game theory aims at developing a scenario that is not only environmentally stable but also provides maximum economical benefit for the supply chain.

APPENDIX

Proof for Case 1 (Equation 10.8):

$$p_{y(i)} = \frac{1}{2}(r^1 + c^1)$$

$$= \frac{1}{2}\left(\left[1 - \sum_{i'=1}^{m} q_{i'} - c_{p1} - t\right] + r_{p1}(\alpha_y - ab) + r_{col1}a + r_{pro1}ab - s(ab)\right)$$

$$= \frac{1}{2}\left((1 - p_x\alpha_x - p_{y(i)}\alpha_y - c_{p1} - t)\left(1 - \sum_{i'=1}^{m} q_{i'} - c_{p1} - t\right)\right.$$

$$\left. + r_{p1}(\alpha_y - ab) + r_{col1}a + r_{pro1}ab - s(ab)\right)$$

Let $\left(\left(1 - \sum_{i'=1}^{m} q_{i'} - c_{p1} - t\right) + r_{p1}(\alpha_y - ab) + r_{col1}a + r_{pro1}ab - s(ab)\right)$ be K

$$\Rightarrow 2(m+1)p_{y(i)} = (1 - p_x\alpha_x - c_{p1} - t)K - p_{y(i)}\alpha_y K \quad (\because s(ab) = t)$$

$$\Rightarrow 2(m+1)p_{y(i)} + p_{y(i)}\alpha_y K = (1 - p_x\alpha_x - c_{p1} - t)K$$

$$\Rightarrow p_{y(i)}(2(m+1) + \alpha_y K) = (1 - p_x\alpha_x - c_{p1} - t)K$$

$$\Rightarrow p_{y(i)} = \frac{(1 - p_x\alpha_x - c_{p1} - t)K}{2(m+1) + \alpha_y K}$$

Replacing K in previous equation,

$$\Rightarrow p_{y(i)} = \frac{(1 - p_x\alpha_x - c_{p1} - t)\left(\begin{array}{c}\left[1 - \sum_{i'=1}^{m} q_{i'} - t\right] - c_{p1} \\ + r_{p1}(\alpha_y - ab) + r_{col1}a + r_{pro1}ab - t\end{array}\right)}{\alpha_y\left(\begin{array}{c}\left[1 - \sum_{i'=1}^{m} q_{i'} - t\right] - c_{p1} + r_{p1}(\alpha_y - ab) \\ + r_{col1}a + r_{pro1}ab - t\end{array}\right) + 2(m+1)}, \quad \forall i$$

Proof for Case 2 (Equation 10.9):

$$p_{y(i)} = \frac{1}{2}(r^1 + r^2)$$

$$= \frac{1}{2}\left(\left(q_i\left[1 - \sum_{i'=1}^{m} q_{i'} - c_{p1} - t\right] + \left(q_i\left[1 - \sum_{i'=1}^{m} q_{i'} - c_{p2} - t\right]\right)\right)\right)$$

$$= \frac{1}{2}\left(q_i\left(\left[1 - \sum_{i'=1}^{m} q_{i'} - c_{p1} - t\right] + \left[1 - \sum_{i'=1}^{m} q_{i'} - c_{p2} - t\right]\right)\right)$$

$$= \frac{1}{2}\left(q_i\left(2\left[1 - \sum_{i'=1}^{m} q_{i'} - t\right] - c_{p1} - c_{p2}\right)\right)$$

$$= \frac{1}{2}\left(\frac{(1 - p_x\alpha_x - p_{y(i)}\alpha_y - c_{p1} - t)}{m+1}\right)\left(2\left[1 - \sum_{i'=1}^{m} q_{i'} - t\right] - c_{p1} - c_{p2}\right)$$

Let $\left(2\left[1 - \sum_{i'=1}^{m} q_{i'} - t\right] - c_{p1} - c_{p2}\right)$ be K

$$\Rightarrow 2(m+1)p_{y(i)} = (1 - p_x\alpha_x - c_{p1} - t)K - p_{y(i)}\alpha_y K$$

$$\Rightarrow 2(m+1)p_{y(i)} + p_{y(i)}\alpha_y K = (1 - p_x\alpha_x - c_{p1} - t)K$$

$$\Rightarrow p_{y(i)}(2(m+1) + \alpha_y K) = (1 - p_x\alpha_x - c_{p1} - t)K$$

$$\Rightarrow p_{y(i)} = \frac{(1 - p_x\alpha_x - c_{p1} - t)K}{(2(m+1) + \alpha_y K)}$$

Replacing K in previous equation,

$$\Rightarrow p_{y(i)} = \frac{1}{2}\frac{\left[(1 - p_x\alpha_x - c_{p1} - t)\left(2\left[1 - \sum_{i'=1}^{m} q_{i'} - t\right] - (c_{p1} + c_{p2})\right)\right]}{\left[m + 1 - \alpha_y\left(\left[1 - \sum_{i'=1}^{m} q_{i'} - t\right] - \frac{1}{2}(c_{p1} + c_{p2})\right)\right]}, \quad \forall i$$

Proof for Case 3 (Equation 10.10):

$$P_{y(i)} = \frac{1}{2}(c^1 + c^2)$$

$$= \frac{1}{2}\left(q_i(r_{p1}(\alpha_y - ab) + r_{col1}a + r_{pro1}ab - s(ab)) + q_i\left(\begin{array}{c} r_{p2}(\alpha_y - ab) + r_{col2}a \\ + r_{pro2}ab - s(ab) \end{array} \right) \right)$$

$$= \frac{1}{2}(q_i((r_{p1} + r_{p2})(\alpha_y - ab) + (r_{col1} + r_{col2})a + (r_{pro1} + r_{pro2})ab - s(ab)))$$

$$= \frac{1}{2}\left(\frac{(1 - p_x\alpha_x - p_{y(i)}\alpha_y - c_{p1} - t)}{m+1} \right)\left(\begin{array}{c} (r_{p1} + r_{p2})(\alpha_y - ab) + (r_{col1} + r_{col2})a \\ + (r_{pro1} + r_{pro2})ab - s(ab) \end{array} \right)$$

Let $(r_{p1} + r_{p2})(\alpha_y - ab) + (r_{col1} + r_{col2})a + (r_{pro1} + r_{pro2})ab - s(ab)$ be K

$$\Rightarrow 2(m+1)p_{y(i)} = (1 - p_x\alpha_x \quad c_{p1} - t)K - p_{y(i)}\alpha_y K \ (\because s(ab) = t)$$

$$\Rightarrow 2(m+1)p_{y(i)} + p_{y(i)}\alpha_y K = (1 - p_x\alpha_x - c_{p1} - t)K$$

$$\Rightarrow p_{y(i)}(2(m+1) + \alpha_y K) = (1 - p_x\alpha_x - c_{p1} - t)K$$

$$\Rightarrow p_{y(i)} = \frac{(1 - p_x\alpha_x - c_{p1} - t)K}{(2(m+1) + \alpha_y K)}$$

Replacing K in previous equation,

$$\Rightarrow p_{y(i)} = \frac{\left[(1 - p_x\alpha_x - c_{p1} - t)\left(\begin{array}{c} (r_{p1} + r_{p2})(\alpha_y - ab) + (r_{col1} + r_{col2})a \\ + (r_{pro1} + r_{pro2})ab - 2t \end{array} \right) \right]}{\left[2m + 2 + \alpha_y\left(\begin{array}{c} (r_{p1} + r_{p2})(\alpha_y - ab) + (r_{col1} + r_{col2})a \\ + (r_{pro1} + r_{pro2})ab - 2t \end{array} \right) \right]}, \quad \forall i$$

Proof for Case 4 (Equation 10.11):

$$P_{y(i)} = \frac{1}{2}(r^2 + c^2)$$

$$= \frac{1}{2}\left(q_i\left[1 - \sum_{i'=1}^{m} q_{i'} - c_{p2} - t \right] + (r_{p2}(\alpha_y - ab) + r_{col2}a + r_{pro2}ab - s(ab)) \right)$$

$$= \frac{1}{2}\left(\frac{(1 - p_x\alpha_x - p_{y(i)}\alpha_y - c_{p1} - t)}{m+1} \right)$$

$$\times \left(\left[1 - \sum_{i'=1}^{m} q_{i'} - c_{p2} - t \right] + (r_{p2}(\alpha_y - ab) + r_{col2}a + r_{pro2}ab - s(ab)) \right)$$

$$\text{Let} \left(\left[1 - \sum_{i'=1}^{m} q_{i'} - c_{p2} - t \right] + \left(r_{p2}(\alpha_y - ab) + r_{col2}a + r_{pro2}ab - s(ab) \right) \right) \text{be } K$$

$$\Rightarrow 2(m+1)p_{y(i)} = (1 - p_x\alpha_x - c_{p1} - t)K - p_{y(i)}\alpha_y K \; (\because s(ab) = t)$$

$$\Rightarrow 2(m+1)p_{y(i)} + p_{y(i)}\alpha_y K = (1 - p_x\alpha_x - c_{p1} - t)K$$

$$\Rightarrow p_{y(i)}(2(m+1) + \alpha_y K) = (1 - p_x\alpha_x - c_{p1} - t)K$$

$$\Rightarrow p_{y(i)} = \frac{(1 - p_x\alpha_x - c_{p1} - t)K}{(2(m+1) + \alpha_y K)}$$

$$\Rightarrow p_{y(i)} = \frac{(1 - p_x\alpha_x - c_{p1} - t)\left(\left[1 - \sum_{i'=1}^{m} q_{i'} - c_{p2} - t \right] + \begin{pmatrix} r_{p2}(\alpha_y - ab) + r_{col2}a \\ + r_{pro2}ab - t \end{pmatrix} \right)}{\left(2(m+1) + \alpha_y \left(\left[1 - \sum_{i'=1}^{m} q_{i'} - c_{p2} - t \right] + \begin{pmatrix} r_{p2}(\alpha_y - ab) + r_{col2}a \\ + r_{pro2}ab - t \end{pmatrix} \right) \right)}, \; \forall i$$

REFERENCES

Agarwal, G., Barari, S., Tiwari, M.K., and Zhang, W.J. 2012. A decision framework for the analysis of green supply chain contracts: An evolutionary game approach. *Expert Systems with Applications* 39(3): 2965–2976.

Allen, D. T. and Allenby, B.R. 1994. *The Greening of Industrial Eco-systems*, Washington, DC: National Academic Press.

Ashley, S. 1993. Designing for the environment. *Mechanical Engineering* 115(3): 52–55.

Atasu, A., Van Wassenhove, L.N., and Sarvary, M. 2009. Efficient take-back legislation. *Production and Operations Management* 18(3): 243–258.

Brass, B. and McIntosh, M.W. 1999. Product, process, and organizational design for remanufacture—An overview of research. *Robotics and Computer-Integrated Manufacturing* 15(3): 167–178.

Cheng, J.-H., Yeh, C.-H., and Tu, C.-W. 2008. Trust and knowledge sharing in green supply chains. *Supply Chain Management: An International Journal* 13(4): 283–295.

Deming, W.E. 1982. *Quality, Productivity, and Competitive Position*, Cambridge, MA: Massachusetts Institute of Technology's Center for Advanced Engineering Study.

Feigenbaum, A.V. 1983. *Total Quality Control*, New York: McGraw-Hill Book Company.

Fortes, J. 2009. Green supply chain management: A literature review. *Otago Management Graduate Review* 7: 51–62.

Grilo, I. and Mertens, J.-F. 2009. Cournot equilibrium without apology: Extension and the Cournot inverse demand function. *Games and Economic Behavior* 65(1): 142–175.

Guide, V.D.R. and Srivastava, R. 1998. Inventory buffers in recoverable manufacturing. *Journal of Operations Management* 16: 551–568.

Gungor, A. and Gupta, S.M. 1999. Issues in environmentally conscious manufacturing and product recovery: A survey. *Computers & Industrial Engineering* 36: 811–853.

Hall, R.W. 1983. *Zero Inventories*, Homewood, IL: Dow-Jones-Irwin.

Ilgin, M.A. and Gupta, S.M. 2010. Environmentally conscious manufacturing and product recovery (ECMPRO): A review of the state of the art. *Journal of Environmental Management* 91(3): 563–591.

Kelle, P. and Silver, E.A. 1989. Forecasting the returns of reusable containers. *Journal of Operations Management* 8(1): 17–35.

Kreps, D.M. 1990. *Game Theory and Economic Modeling*, Oxford, U.K.: Clarendon Press.

Laan, E.A. van der, Salomon, M., and Dekker, R. 1996. Product remanufacturing and disposal: A numerical comparison of alternative control strategies. *International Journal of Production Economics* 45: 489–498.

Lovejoy, W.S. 2010. Bargaining chains. *Management Science* 56: 2282–2301.

Navin-Chandra, D. 1991. Design for environment ability. *Design Theory and Methodology* 31: 99–124.

Sarkis, J. and Cordeiro, J. 2001. An empirical evaluation of environmental efficiencies and firm performance: Pollution prevention versus end-of-pipe practice. *European Journal of Operational Research* 135: 102–113.

Schonberger, R.J. 1982. Japanese manufacturing techniques: Nine hidden lessons in simplicity. In *World Class Manufacturing*, pp. 157–179, New York: Free Press.

Sheu, J.-B. 2011. Bargaining framework for competitive green supply chains under governmental financial intervention. *Transportation Research Part E* 47: 573–592.

Simpson, D. and Samson, D. 2008. Developing strategies for green supply chain management. *Decision Line* 39(4): 12–15.

Srivastava, S. 2007. Green supply-chain management: A state-of-the-art literature review. *International Journal of Management Reviews* 9(1): 53–80.

Treleven, M. and Schweikhart, S.B. 1988. A risk/benefit analysis of sourcing strategies: Single Vs multiple sourcing. *Journal of Operations Management* 7: 93–114.

Wang, H.-F. and Gupta, S.M. 2011. *Green Supply Chain Management: Product Life Cycle Approach*, New York: McGraw Hill.

Zhang, H.C., Kuo, T.C., Lu, H., and Huang, S.H. 1997. Environmentally conscious design and manufacturing: A state of the art survey. *Journal of Manufacturing Systems* 16: 352–371.

11 Barrier Analysis to Improve Green in Existing Supply Chain Management

Mathiyazhagan Kaliyan, Kannan Govindan, and Noorul Haq

CONTENTS

11.1 INTRODUCTION

In the present manufacturing scenario, the manufacturing sectors around the globe have started to emphasize more on environmental issues for the products they manufacture to compete in the international market (Zhu and Sarkis, 2006). Nowadays, integrating environmental, social, and economic concerns of the region is one of the most crucial causes to improve green in manufacturing industries (Benoit and Comeau, 2005). The manufacturing sectors have realized that disregarding the final product they manufacture, they need to concentrate more on the entire supply chain of the product, which embraces activities from initial raw material procurement to the final disposal of the product. They should make sure that it should be well disposed into the environment.

This has led to the emergence of the concept called green supply chain management (GSCM; Srivastava, 2007).

The manufacturing product may be of any type, but the entire activity to produce the main products must be environment friendly, which will preserve and maintain the green supply chain. Even though various activities have been adopted by the manufacturers to improve the GSCM, some industries need to alter their technology or activity through which they will be able to improve GSCM to some more percentage. The main objective of the present work is to carry out a survey in various departments of different industries (30 industries have been taken) and considering 10 barriers. The data obtained from the survey were analyzed to find out the impacts of barriers for the implementation of GSCM in the existing supply chain management scenario.

11.2   LITERATURE REVIEW

Various researchers have gone in the field of green supply chain and created key factors related to green purchasing and developed a validation scale to measure the green procurement (Carter et al., 2000). Unlike the traditional environmental management, the concept of green supply chain assumes full responsibility of a firm toward its products from the extraction or acquisition of raw materials up to final use and disposal of products (Hart, 1995). A design should facilitate reuse of a product or part of it with minimal treatment of the used product (Sarkis, 1995). Hence, an integrated research has to be carried out in the field of supply chain management (Theyel, 2001) and various environmental indicators were found to evaluate the GSCM (Michelse et al., 2006).

The major identification of information flow demand in top management of GSCM and available opportunities for a GSCM for the small and medium enterprises have been discussed (Raymond et al., 2008). The co-operation among the internal and external systems has helped to improve the GSCM (Azzone et al., 1997). Certain regulations in industries were found to be a hindrance in the implementation of the GSCM (Porter and van der Linde, 1995). Every company has its own barriers and drivers and only a few barriers were found to be common among themselves (Zhu and Sarkis, 2006).

Carter and Dresner (2001) mentioned proper training has to be provided to every higher official of the department so that they may help the improvement of GSCM by implementing these activities in supply chain management and examine the positive or negative variation occurring in the industrial metabolism (Xu, 2007). Researchers who have studied the cases of integrated chain management have been rather optimistic about the possibilities of co-operation. The main barrier for implementing the GSCM is the mindset of the top-level management on sacrificing the cost (Srivastava, 2007).

The GSCM can reduce the ecological impact of industrial activity without sacrificing the cost, quality, and energy utilization (Montabon et al., 2007). Wang and Gupta (2011) pointed out a stage-by-stage production methodology for integrating environmental themes into supply chain management. This includes applying logical techniques to quantify the environmental impacts of supply chains and identifying opportunities for making improvements that lead to green engineering of a product. A thorough mining of the ingredients of the various literatures pertaining to GSCM reveals that there are various factors in industries that affect the implementation of GSCM. Hence, it is important to conduct a survey to identify the crucial barriers that obstruct the industries in implementing the GSCM and suggest possible ways to eradicate the barriers.

11.3   GREEN SUPPLY CHAIN MANAGEMENT

A number of authors have referred to the green supply chain over the past decade due to the emerging environmental management topics. Penfield (2007) defined green supply chain as "the process of using environmental friendly inputs and transforming these inputs through change agents—whose by products can be improved or recycled within the existing environment." This action moves up end product that can be regenerated and reprocessed at the end of their life-cycle and is creating a green supply chain. Ali and Kannan (2010) mentioned GSCM has emerged as an important organizational philosophy to reduce environmental risks. There are outwardly lots of benefits accruable to an organization from reverse logistics. It has been posited that despite being environmentally responsible, an effective management of reverse logistics operations will result in higher profitability through reduction in transportation, inventory, and warehousing costs (Ilgin and Gupta, 2010). The unanimous idea of a green supply chain is to bring down the prices, while aiding the environment.

The Chinese legal philosophy restricts all of the same contents that the European law does, but it allows no immunities and enforces laboratory testing and labeling requirements (Penfield, 2007). The advantages of GSCM are (1) positive impact on financial performance; (2) effective management of suppliers; (3) dissemination of technology, advanced techniques, capital, and knowledge among the chain partners; (4) transparency of the supply chain; (5) large investments and risks getting shared among partners in the chain; (6) better control of product safety and quality; (7) increase of sales; and (8) finding beneficial uses of waste.

Zhu et al. (2005) mentions that GSCM is strongly related to interorganizational environmental topics such as industrial ecosystems, industrial ecology, product life cycle analysis, extended producer responsibility, and product stewardship. In a broader sense, GSCM also falls within the purview of the burgeoning literature of ethics and sustainability, which incorporates other social and economic influences.

Srivastava (2007) points out that "green" component to supply chain management involves addressing the influence and correlation between supply chain management and natural environment. Organizations implementing GSCM practices have to improve their environmental and financial performance, investment recovery, and ecodesign or design for environmental practices (Zhu and Sarkis, 2004). The literature in GSCM has been growing as organizations and researchers begin to realize that the management of environmental programs and operations do not end at the boundaries of the organization (Zhu et al., 2005).

11.3.1 Benefits of GSCM

Greening the supply chain provides a range of benefits from cost reduction to integrating suppliers in a participative decision-making process that promotes environmental innovation in a firm (Rao, 2002).

The benefits or positive outcomes of adopting GSCM are classified into four categories based on an extensive literature survey: environmental outcomes, economic outcomes, operational outcomes, and intangible outcomes (Eltayeb et al., 2010). Environmental outcomes include the effects of green supply chain initiatives on the natural environment inside and outside the firm. They include reduction of solid/liquid wastes, reduction of emissions and decrease of consumption for hazardous materials, decrease of frequency of environmental accidents, and improved employee and community health (Eltayeb et al., 2010; Zhu and Sarkis, 2004).

Economic outcomes include financial benefits that relate to the whole organization such as profitability, sales, market share, and productivity (Fuentes et al., 2004). They also include the reduction in the cost of material purchasing and energy consumption, reduced cost of waste treatment and discharge, and avoiding a fine in the case of environmental accidents (Zhu and Sarkis, 2004). Operational outcomes include benefits that reflect on the operational level of the organization such as cost reductions, quality, flexibility, and delivery. Intangible outcomes include benefits that are conceptual or difficult to quantify outcomes such as organizational image and customer satisfaction (Eltayeb et al.. 2010; Smith, 2005). Ten reasons have been identified for a company to adopt green supply chain such as target marketing, sustainability of resources, lowered costs/increased efficiency, product differentiation and competitive advantage, competitive and supply chain pressures, adapting to regulation and reducing risk, brand reputation, return on investment, employee morale, and the ethical imperative (Duber-Smith, 2005).

Stevels (2002) demonstrated the benefits of GSCM to different roles of supply chain, including environment and society in terms of different categories: material, immaterial, and emotion. For material, GSCM helps lower environmental defilement, lower cost prices for supplier, lower cost for producer, lower cost of ownership for customer, and less consumption of resources for society. In terms of "immaterial," GSCM helps in overcoming prejudice and cynicism toward environment, less rejects for supplier, easier manufacturing for producer, convenience for customer, and better compliance for society. For "emotion," GSCM helps the motivation of stakeholder for environment, better image for supplier and producer, gives better quality of life for customer, and makes the industry go on the right track for society.

11.4 INTEGRATING ISO 14001 AND GSCM

ISO14001 is an international standard developed by the International Organization for Standardization to preserve and safeguard the environment. It is applicable to any business or organization, regardless of size, location, or income (Andreas and Suji, 1998). The ISO 14001 certified companies should employ environmental aspects both in internal operations and the supply chain (Green et al., 1998).

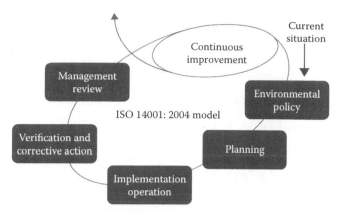

FIGURE 11.1   Frame work of ISO 14001: 2004.

The standard model depicted in Figure 11.1 strongly follows a sequence of operations that efficiently monitors the environmental safety and performance. It is one of the illustrations of ISO 14000 family.

The primary step is to frame an environmental policy. Successive planning is undertaken to calculate the methods of implementing it and the barriers for the policy. Then follows the implementation operation. It is the vital step since the planning is put to practical purpose. The verification of the functioning and any needed corrective actions are done. Finally, a management review is stipulated for a follow up. The ISO 14001: 2004 program does not end there. It gives room for continuous improvement, thus closing the loop and continuously improving the whole chain.

The GSCM mainly involves the following three activities:

1. Prevention of pollution
2. Continuously improving the product quality and service with green
3. Recycling, reusing, and minimizing the waste

To find out the extent up to which the aforementioned activities have been effectively undertaken, there is a need for a single standard tool. The ISO 14001 obviously solves the purpose. Hence, the companies certified with this standard are near to Green implementation, and their products and services, help to implement GSCM with minimum cost investment (Arimura et al., 2009). The GSCM and ISO 14001 are both concerned about the environment performance goals; hence, their mutuality and coupling are important.

This standard is the reason for the increase in focus on environmental management system (EMS) reducing waste and pollution, while simultaneously improving overall performance. ISO 14001 is concerned with processes involving the creation, management, and elimination of pollution (Melnyk et al., 2003). It is believed that the facilities that have certified ISO 14001 may be able to implement GSCM at a lower cost. The skills, management practices, and overall effort required to certify ISO 14001 complement the skills, practices, and effort required for GSCM. The standard includes EMS, auditing, performance evaluation, labeling, life cycle assessment, and product standards (Tibor and Feldman, 1996).

11.5 BARRIERS IN GREEN SUPPLY CHAIN MANAGEMENT

Zhang et al. (2009) summarized SMEs are less active in adopting environmental management initiatives than larger companies. The identified list of barriers to the implementation of environmental initiatives based on some aspects of financial resources include lack of funding for environmental projects or very long return on investment periods (Vernon et al., 2003). SMEs are lacking human resources both in quantity and in technical knowledge to pursue environmental management (Perron, 2005).

Van Hemel and Cramer (2002) mentioned the following barriers for the implementation of GSCM in SMEs: "no clear environmental benefit," "not perceived as responsibility," and "no alternative solution is available." They are difficult to deal with through a regulatory approach due to their diverse nature, geographical spread, and often their lack of formal organization (Perron, 2005). Green supply chain initiatives lack legitimacy to some managers and are often dismissed as the ploy to appease customers. A questionnaire-based survey was conducted to rank these enablers. They found contextual relationships among enablers and developed a hierarchy-based model for the enablers by using ISM (Luthra et al., 2010; Mudgal et al., 2009). From the literature, it is evident that there is no work on barrier impact analysis. This chapter addresses the gap of barriers impact analysis for the implementation of GSCM found from the literature.

11.6 PROBLEM DESCRIPTION

From the south Indian region, 70 industries have been chosen for the research. After thoroughly surveying, it was found that only 30 varieties of industries from the Electrical, Electronic, Textiles, Computer, Automobile with the certification of ISO14001 were shortlisted and found suited for the research work, because these industries are leading and well known in Indian market and close to green and have the awareness of environmental issues.

Due to environmental legislation and customer pressure, industries have started to implement green supply chain concept in their industries. From the literature, there is no work in analysis of barrier impacts (eradication of barrier). Analysis of barriers will help them to achieve the green supply chain concept in their industries. Barrier analysis is done based on the expert's opinion from the 6 departments that were classified according to their functional areas in these 30 industries. Ten barriers were selected from literature (Fawcett and Gregory, 2001). Table 11.1 shows the 10 barriers taken for the work and a succinct description about their importance. These 10 barriers play a crucial role in affecting GSCM implementation and effective operation in existing supply chain condition.

TABLE 11.1
Barriers Description

Barrier	Definition
1. Lack of top management support	The existing system can be modified only with the support of top management from all points of consideration, but the main objective is that it should not have any negative impact on the system
2. Non-aligned strategic and operating philosophies	The proper system definition that the company adopts should be clear to all the levels of the people involved in the system
3. Inability or unwillingness to share information	Communication lag is a potential problem. When the flow of information is affected due to carelessness or unwillingness, there is big disturbance in the process flow and no improvement will be possible
4. Lack of trust among supply chain members	Lack of trust is a psychological barrier. It is very harmful since it is concerned with a single individualistic attitude. Lack of trust decelerates the progress
5. An unwillingness to share risks and rewards	Risk taking is a common trait for success. Failing to take risks means failing to improve and sometimes even failing to progress. Calculated risks should be taken at all levels. The rewards should also be shared, not enjoyed selfishly. Rewards motivate workers
6. Inflexible organizational systems and processes	The organization structure must be designed in a manner to rearrange its form to any forthcoming changes
7. Cross-functional conflicts and "turf" protection	Each one should share the knowledge openly in all conditions. It should not be postponed to any further stages
8. Inconsistent/inadequate performance measures	Performance measurement is an important criterion to evaluate performance. Improper tools and methods result in poor measures
9. Resistance to change	Change is inevitable for progress. Reluctance to change is a negative mindset that curbs the interest and improvement potential of an industry
10. Lack of training for new mindsets and skills	Continuous improvement is a compulsion. It is achieved by means of training. Training for skill and attitude is essential to survive and excel Failing so will render one unfit in an organization

11.7 SOLUTION METHODOLOGY

The aim of this research work is to identify the barriers that have major and minor impacts in the implementation of GSCM. Initially, the survey was carried out among the experts of 6 departments, which were classified according to their functional areas in these 30 industries. Every industrial setup embraces its own set of departments based on its requirements, and it is quite evident that all the industries will have similar departments with a common agenda.

Hence the following departments alone have been chosen for the survey: Design, Planning, Procurement, Manufacturing, Quality, Health, and Safety Environmental

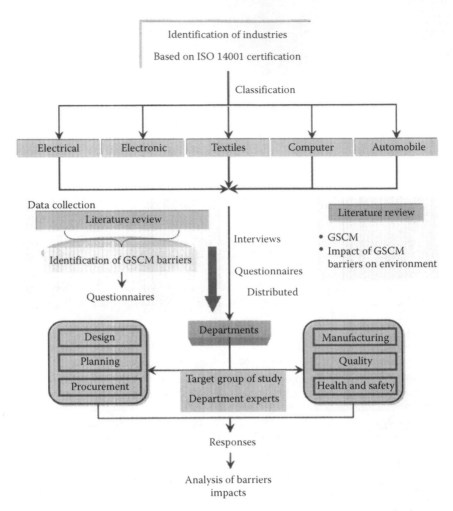

FIGURE 11.2 Schematic representation of GSCM barrier analysis (own contribution).

Department (HSED). Experts from the department were given Barriers validation card to score the barrier. Each department was interlinked to adopt GSCM and to make the entire chain green.

Figure 11.2 shows the methodology of this article in brief. Six different departments, namely, design, planning, procurement, manufacturing, quality, and HSED from 30 ISO 14001 certified companies are taken. Ten barriers that influence the implementation of GSCM are considered.

11.8   DATA COLLECTION

The managers were asked to provide a barrier score value between 1 and 10. If the value of the barrier lies between 1 and 4, it means the barrier can be easily changed and if the value lies between 5 and 7, it can be eradicated within a short period of time. If the value lies between 8 and 10, it is a strong barrier. The industries always concentrate to eradicate maximum impact barriers. When the maximum impact barriers are eradicated, it will be easily possible to eradicate average impact barriers in the value of 5–7. For this study, only minimum and maximum values of barrier level 1–4 and 8–10 have been considered. In this work, 10 barriers were taken into account and a survey was conducted in each of the 6 Departments of the 30 industries and hence a total of 1800 values {10 barriers × 6 Departments × 30 industries = 1800} were recorded.

11.9   RESULTS AND DISCUSSION

In this chapter, the results are grouped in accordance with three aspects mainly:

1. Possibilities to improve a department to be green
2. Minimum impact of barriers in the departments
3. Maximum impact of barriers in the departments

Initially, a barrier's validation card as in Table 11.2 was prepared and distributed to the experts of 6 departments in 30 industries. By adding the score of each barrier per department for 30 industries, the cumulative values are found and tabulated in Table 11.3.

From Table 11.3, the bar graph of the Sum of Barriers in each barrier is drawn and the same is shown in Figure 11.3, which shows the extent of each barrier per department in the 30 industries.

To find out the potentiality of each department to becoming green, the mean values of barriers are calculated from the cumulated values of Table 11.3 and the same is shown in Figure 11.4.

The department with more count of mean values between 1 and 4 has more possibility of getting green, since these barriers may be eradicated easily. From the Figure 11.4 it can be noted that the implementation of GSCM based on the impact of barriers can be discussed as; the first barrier B-1 influences the design and HSED departments on an average level and its impact is medium, because the top managements accept the GSCM needs. Planning, manufacturing, quality, and HSED departments have been strongly impacted by barrier B-2 because of poor operating philosophies. It cannot be eradicated without stringent action.

The Quality and HSED departments have high impact, whereas the planning and procurement departments have average impact of barrier on implementing GSCM as far as barrier B-3 is concerned. This is probably due to the lack of exchanging the information among the industrial people. Considering barrier B-4, the design and manufacturing departments have high-barrier impact because of lack of confidence within the supply chain members.

In the case of barrier B-5, design, procurement, quality, and HSED have high-barrier impact. This is considered to be due to the lack of motivation among the subordinates and less encouragement for their innovations. As far as barrier B-6 is concerned, the design department has the most impact and the four departments, namely, planning, procurement, manufacturing, and quality have the average impact

TABLE 11.2
Barrier Validation Card

	Departments					
Barrier	**Design**	**Planning**	**Procurement**	**Manufacturing**	**Quality**	**(HSED)**
1. Lack of top management support						
2. Nonaligned strategic and operating philosophies						
3. Inability or unwillingness to share information						
4. Lack of trust among supply chain members						
5. An unwillingness to share risks and rewards						
6. Inflexible organizational systems and processes						
7. Cross-functional conflicts and "turf" protection						
8. Inconsistent/ inadequate performance measures						
9. Resistance to change						
10. Lack of training for new mindsets and skills						

on implementing the GSCM. This is expected to be due to not adopting the new trends and technologies. The departments of planning and quality are having high impact, whereas procurement, manufacturing, and HSED are observed to be having the average impact of B-7. The reason is that the departments sometimes infringe on the concept of green.

Regarding the barrier B-8, the design, planning, and quality departments have the average effect, whereas the departments of manufacturing and HSED possess high effect. Inadequate/insufficient tools/systems/methodologies present in the industries

TABLE 11.3
Cumulative Value of Barriers

Departments	Barriers									
	B[b]-1	B-2	B-3	B-4	B-5	B-6	B-7	B-8	B-9	B-10
Design	196	159	128	297	292	277	91	201	81	207
Planning	112	262	192	111	121	221	252	214	241	88
Procurement	98	190	216	216	234	161	224	126	284	147
Manufacturing	98	299	81	258	117	172	163	242	173	108
Quality	90	228	233	151	229	171	250	213	104	290
HSED[a]	213	270	274	219	277	126	184	274	104	199

[a] Health and Safety Environmental Department.
[b] Represents the barrier.

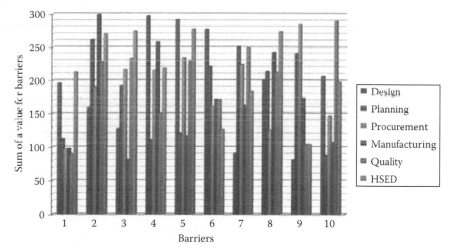

FIGURE 11.3   Sum of barriers in each barrier.

Departments	Barriers										Mean Value
	B-1	B-2	B-3	B-4	B-5	B-6	B-7	B-8	B-9	B-10	
Design	7	5	4	10	10	9	3	7	3	7	6.5
Planning	4	9	6	4	4	7	8	7	8	3	6.3
Procurement	3	6	7	7	8	5	7	4	9	5	6.1
Manufacturing	3	10	3	9	4	6	5	8	6	4	5.8
Quality	3	8	8	5	8	6	8	7	3	10	6.6
HSED	7	9	9	7	9	4	6	9	3	7	7.0

Note: Shading indicates the significance of the difference.

FIGURE 11.4   Mean value.

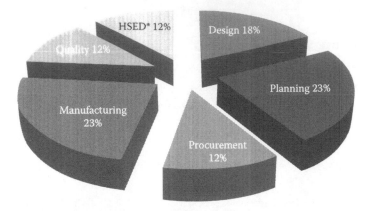

FIGURE 11.5   Department's possibilities to be green.

are responsible for barrier B-8. Considering barrier B-9, the planning and procurement departments have the high impact because of high resistance from the people to the adoption for progress. In barrier B-10, the department of quality is having high impact, and design, procurement, and HSED departments have the average impact in implementing the GSCM. This may be due to insufficient training related to green environment.

Further, it can be clearly observed from Figure 11.4 that all the departments have the average impact of barriers for implementation of GSCM. On improvement of the enterprise resource planning system (ERP), there will be a flexible mode of communication, which is encouraged by the top management that will be highly helpful for eradicating the barriers B-3, B-7, and B-9 for the design department. From prediction through the Figure 11.5, it can be seen that the design department has 18% possibilities to become green and B-1, B-2, B-8, and B-10 also can be eradicated within a short time by the planning department.

The minimum contributing barriers are B-1, B-4, B-5, and B-10 because the planning has a very good relationship among the departments. On the other hand, the procurement department plays a vital role in the organization justifying the organization's profit. That's why it has minimum number of strong barriers and 12% possibilities to become green. The quality and HSED also have 12% possibilities of getting green but have a lot of strong impacting barriers because these two involve significant initial changes. But once introduced into the system, it becomes highly flexible, but still a lot of training and expenditure are required and adopting those regulations is a competitive work.

If we consider the order of rank for the possibility of the departments going green, manufacturing and planning departments share the first rank. Hence the remaining departments start to adopt the system; the manufacturing will automatically comply with them. But the working environment is going to be tough for a while till adopting and practicing green become well applied and settled.

In Figure 11.6, the barrier B-1 has maximum possibilities to eradicate and convert the system into green. Among the rest of the barriers B-3, B-5, B-9, and B-10 also have the flexibility to convert the system into green. Further, the departments need some more alteration to reach these possibilities.

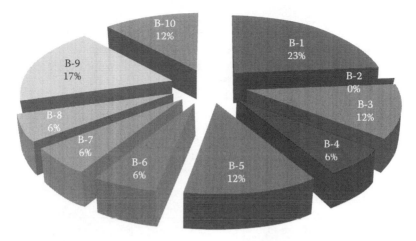

FIGURE 11.6   Minimum impacts of barriers in departments.

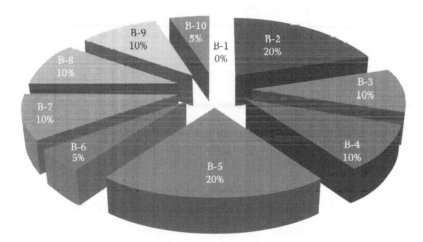

FIGURE 11.7   Maximum impacts of barriers in departments.

In Figure 11.7, it is clearly graphed that B-2 and B-5 act as a strong barrier for the implementation of GSCM. To eradicate this barrier, the organizations may require a lot of changes in the system and it is not possible within a short time. But it can be overcome with continual improvement over a long period.

11.10   CONCLUSIONS

From this study, the barriers impacting the implementation of GSCM at maximum and minimum level are shown in Figures 11.6 and 11.7. From the figures the possibilities of industries close to green can be seen based on department's expert's opinion. Still some industries are struggling to implement the GSCM concept from the existing supply chain condition, and also from this chapter it is evident that industries are not speeding up the GSCM concept in their industries. It evaluates the scope for improvement of each department toward going green. Further, it is recommended to examine these barriers with the environmental evidence to confirm their real impact on the ecological condition of an industry to make the system green.

ACKNOWLEDGMENT

The second author is supported by a grant from Forsknings- og Innovationsstyrelsen for "The International Network programme" (1681448).

REFERENCES

Ali, D and Kannan, G, 2010. An analysis of drivers affecting the implementation of green supply chain management, *Resources Conservation and Recycling*, 55(6), 659–667.

Andreas, S and Suji U, 1998. ISO 14001—Implementing an Environmental Management System (Version 2.02), ELLIPSON, Management Consultants Leonhardsgraben 52 CH-4051 Basel Switzerland.

Arimura, TH, Hibiki, A, and Katayama, H, 2008. Is a voluntary approach an effective environmental policy instrument? A case for environmental management systems. *Journal of Environmental Economics and Management*, 55, 281–295.

Azzone, G, Bertelè, U, and Noci, G, 1997. At last we are creating environmental strategies which work, *Long Range Planning*, 30(4), 478–479, 562–571.

Benoit, G and Comeau, A, 2005. A sustainable future for the Mediterranean: The blue plan's environment and development outlook, *Routledge*, pp. 1–462.

Carter, CR and Dresner, M, 2001. Purchasing's role in environmental: Cross-functional development of grounded theory, *Journal of Supply Chain Management*, 37(3), 12–27.

Carter, CR, Kale, R, and Grimm, CM, 2000. Environmental purchasing and firm performance: An empirical investigation, *Transportation Research Part E: Logistics and Transportation Review*, 36(3), 219–228.

Duber-Smith, DC, 2005. The green imperative, *Soap, Perfumery, and Cosmetics*, 78(8), 24–26.

Eltayeb, TK, Zailani, S, and Ramayah, T, 2010. Green supply chain initiatives among certified companies in Malaysia and environmental sustainability: Investigating the outcomes, *Resources, Conservation and Recycling*, 55(5), 495–506.

Fawcett, SE and Gregory, MM, 2001. Achieving world-class supply chain alignment: Benefits, barriers, and bridges, Center for Advanced Purchasing Studies Arizona State University Research Park, pp. 1–159.

Fuentes, MM, Albacete-Saez, CA, and Llorens-Montes, FJ, 2004. The impact of environmental characteristics on TQM principles and organizational performance, *International Journal of Management Science*, 32, 425–442.

Green, K, Morton, B, and New, S, 1998. Purchasing and environmental management: Interactions, policies and opportunities, *Business Strategy and the Environment 1996*, 5(3), 188–197.

Hart, SL, 1995. A natural-resource-based view of the firm, *The Academy of Management Review*, 20(4), 986–1014.

Ilgin, MA and Gupta, SM, 2010. Environmentally conscious manufacturing and product recovery (ECMPRO): A review of the state of the art, *Journal of Environmental Management*, 91(3), 563–591.

Luc, H, 2010. *A Sustainable Future for the Mediterranean: The Blue Plan's Environment and Development Outlook* by Guillaume Benoit and Aline Comea, *International Journal of Environment and Pollution*, 40(1/2/3), 299–300.

Luthra, S, Manju, KS, and Haleem, A, 2010. Suggested implementation of the green supply chain management in automobile industry of India: A review. *Proceedings of National Conference on Advancements and Futuristic Trends of Mechanical and Industrial Engineering*, November, GITM, Bilaspur, India, pp. 12–13.

Melnyk, SA, Sroufe, RP, and Calantone, R, 2003. Assessing the impact of environmental management systems on corporate and environmental performance, *Journal of Operations Management*, 21, 329–351.

Michelse, O, Fet, AM, and Dahlsrud, A, 2006. Eco-efficiency in extended supply chains: A case study of furniture production, *Journal of Environmental Management*, 79(3), 290–297.

Montabon, F, Sroufe, R, and Narasimhan, R, 2007. An examination of corporate reporting, environmental management practices and firm performance, *Journal of Operations Management*, 25(5), 998–1014.

Mudgal, RK, Shankar, R, Talib, P, and Raj, T, 2009. Greening the supply chain practices: An Indian perspective of enablers' relationship, *International Journal of Advanced Operations Management*, 1(2 and 3), 151–176.

Penfield, PW, 2007. *The Green Supply Chain—Sustainability Can Be a Competitive Advantage*, Material Handling Industry of America, School of Management, Syracuse University, New York.

Perron, GM, 2005. *Barriers to Environmental Performance Improvements in Canadian SMEs*, Dalhousie University, Nova Scotia, CA.

Porter, ME and van der Linde, C, 1995. Green and competitive: Ending the stalemate, *Harvard Business Review*, 73(5), 120–134.

Rao, P, 2002. Greening of the supply chain: A new initiative in South East Asia, *International Journal of Operations and Production Management*, 22(6), 632–655.

Raymond, P, Côté, LJ, Marche, S, Geneviève, MP, and Ramsey, W, 2008. Influences, practices and opportunities for environmental supply chain management in Nova Scotia SMEs, *Journal of Cleaner Production*, 16(15), 1561–1570.

Sarkis, J, 1995. Supply chain management and environmentally conscious design and manufacturing, *International Journal of Environmentally Conscious Design and Manufacturing*, 4(2), 43–52.

Smith, AD, 2005. Reverse logistics programs: Gauging their effects on CRM and online behavior, *VINE: The Journal of Information and Knowledge Management Systems*, 35(3), 166–181.

Srivastava, SK, 2007. Green supply-chain management: A state-of-the-art literature review, *International Journal of Management Review*, 9(1), 53–80.

Stevels, A, 2002. Green supply chain management much more than questionnaires and ISO 14001, *IEEE International Symposium on Electronics and the Environment*, 96–100.

Theyel, G, 2001. Customer and supplier relations for environmental performance, *Greener Management International*, 35, 61–69.

Tibor, T and Feldman, I, 1996. *ISO 14001: A Guide to the New Environmental Management Standards*, Irwin, Burr Ridge, IL.

Van Hemel, C and Kramer, J, 2002. Barriers and stimuli for eco-design in SMEs, *Journal of Cleaner Production*, 10, 439–453.

Vernon, J, Essex, S, Pinder, D, and Curry, K, 2003. The 'greening' of tourism micro-businesses: Outcomes of focus group investigations in South East Cornwall, *Business Strategy and the Environment*, 12, 49–69.

Wang, H-F and Gupta, SM, 2011. *Green Supply Chain Management: Product Life Cycle Approach*, McGraw Hill, New York, ISBN: 978-0-07-162283-7.

Xu, D-w, 2007. An evaluation of the influence of enterprises with green supply chain management on the environment, *Journal of Shanxi Datong University (Natural Science)*, 1(2), 48–51.

Zhang, B, Bi, J, and Liu, B, 2009. Drivers and barriers to engage enterprises in environmental management initiatives in Suzhou Industrial Park, China, *Frontiers of Environmental Science and Engineering in China*, 3(2), 210–220.

Zhu, Q and Sarkis, J, 2004. Relationships between operational practices and performance among early adopters of green supply chain management practices in Chinese manufacturing enterprises, *Journal of Operations Management*, 22, 265–289.

Zhu, Q and Sarkis, J, 2006. An inter-sectoral comparison of green supply chain management in China: Drivers and practices, *Journal of Cleaner Production*, 14(5), 472–486.

Zhu, Q, Sarkis, J, and Geng, Y, 2005. Green supply chain management in China: Pressures, practices and performance, *International Journal of Operations and Production Management*, 25, 449–468.

12 River Formation Dynamics Approach for Sequence-Dependent Disassembly Line Balancing Problem

Can B. Kalayci and Surendra M. Gupta

CONTENTS

12.1 INTRODUCTION

Product recovery seeks to obtain materials and parts from old or outdated products through recycling and remanufacturing in order to minimize the amount of waste sent to landfills. Gungor and Gupta (1999) and Ilgin and Gupta (2010) provide an extensive review of product recovery. See Wang and Gupta (2011) and Ilgin and Gupta (2012) for more information on remanufacturing and green supply chain management. The first crucial and the most time-consuming step of product recovery is disassembly. Disassembly is defined as the systematic extraction of valuable parts and materials from discarded products through a series of operations to use in remanufacturing or recycling after appropriate cleaning and testing operations. Disassembly operations can be performed at a single workstation, in a disassembly cell or on a disassembly line. Although a single workstation and disassembly cell are more flexible, the highest productivity rate is provided by a disassembly line and hence is the best choice for automated disassembly processes, a feature that will be essential in the future disassembly systems (Gungor and Gupta 2001, 2002). Disassembly operations have unique characteristics and cannot be considered as the reverse of assembly operations. The quality and quantity of components used in the stations of an assembly line can be controlled by imposing strict conditions. However, there are no such conditions of EOL products moving on a disassembly line. In a disassembly environment, the flow process is divergent; a single product is broken down into many subassemblies and parts, while the flow process is convergent in an assembly environment. There is also a high degree of uncertainty in the structure, quality, reliability, and condition of the returned products in disassembly. Additionally, some parts of the product may be hazardous and may require special handling that will affect the utilization of disassembly workstations. Since disassembly tends to be expensive, disassembly line balancing becomes significant in minimizing resources invested in disassembly and maximizing the level of automation. Disassembly line balancing problem (DLBP) is a multiobjective problem that is described in Güngör and Gupta (2002) and has mathematically been proven to be NP-complete (Tovey 2002) in (McGovern and Gupta 2007a) making the goal to achieve the optimal balance computationally expensive. Exhaustive search works well enough in obtaining optimal solutions for small-sized instances; however, its exponential time complexity limits its application on the large-sized instances. An efficient search method needs to be employed to attain a (near) optimal condition with respect to objective functions. Although some researchers have formulated the DLBP using mathematical programming techniques (Altekin 2005, Koc et al. 2009, Altekin and Akkan 2012), it quickly becomes unsolvable for a practical sized problem due to its combinatorial nature. For this reason, there is an increasing need to use metaheuristic techniques such as genetic algorithms (GA) (McGovern and Gupta 2007a, Kalayci and Gupta 2011a), ant colony optimization (ACO) (McGovern and Gupta 2005, 2006, 2007b, Agrawal and Tiwari 2008, Tripathi et al. 2009, Ding et al. 2010), simulated annealing (SA) (Kalayci et al. 2011b, 2012), tabu search (TS) (Kalayci and Gupta 2011b), and artificial bee colony (ABC) (Kalayci et al. 2011a). See McGovern and Gupta (2011) for more information on DLBP. In this chapter, we consider a sequence-dependent DLBP and solve it by a river formation dynamics (RFD) approach. In the literature, RFD approach has been used to design heuristic algorithms (Rabanal et al. 2007); to find minimum spanning/distance trees (Rabanal et al. 2008b); to solve dynamic traveling salesman problem (Rabanal et al. 2008a), NP-complete problems (Rabanal et al. 2009), steiner tree problem (Rabanal et al. 2011); and for testing restorable systems (Rabanal et al. 2012).

The rest of the chapter is organized as follows. In Section 12.2, notation used in this chapter is presented. Problem definition and formulation is given in Section 12.3. Section 12.4 describes the proposed RFD algorithm for the multiobjective SDDLBP. The computational experience to evaluate its performance on a numerical example is provided in Section 12.5. Finally, some conclusions are pointed out in Section 12.6.

12.2 NOTATIONS

α	relative importance of the pheromone trail in path selection
β	relative importance of heuristic information in path selection
η	heuristic information (visibility) of task j (i.e., the priority rule value for task j)
ω	up gradient coefficient for climbing drops
ρ	evaporation coefficient
τ_0	initial pheromone level
τ_{ij}	pheromone trail intensity in the path "selecting task j after selecting task i"
θ	flat gradient coefficient for climbing drops
a	ant count $(1, \ldots, an)$
$altitude_i$	altitude of part i
an	number of ants
AS_i^a	the set of assignable tasks for ant a after the selection of task i
AV	available task list
c	cycle time (Maximum time available at each workstation)
$cumulatedSediment$	the amount of sediment carried by the drop dr
d_i	demand; quantity of part i requested
$decreasingGradient_{ij}$	gradient value between part i and j
$DOWN_i^{dr}$	the set of parts that are neighbors of part i that can be visited by drop dr and connected through a down gradient
dn	number of drops
dr	drop count $(1, \ldots, dn)$
$erosion_{ij}$	erosion created between part i and part j
$erosionProduced$	the sum of erosions introduced in all graph in the previous phase
$fitness_{ij}$	fitness value of the edge connecting part i and part j
$FLAT_i^{dr}$	the set of parts that are neighbors of part i that can be visited by drop dr and connected through an up gradient
h_i	binary value; 1 if part i is hazardous, else 0
i	part identification, task count $(1, \ldots, n)$
IP	set (i,j) of parts such that task i must precede task j
j	part identification, task count $(1, \ldots, n)$
j_1	part identification, task count $(1, \ldots, n)$
j_2	part identification, task count $(1, \ldots, n)$
j_3	part identification, task count $(1, \ldots, n)$
k	workstation count $(1, \ldots, m)$
LB	line balance solution
m	number of workstations required for a given solution sequence
m^*	minimum possible number of workstations
M	sufficiently large number
n	number of parts for removal
N	the set of natural numbers

notClimbingFactor	the variable used to decide whether drop *dr* can climb upward gradients
paramBlockedDrop	parameter to deposit sediment whenever a drop *dr* is blocked to climb
paramErosion	parameter of the erosion process
P_{ij}	transition rule: Probability of a part *j* being selected after part *i* from available task set (*AV*)
PD_{dr}	probability of a drop *dr* can climb toward upward gradients
PS_i	ith part in a solution sequence
q_0	selection probability parameter for the use of heuristic information
q_1	selection probability parameter for the use of pheromone trail information
Q	constant parameter (amount of pheromone added if a path is selected)
r	uniformly distributed random number between 0 and 1
sd_{ij}	sequence dependent time increment influence of *i* on *j*
$|SUC_j|$	the number of all successors of task *j*
t_i	part removal time of part *i*
t_i'	part removal time of part *i* considering sequence dependent time increment
totalGradient	sum of the values of down, up and flat gradients
UP_i^{dr}	the set of parts that are neighbors of part *i* that can be visited by drop *dr* and connected through a up gradient

12.3 PROBLEM DEFINITION AND FORMULATION

The sequence-dependent disassembly line balancing problem (SDDLBP) investigated in this chapter is concerned with the paced disassembly line for a single model of product that undergoes complete disassembly. Model assumptions include the following: a single product type is to be disassembled on a disassembly line; the supply of the end-of-life product is infinite; the exact quantity of each part available in the product is known and constant; a disassembly task cannot be divided between two workstations; each part has an assumed associated resale value, which includes its market value and recycled material value; the line is paced; part removal times are deterministic, constant, and discrete; each product undergoes complete disassembly even if the demand is zero; all products contain all parts with no additions, deletions, modifications, or physical defects; each part is assigned to one and only one workstation; the sum of part removal times of all the parts assigned to a workstation must not exceed cycle time; and the precedence relationships among the parts must not be violated.

The difference between DLBP and SDDLP is task time interactions. As opposed to the DLBP, in SDDLBP whenever a task interacts with another task, their task times may be influenced. For example, consider the disassembly of a personal computer, where several components have to be disassembled at the same workstation or neighboring ones. Disassembling a particular component before another component from the same motherboard may prolong (or curtail) the task time, as opposed to disassembling them in reverse order, because one component could hinder the other because it requires additional movements and/or prevents it from using the most efficient disassembly process.

In order to model this very typical situation adequately, the concept of sequence-dependent task time increments is introduced. If task j is performed before task i, its standard time t_j is incremented by sd_{ij}. This sequence-dependent increment measures the prolongation of task j forced by the interference of already waiting task i. Obviously, tasks i and j only can interact in the described manner if do not have any precedence relationships, i.e., there is no path in the precedence graph directly connecting i and j.

Illustrative example: The precedence relationships (solid line arrows) and sequence-dependent time increments (dashed line arrows) for an eight part PC disassembly process are illustrated in Figure 12.1 and their knowledge database is given in Table 12.1. This example is modified from Güngör and Gupta (2002).

Sequence dependencies for the PC example are given as follows: $sd_{23} = 2$, $sd_{32} = 4$, $sd_{56} = 1$, $sd_{65} = 3$.

For a feasible sequence $\langle 1,2,3,6,5,8,7,4 \rangle$; since part 2 is disassembled before part 3, sequence dependency $sd_{32} = 4$ takes place because when part 2 is disassembled, the

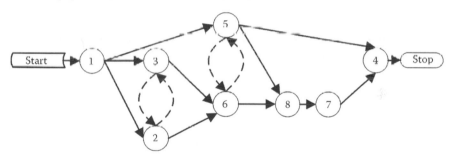

FIGURE 12.1 Precedence relationships (solid line arrows) and sequence dependent time increments (dashed line arrows) for the PC example.

TABLE 12.1
Knowledge Database for the PC Example

Part	Task	Time	Hazardous	Demand
PC top cover	1	14	No	360
Floppy drive	2	10	No	500
Hard drive	3	12	No	620
Back plane	4	18	No	480
PCI cards	5	23	No	540
RAM modules	6	16	No	750
Power supply	7	20	No	295
Motherboard	8	36	No	720

obstacle part 3 is still not taken out, i.e., the part removal time for part 2 is increased, which results in $t_2' = t_2 + sd_{32} = 14$; similarly, since part 6 is disassembled before part 5, sequence dependency $sd_{56} = 1$ takes place, because when part 6 is disassembled, the obstacle part 5 is still not taken out, i.e., the part removal time for part 6 is increased, which results in $t_6' = t_6 + sd_{56} = 17$.

For another feasible sequence $\langle 1,3,2,5,6,8,7,4 \rangle$ with the same PC example; since part 3 is disassembled before part 2, sequence dependency $sd_{23} = 2$ takes place because when part 3 is disassembled, the obstacle part 2 is still not taken out, i.e., the part removal time for part 3 is increased, which results in $t_3' = t_3 + sd_{23} = 14$; since part 5 is disassembled before part 6, sequence dependency $sd_{65} = 3$ takes place because when part 5 is disassembled, the obstacle part 6 is still not taken out, i.e., the part removal time for part 5 is increased, which results in $t_5' = t_5 + sd_{65} = 26$.

The mathematical formulation of our SDDLBP is given as follows:

In this chapter, the precedence relationships considered are of AND type and are represented using the immediately preceding matrix $[y_{ij}]_{n \times n}$, where

$$
y_{ij} = \begin{cases} 1 & \text{if task } i \text{ is executed after task } j \\ 0 & \text{if task } i \text{ is executed before task } j \end{cases} \tag{12.1}
$$

In order to state the partition of total tasks, we use the assignment matrix $[x_{jk}]_{n \times m}$, where

$$
x_{jk} = \begin{cases} 1 & \text{if part } j \text{ is assigned to station } k \\ 0 & \text{otherwise} \end{cases} \tag{12.2}
$$

The mathematical formulation of SDDLBP is given as follows:

$$
\min f_1 = m \tag{12.3}
$$

$$
\min f_2 = \sum_{i=1}^{m} (c - t_i')^2 \tag{12.4}
$$

$$
\min f_3 = \sum_{i=1}^{n} i \times h_{PS_i}, \quad h_{PS_i} = \begin{cases} 1 & \text{hazardous} \\ 0 & \text{otherwise} \end{cases} \tag{12.5}
$$

$$
\min f_4 = \sum_{i=1}^{n} i \times d_{PS_i}, \quad d_{PS_i} \in N, \forall PS_i \tag{12.6}
$$

Subject to

$$\sum_{k=1}^{m} x_{jk} = 1, \quad j = 1,\ldots,n \tag{12.7}$$

$$\left\lceil \frac{\sum_{i=1}^{n} t_i}{c} \right\rceil \leq m^* \leq n \tag{12.8}$$

$$\sum_{j=1}^{n} \left(t_j + \sum_{i=1}^{n} sd_{ij} \times y_{ij} \right) \times x_{jk} \leq c \tag{12.9}$$

$$x_{ik} \leq \sum_{k=1}^{m} x_{jk}, \quad \forall (i,j) \in IP \tag{12.10}$$

The first objective given in Equation 12.3 is to minimize the number of workstations for a given cycle time (the maximum time available at each workstation) (Baybars 1986). It rewards the minimum number of workstations, but allows the unlimited variance in the idle times between workstations because no comparison is made between station times. It also does not force to minimize the total idle time of workstations.

The second objective given in Equation 12.4 is to aggressively ensure that idle times at each workstation are similar, though at the expense of the generation of a nonlinear objective function (McGovern and Gupta 2007a). The method is computed based on the minimum number of workstations required as well as the sum of the square of the idle times for all the workstations. This penalizes solutions where, even though the number of workstations may be minimized, one or more have an exorbitant amount of idle time when compared to the other workstations. It also provides for leveling the workload between different workstations on the disassembly line. Therefore, a resulting minimum performance value is the more desirable solution indicating both a minimum number of workstations and similar idle times across all workstations.

As the third objective (see Equation 12.5), a hazard measure is developed to quantify each solution sequence's performance, with a lower calculated value being more desirable (McGovern and Gupta 2007a). This measure is based on binary variables that indicate whether a part is considered to contain hazardous material (the binary variable is equal to 1 if the part is hazardous, else 0) and its position in the sequence. A given solution sequence hazard measure is defined as the sum of hazard binary flags multiplied by their position number in the solution sequence, thereby rewarding the removal of hazardous parts early in the part removal sequence.

As the fourth objective (Equation 12.6), a demand measure was developed to quantify each solution sequence's performance, with a lower calculated value being more desirable (McGovern and Gupta 2007a). This measure is based on positive integer values that indicate the quantity required of a given part after it is removed (or 0 if it is not desired) and its position in the sequence. A solution sequence demand measure is then defined as the sum of the demand value multiplied by the position of the part in the sequence, thereby rewarding the removal of high-demand parts early in the part removal sequence.

The constraints given in Equation 12.7 ensure that all tasks are assigned to at least and at most one workstation (the complete assignment of each task), Equation 12.8 guarantees that the number of work stations with a workload does not exceed the permitted number, Equation 12.9 ensures that the work content of a workstation cannot exceed the cycle time, and Equation 12.10 imposes the restriction that all the disassembly precedence relationships between tasks should be satisfied.

12.4 PROPOSED RIVER FORMATION DYNAMICS APPROACH

The decision version of DLBP was proven to be NP-complete (and hence, the optimization version is NP-hard) (McGovern and Gupta 2007a). Since SDDLBP is a generalization of DLBP (setting all sequence-dependent time increments to zero, SDDLBP reduces to DLBP), the decision version of SDALBP is NP-complete and its optimization versions are NP-hard, too.

Since SDDLBP falls into the NP-Complete class of combinatorial optimization problems, when the problem size increases, the solution space is exponentially increased and an optimal solution in polynomial time cannot be found as it can be time consuming for optimum seeking methods to obtain an optimal solution within this vast search space. Therefore, it is necessary to use alternative methods in order to reach (near) optimal solutions faster. In this regard, nature has inspired many heuristic algorithms to obtain reasonable solutions to complex problems. For this reason, a recent metaheuristic, RFD approach is used to solve the problem.

12.4.1 SOLUTION REPRESENTATION

One of the most important decisions in designing a metaheuristic lies in deciding how to represent solutions and relate them in an efficient way to the searching space. Also, solution representation should be easy to decode to reduce the cost of the algorithm. In the proposed ACO algorithm, permutation-based representation is used, so elements of a solution string are integers. Each element represents a task assignment to work station. The value of the first element of the array shows which task is assigned to workstations first, the second value shows which task is assigned second, and so on. For example, if there are eight tasks to be assigned to workstations, then the length of the solution string is 8. Figure 12.2 illustrates assignment of tasks to workstations as an example.

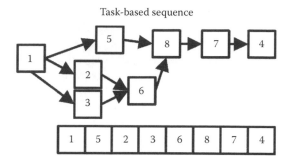

FIGURE 12.2 Assignment of tasks to workstations.

12.4.2 Feasible Solution Construction Strategy

The strategy of building a feasible balancing solution is the key issue to solve the line balancing problems. We use station-oriented procedure for a solution constructing strategy in which solutions are generated by filling workstations successively one after the other (Ding et al. 2010). The procedure is initiated by the opening of a first station. Then, tasks are successively assigned to this station until more tasks cannot be assigned and a new station is opened. In each iteration, a task is selected according to the probabilistic selection criteria from the set of candidate tasks to assign to the current station. When no more tasks may be assigned to the open station, this is closed and the following station is opened. The procedure finalizes when there are no more tasks left to assign. In order to describe the process to build a feasible balancing solution, available task and assignable task are defined as follows: A task is an available task if and only if it has not already been assigned to a workstation and all of its predecessors have already been assigned to a workstation. A task is an assignable task if and only if it belongs to the set of available task and the idle time of current workstation is higher than or equal to the processing time of the task.

The generation procedure of a feasible balancing solution is given as follows:

Step 0: Initialization; open the first workstation for task assignment.

Step 1: According to the precedence constraints, construct the available task set.

Step 2: According to the cycle time, construct the assignable task set.

Step 3: If the set of candidate task is null, go to Step 5.

Step 4: Select a task from the available task set according to the transition rule and assign the task to the current workstation; go back to Step 1.

Step 5: If the set of available task is null, go to Step 7.

Step 6: Open a new workstation, go back to Step 1.

Step 7: Stop the procedure.

Although this is a multicriteria problem, only a single criterion (the measure of balance, f_2) is being used in the basic RFD calculations and trail selection. The other objectives are only considered after balance and then only at the completion of each cycle, not as part of the probabilistic node selection. That is, a drop's tour solution is produced based on f_2, while at the end of each cycle the best overall solution is then updated based on f_1, f_2, f_3, f_4 in that order. This is done because for the purpose of this study, the balance is considered to be the dominant requirement, as well as to be consistent with previous multicriteria DLBP studies by the authors (McGovern and Gupta 2006).

12.4.3 Main Steps of the Proposed Method

RFD is a recent swarm-based metaheuristic optimization algorithm introduced by Rabanal et al. (2007). It is based on copying how water forms rivers by eroding the ground and depositing sediments. The basic scheme of the modified RFD method is given as follows:

Start

Step 1: Read disassembly data, initialize parameters, construct vectors and matrices.

Iterative loop:

Step 2:

    2.1 Construct a complete assignment for each drop:

    Repeat

        Move drops (Apply state transition rule to select the next task)

        Erode paths

        Deposit sediments

    Until a complete assignment is constructed

    2.2 Analyze paths and record the best solution found so far

Step 3: If the maximum number of iterations is realized, then STOP; else go to Step 2

End

The (move drops) phase of RFD consists of moving the drops across the nodes of the graph in a partially random way. The following transition rule defines the probability that a drop dr at a part i chooses the part j to move next:

$$P_{ij} = \begin{cases} \dfrac{decreasingGradient_{ij}}{totalGradient} & \text{if } j \in DOWN_i^{dr} \\[2ex] \dfrac{\omega / \left| decreasingGradient_{ij} \right|}{totalGradient} & \text{if } j \in UP_i^{dr} \\[2ex] \dfrac{\theta}{totalGradient} & \text{if } j \in FLAT_i^{dr} \\[2ex] 0 & \text{otherwise} \end{cases} \qquad (12.11)$$

where

$$totalGradient = \left(\sum_{j_1 \in DOWN_i^{dr}} decreasingGradient_{ij_1} \right) + \left(\sum_{j_1 \in UP_i^{dr}} decreasingGradient_{ij_1} \right)$$

$$+ \left(\sum_{j_1 \in FLAT_i^{dr}} \theta \right) \tag{12.12}$$

$DOWN_i^{dr}$, UP_i^{dr}, and $FLAT_i^{dr}$ are the sets of parts that are neighbors of part i that can be visited by the drop dr and are connected through a down, up, and flat gradient, respectively, ω and θ are considered to be the parameters of the algorithm and $decreasingGradient_{ij}$ represents the gradient between parts i and j and is defined as follows:

$$decreasingGradient_{ij} = \frac{altitude_j - altitude_i}{fitness_{ij}} \tag{12.13}$$

where
  $altitude_i$ is the altitude of the part i
  $fitness_{ij}$ is the length of the edge connecting part i and part j

Drops climb increasing slopes with a low probability. This probability will be inversely proportional to the increasing gradient, and it will be reduced during the execution of the algorithm to improve the search of good paths. Enabling drops to climb increasing slopes with some low probability allows us to find alternative choices and enables the exploration of other paths in the graph. If a drop dr fails to be considered as a climbing drop then it only considers taking down or flat gradients. Given a drop dr located at part i, we randomly decide whether dr can climb upward gradients according to the following probability:

$$PD_{dr} = \frac{1}{notClimbingFactor} \tag{12.14}$$

At the beginning of the algorithm, the altitude of all parts is the same, so *totalGradient* is 0, so the probability of a drop moving through an edge with zero gradient is set to some (non null) value. This enables drops to spread around a flat environment, which is required, in particular, at the beginning of the algorithm.

In the next (erode paths) phase of the algorithm, paths are eroded according to the movements of drops in the previous phase. In particular, if a drop moves from part i to part j, then we erode i. The reduction of the altitude of this node depends on the current gradient between i and j. In particular, the erosion is higher if the downward gradient between i and j is high. The altitude of the eroded node i is changed as follows:

$$altitude_i = altitude_j - erosion_{ij} \qquad (12.15)$$

where

$$erosion_{ij} = \frac{paramErosion}{(n-1)dn} \cdot decreasingGradient_{ij} \qquad (12.16)$$

where
 paramErosion is a parameter of the erosion process
 n is the number of parts of the graph
 dn is the number of drops used in the algorithm

If the edge is flat or increasing, then less erosion is performed.

Once the erosion process finishes, the altitude of all nodes of the graph is slightly increased (deposit sediments). The objective is to avoid, after some iterations, the erosion process leading to a situation where all altitudes are close to 0, which would make gradients negligible and would ruin all formed paths. In particular, the altitude of a part i is increased according to the following expression:

$$altitude_i = altitude_i + \frac{erosionProduced}{n-1} \qquad (12.17)$$

where
 erosionProduced is the sum of erosions introduced in all graph nodes in the previous phase
 n is the number of parts

Individual drops deposit sediment when all movements available for a drop imply climbing an increasing slope and the drop fails to climb any edge (according to the probability assigned to it). In this case, the drop is blocked and it deposits the sediments it is transporting. This increases the altitude of the current node in proportion to the cumulated sediment carried by the drop, which in turn is proportional to the erosions produced by the drop in previous movements. If a drop gets blocked at part i, then the altitude of i is increased as follows:

$$altitude_i = altitude_i + paramBlockedDrop \cdot cumulatedSediment \qquad (12.18)$$

where
 paramBlockedDrop is a parameter
 cumulatedSediment is the amount of sediment carried by the drop

Finally, all solutions are found by drops and stores the best solution found so far (analyze paths).

Proposed RFD algorithm is compared to ACO algorithm, which is a popular approach for solving combinatorial optimization problems in the literature. It was first introduced by Dorigo et al. (1996).

The structure of the ACO algorithm is given in the following steps:

Step 1: Start.

Step 2: Read disassembly data, initialize parameters, construct vectors and matrices, and start iteration.

Step 3: Create a new ant $(a = a + 1)$.

Step 4: Open kth station $(k = k + 1)$.

Step 5: Form the assignable task list.

Step 6: Form the available task list.

Step 7: Determine all task(s) selection probability in the available task list according to the global pheromone quantities, positional weight values of the tasks. Generate a random number (r) and choose the jth task randomly whose cumulative probability satisfies the following rule:

$$
j = \begin{cases}
j_1 : \arg\max_{j \in AS_i^a} \left\{ (\tau_{ij})^\alpha (\eta_j)^\beta \right\} & \text{if } 0 \leq r \leq q_0 \quad \text{(exploitation)} \\[2ex]
j_2 : P_{ij} = \dfrac{(\tau_{ij})^\alpha (\eta_j)^\beta}{\sum\limits_{j_2 \in AS_i^a} (\tau_{ij_2})^\alpha (\eta_{j_2})^\beta} & \text{else if } q_0 < r < q_1 \quad \text{(biased exploration)} \\[2ex]
j_3 : \text{Random select } j \in AS_i^a & \text{else if } q_1 < r \leq 1 \quad \text{(random selection)}
\end{cases}
$$

$$(12.19)$$

where

$$
\eta_j = \frac{t_j}{c} + \frac{|SUC_j|}{\text{Max}_{1 \leq i \leq n} \left\{ |SUC_i| \right\}}
\tag{12.20}
$$

Step 8: Assign the selected task to the kth station.

Step 9: Update local pheromone level for the chosen task and assigned station by the following rule:

$$
\tau_{ij} = (1 - \rho)\tau_{ij} + \rho\tau_0
\tag{12.21}
$$

Step 10: If all tasks are not assigned to workstations, go to Step 5.

Step 11: If all ants are not created, go to Step 3.

Step 12: Calculate the objective function values. If better than the best solution is found then update the best solution.

Step 13: Calculate remaining time. If finished, go to Step 15.

Step 14: Update global pheromone level by the following rule:

$$\tau_{ij} = (1-\rho)\tau_{ij} + \rho\Delta\tau_{ij} \qquad (12.22)$$

where

$$\Delta\tau_{ij} = \begin{cases} Q/f_2 & \text{if } (i,j) \in \text{best schedule} \\ 0 & \text{otherwise} \end{cases} \qquad (12.23)$$

Step 15: Stop.

Parameters used (best found according to the full factorial experimental results):

$an = n$, $Q = 1$, $\alpha = 1$, $\beta = 0.2$, $\rho = 0.5$, $q_0 = 0.1$, $q_1 = 0.9$

12.5   NUMERICAL RESULTS

The proposed algorithm was coded in MATLAB® and tested on Intel Core2 1.79 GHz processor with 3 GB RAM. After engineering, the program is investigated on two different scenarios for verification and validation purposes. Best parameters set for the proposed RFD approach are as the following: $\omega = 0.1$, $\theta = 1$, $dn = n$, $notClimbingFactor = 1$, $paramBlockedDrop = 1$, $paramErosion = 1$.

The first scenario for a given product consists of $n = 10$ components. The knowledge database and precedence relationships for the components are given in Table 12.2 and Figure 12.3, respectively. The problem and its data were modified

TABLE 12.2
Knowledge Database for the 10 Part Product

Task	Time	Hazardous	Demand
1	14	No	0
2	10	No	500
3	12	No	0
4	17	No	0
5	23	No	0
6	14	No	750
7	19	Yes	295
8	36	No	0
9	14	No	360
10	10	No	0

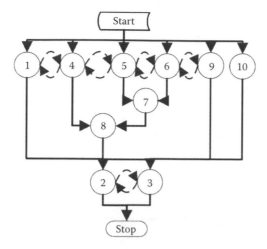

FIGURE 12.3 Precedence relationships (solid line arrows) and sequence dependent time increments (dashed line arrows) for the 10 part product.

from McGovern and Gupta (2006) with a disassembly line operating at a speed that allows $c = 40$ s for each workstation to perform its required disassembly tasks.

The sequence dependencies for the 10 part product are given as the following: $sd_{14} = 1$, $sd_{23} = 2$, $sd_{32} = 3$, $sd_{41} = 4$, $sd_{45} = 4$, $sd_{54} = 2$, $sd_{56} = 2$, $sd_{65} = 4$, $sd_{69} = 3$, $sd_{96} = 1$.

Here the objective is to create feasible solutions for the complete disassembly of the product. The proposed approach rapidly found optimal solutions in an exponentially large search space (as large as $10! = 3,628,800$). Table 12.3 depicts an optimal

TABLE 12.3

Optimal Solution Sequence for 10-Part Product Disassembly

		Workstations				
		I	II	III	IV	V
↑ Part removal sequence	6	14(+2+1)				
	1	14(+4)				
	10		10			
	5		23(+4)			
	7			19		
	4			17		
	8				36	
	9					14
	2					10(+3)
	3					12
Total time		35	37	36	36	39
Idle time		5	3	4	4	1

(Time to remove parts (s))

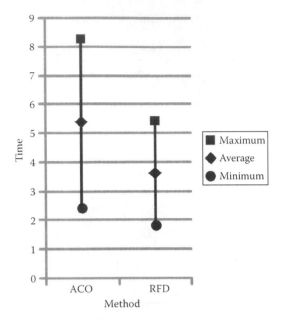

FIGURE 12.4 Time performance comparison for 10 part disassembly within 90% confidence interval.

solution sequence. The fitness function values of the optimal solution are found to be $f_1 = 5, f_2 = 67, f_3 = 5, f_4 = 9605$. According to this sequence, sequence-dependent time increments sd_{56}, sd_{96}, sd_{41}, sd_{45}, sd_{32} are added to the part removal times of part 6, 6, 1, 5, 2, respectively.

While the exhaustive search method was able to find optimal solution in $215t$ time, RFD approach reaches $<4t$ time on average. As it can be seen in Figure 12.4, although RFD tends to reach optimal solutions faster than ACO on average, there is not a statistically significant difference in terms of time performance within 90% confidence interval.

The second scenario is for a cellular telephone instance that consists of $n = 25$ components. The knowledge database and precedence relationships for the components are given in Table 12.4 and Figure 12.5, respectively. The problem and its data were modified from Gupta et al. (2004) with a disassembly line operating at a speed that allows $c = 18$ for each workstation to perform its required disassembly tasks.

The sequence dependencies for the 25 part product are given as the following:

$$sd_{45} = 2, sd_{54} = 1, \ sd_{67} = 1, \ sd_{69} = 2, \ sd_{76} = 2, \ sd_{78} = 1, \ sd_{87} = 2, \ sd_{96} = 1,$$

$$sd_{13,14} = 1, \ sd_{14,13} = 2$$

$$sd_{14,15} = 2, \ sd_{15,14} = 1, \ sd_{20,21} = 1, \ sd_{21,20} = 2, \ sd_{22,25} = 1, \ sd_{25,22} = 2$$

TABLE 12.4

Knowledge Base of Cellular Telephone Instance

Part	Task	Part Removal Time	Hazardous	Demand
Antenna	1	3	Yes	4
Battery	2	2	Yes	7
Antenna guide	3	3	No	1
Bolt (type 1) A	4	10	No	1
Bolt (type 1) B	5	10	No	1
Bolt (type 2) 1	6	15	No	1
Bolt (type 2) 2	7	15	No	1
Bolt (type 2) 3	8	15	No	1
Bolt (type 2) 4	9	15	No	1
Clip	10	2	No	2
Rubber seal	11	2	No	1
Speaker	12	2	Yes	4
White cable	13	2	No	1
Red/blue cable	14	2	No	1
Orange cable	15	2	No	1
Metal top	16	2	No	1
Front cover	17	2	No	2
Back cover	18	3	No	2
Circuit board	19	18	Yes	8
Plastic screen	20	5	No	1
Keyboard	21	1	No	4
LCD	22	5	No	6
Sub-keyboard	23	15	Yes	7
Internal IC board	24	2	No	1
Microphone	25	2	Yes	4

Since within the vast search space (25!), the exhaustive search is useless due to the exponential growth of the time complexity, i.e., the optimal solution is unknown. A typical solution found by RFD and ACO is given in Figure 12.6 as the best performance.

Given $500t$ time for both algorithms, the fitness performance comparison of RFD and ACO is given in Figures 12.7 through 12.10, respectively. As can be seen in Figure 12.7, minimum, average, and maximum values of first objective overlap each other. Both algorithms found 10 as the minimum number of workstations that has the biggest priority among all other objectives. Since the second objective value is the next best performance criterion, as it can be seen from Figure 12.8, RFD algorithm finds statistically significantly better results than ACO algorithm within 90% confidence interval. It is not necessary to check the next two objectives for a better understanding of the fitness performance since second objective can

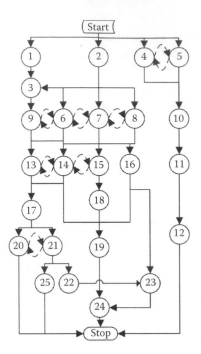

FIGURE 12.5 Precedence relationships (solid line arrows) and sequence dependent time increments (dashed line arrows) for the cellular telephone instance.

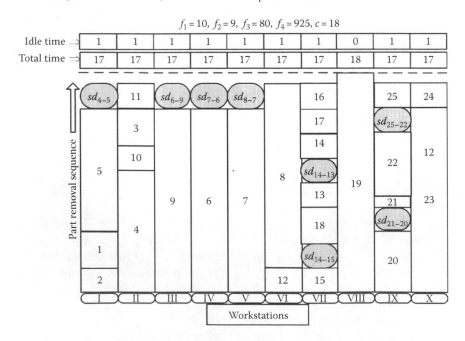

FIGURE 12.6 A typical solution found using the cellular telephone instance.

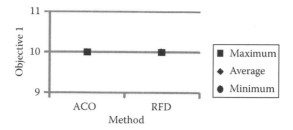

FIGURE 12.7 Objective 1 performance comparison for 25 part disassembly within 90% confidence interval.

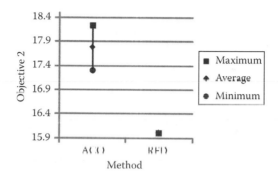

FIGURE 12.8 Objective 2 performance comparison for 25 part disassembly within 90% confidence interval.

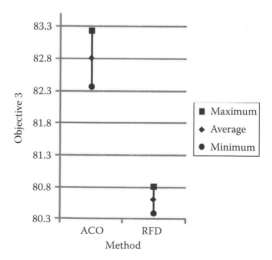

FIGURE 12.9 Objective 3 performance comparison for 25 part disassembly within 90% confidence interval.

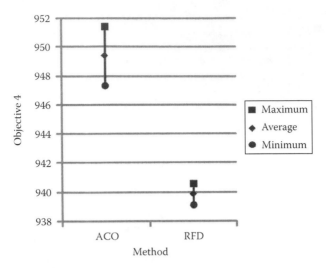

FIGURE 12.10 Objective 4 performance comparison for 25 part disassembly within 90% confidence interval.

TABLE 12.5
Detailed Average and Standard Deviation Values for Two Scenarios

Objective	Method	Average		Standard Deviation	
		p10	p25	p10	p25
f_1	ACO	5	10	0	0
	RFD	5	10	0	0
f_2	ACO	67	17.77	0	1.41
	RFD	67	16	0	0
f_3	ACO	5	82.8	0	1.32
	RFD	5	80.6	0	0.62
f_4	ACO	9605	949.37	0	6.31
	RFD	9605	939.83	0	2.29
t	ACO	5.36	244.67	5.61	161.49
	RFD	3.62	222.25	3.45	119.24

TABLE 12.6
Detailed Standard Error and Confidence Interval Values for Two Scenarios

		p10			p25		
Objective		Standard Error	90% Confidence Interval		Standard Error	90% Confidence Interval	
f_1	ACO	0	5	5	0	10	10
	RFD	0	5	5	0	10	10
f_2	ACO	0	67	67	−0.46	17.30	18.23
	RFD	0	67	67	0	16	16
f_3	ACO	0	5	5	−0.44	82.36	83.24
	RFD	0	5	5	−0.20	80.40	80.80
f_4	ACO	0	9605	9605	−2.07	947.29	951.44
	RFD	0	9605	9605	−0.75	939.08	940.59
t	ACO	−2.92	2.44	8.27	−53.12	191.55	297.80
	RFD	−1.79	1.83	5.41	−39.23	183.02	261.47

clearly show the performance difference. Finally: according to our test results, it can be said that RFD algorithm showed better performance than ACO algorithm for solving the SDDLBP (Tables 12.5 and 12.6).

12.6   CONCLUSIONS

The main objective of this chapter was to solve SDDLBP, which aimed to minimize the number of disassembly workstations, minimize the total idle time of all workstations by ensuring similar idle time at each workstation considering sequence-dependent time increments, maximize the removal of hazardous components as early as possible in the disassembly sequence, and maximize the removal of high-demand components before low-demand components. A fast, near-optimal, RFD approach was modified, developed, and presented in this chapter to solve multiobjective SDDLBP. Also, the performance comparison with ACO method had been provided in terms of time complexity and fitness measures. It was clearly shown that RFD approach showed better performance than ant colony optimization.

REFERENCES

Agrawal, S. and Tiwari, M. K. 2008. A collaborative ant colony algorithm to stochastic mixed-model U-shaped disassembly line balancing and sequencing problem. *International Journal of Production Research*, 46(6), 1405–1429.

Altekin, F. T. 2005. Profit oriented disassembly line balancing. PhD dissertation, Middle East Technical University, Ankara, Turkey.

Altekin, F. T. and Akkan, C. 2012. Task-failure-driven rebalancing of disassembly lines. *International Journal of Production Research*, 46(10), 1–22.

Baybars, I. 1986. A survey of exact algorithms for the simple assembly line balancing problem. *Management Science*, 32(8), 909–932.

Ding, L.-P. et al. 2010. A new multi-objective ant colony algorithm for solving the disassembly line balancing problem. *The International Journal of Advanced Manufacturing Technology*, 48(5–8), 761–771.

Dorigo, M., Maniezzo, V., and Colorni, A. 1996. Ant system: Optimization by a colony of cooperating agents. *IEEE Transactions on Systems Man and Cybernetics Part B: Cybernetics*, 26(1), 29–41.

Gungor, A. and Gupta, S. M. 1999. Issues in environmentally conscious manufacturing and product recovery: A survey. *Computers & Industrial Engineering*, 36(4), 811–853.

Gungor, A. and Gupta, S. M. 2001. A solution approach to the disassembly line balancing problem in the presence of task failures. *International Journal of Production Research*, 39(7), 1427–1467.

Güngör, A. and Gupta, S. M. 2002. Disassembly line in product recovery. *International Journal of Production Research*, 40(11), 2569–2589.

Gupta, S. M., Erbis, E., and McGovern, S. M. 2004. Disassembly sequencing problem: A case study of a cell phone. In: Gupta, S. M. (Ed.), *Environmentally Conscious Manufacturing IV*. Bellingham, WA: SPIE-International Society for Optical Engineering, pp. 43–52.

Ilgin, M. A. and Gupta, S. M. 2010. Environmentally conscious manufacturing and product recovery (ECMPRO): A review of the state of the art. *Journal of Environmental Management*, 91(3), 563–591.

Ilgin, M. A. and Gupta, S. M. 2012. *Remanufacturing Modeling and Analysis*. Boca Raton, FL: CRC Press.

Kalayci, C. B. and Gupta, S. M. 2011a. A hybrid genetic algorithm approach for disassembly line balancing. *Proceedings of the 42nd Annual Meeting of Decision Science Institute (DSI 2011)*, November 19–22, Boston, MA, pp. 2142–2148.

Kalayci, C. B. and Gupta, S. M. 2011b. Tabu search for disassembly line balancing with multiple objectives. *41st International Conference on Computers and Industrial Engineering (CIE41)*, October 23–26, University of Southern California, Los Angeles, CA, pp. 477–482.

Kalayci, C. B., Gupta, S. M., and Nakashima, K. 2011a. Bees colony intelligence in solving disassembly line balancing problem. *Proceedings of the 2011 Asian Conference of Management Science and Applications (ACMSA2011)*, December 21–22, Sanya, Hainan, China, pp. 34–41.

Kalayci, C. B., Gupta, S. M., and Nakashima, K. 2011b. A simulated annealing algorithm for balancing a disassembly line. *Proceedings of the Seventh International Symposium on Environmentally Conscious Design and Inverse Manufacturing (EcoDesign 2011)*, November 30–December 2, Kyoto, Japan, pp. 713–718.

Kalayci, C. B., Gupta, S. M., and Nakashima, K. 2012. A simulated annealing algorithm for balancing a disassembly line. In: Umeda, Y. (Ed.), *Design for Innovative Value towards a Sustainable Society*. Dordrecht, the Netherlands: Springer, pp. 714–719.

Koc, A., Sabuncuoglu, I., and Erel, E. 2009. Two exact formulations for disassembly line balancing problems with task precedence diagram construction using an AND/OR graph. *IIE Transactions*, 41(10), 866–881.

McGovern, S. M. and Gupta, S. M. 2005. Uninformed and probabilistic distributed agent combinatorial searches for the unary NP-complete disassembly line balancing problem. *Environmentally Conscious Manufacturing V*, Boston, MA, pp. 81–92.

McGovern, S. M. and Gupta, S. M. 2006. Ant colony optimization for disassembly sequencing with multiple objectives. *The International Journal of Advanced Manufacturing Technology*, 30(5), 481–496.

McGovern, S. M. and Gupta, S. M. 2007a. A balancing method and genetic algorithm for disassembly line balancing. *European Journal of Operational Research*, 179(3), 692–708.

McGovern, S. M. and Gupta, S. M. 2007b. Combinatorial optimization analysis of the unary NP-complete disassembly line balancing problem. *International Journal of Production Research*, 45(18–19), 4485–4511.

McGovern, S. M. and Gupta, S. M. 2011. *The Disassembly Line: Balancing and Modeling*. New York: McGraw Hill.

Rabanal, P., Rodríguez, I., and Rubio, F. 2007. Using river formation dynamics to design heuristic algorithms. In: Akl, S. et al. (Eds.), *Unconventional Computation*. Berlin/Heidelberg, Germany: Springer, pp. 163–177.

Rabanal, P., Rodriguez, I., and Rubio, F. 2008a. Solving dynamic TSP by using river formation dynamics. *ICNC '08, Fourth International Conference on Natural Computation*, October 18–20, Jinan, China, pp. 246–250.

Rabanal, P., Rodríguez, I., and Rubio, F. 2008b. Finding minimum spanning/distances trees by using river formation dynamics. In: Dorigo, M. et al. (Eds.), *Ant Colony Optimization and Swarm Intelligence*. Berlin/Heidelberg, Germany: Springer, pp. 60–71.

Rabanal, P., Rodríguez, I., and Rubio, F. 2009. Applying river formation dynamics to solve NP-complete problems. In: Chiong, R. (Ed.), *Nature-Inspired Algorithms for Optimisation*. Berlin/Heidelberg, Germany: Springer, pp. 333–368.

Rabanal, P., Rodríguez, I., and Rubio, F. 2011. Studying the application of ant colony optimization and river formation dynamics to the steiner tree problem. *Evolutionary Intelligence*, 4(1), 51–65.

Rabanal, P., Rodríguez, I., and Rubio, F. 2012. Testing restorable systems: Formal definition and heuristic solution based on river formation dynamics. *Formal Aspects of Computing*, January, 1–26.

Tovey, C. A. 2002. Tutorial on computational complexity. *Interfaces*, 32(3), 30–61.

Tripathi, M. et al. 2009. Real world disassembly modeling and sequencing problem: Optimization by algorithm of self-guided ants (ASGA). *Robotics and Computer-Integrated Manufacturing*, 25(3), 483–496.

Wang, H.-F. and Gupta, S. M. 2011. *Green Supply Chain Management: Product Life Cycle Approach*. New York: McGraw Hill.

13 Graph-Based Approach for Modeling, Simulation, and Optimization of Life Cycle Resource Flows

Fabio Giudice

CONTENTS

13.1   INTRODUCTION

For a complete analysis directed at evaluating and reducing the environmental impact of a product, it is necessary to consider, together with the phases of development and production, those of use, recovery, and retirement of product at end of life. Further, all these phases must be understood not in relation to the specific actors involved (manufacturer, consumer, etc.), but rather from a wider perspective, according to which it is possible to speak of *product-system*, where the product is considered integral with its life cycle and within the environmental, technological, economic, social context in which the life cycle develops.

From the specific viewpoint of environmental analysis and planning, this system is characterized by physical flows of resources transformed through the various processes making up the life cycle and by interactions with the ecosphere. The impact this product-system has on the environment is the result of life cycle processes that exchange substances and resources with the ecosphere.

Particularly, the flows of material resources that run through the life cycle of product, fully expressing its physical dimension, outline a key perspective in environmental protection approaches, since they represent one of the main aspects of a product's impact on the environment, that of the use and consumption of material resources. This partial view of the environmental problem may seem limited, but in reality it is very wide-ranging. It does not exclude, in fact, the possibility of taking into account the other aspects of impact (mainly energy consumption and emissions) in environmental analysis. With regard to the contributions to the impact due to the energy and emission content of the material resources in play, these are clearly ascribable to the volumes of the material flows. Regarding the contributions to the impact due to the energy fueling the processes and to the direct emissions from it, also these can generally be ascribed to the volumes being processed or to specific process parameters dependent on the physical properties of the material resources.

In this perspective, the analysis of material resource flows through the life cycle becomes a powerful approach to the environmental issues in product design, production planning, and whole life cycle management (LCM). The following directives emerge as factors of success in containing the environmental impacts:

- The use of resources must be planned taking account of the environmental efficiency of the distribution of all the resource flows in the entire life cycle.
- The production system, from supply chain to manufacturing, must be organized to take account of the environmental efficiency of the production cycle, minimizing discards and waste.
- All the systems that meet the necessities arising in the implementation of key environmental strategies (extension of useful life, recovery at end of life) such as service, reverse logistics, recovery systems, must be arranged and configured to support the related phases of the life cycle, allowing to exploit efficiently the potentialities of these strategies, in terms of improvement of environmental behavior of the life cycle.

With these premises, approaches to analysis and management of distribution of flows of material resources through the life cycle of a product, must be outlined, and suitable instruments are required. System modeling based on graph theory can meet these requests. Particularly, network modeling does look a powerful resource, since it is based on the concept of *flow*, and of other entities that are fit for supporting various type of analysis on flows. This chapter would outline a survey with preliminary statements, theoretical developments, prescriptions, and reflections on direct application to product life cycle reference model, identification of potentialities, with regard to the recourse to graph-based modeling for the analysis, simulation, and optimization of flows of material resources through a product life cycle. With this aim, the investigation has been structured and reported as follows. In Section 13.2, the life cycle approach in product design and management has been described, defining life cycle concept, outlining the basic premises for life cycle modeling and simulation, and introducing the main methods and techniques used in life cycle approach for environmental impact evaluation and reduction and the most effective strategies to improve environmental behavior of the product in its life cycle. Section 13.3 introduces graph theory and synthesizes the concepts and mathematical formulation related to the issues considered the most suitable to the aims of the investigation, with particular regard to directed graph and network flows. Finally, in Sections 13.4 and 13.5, the investigation on application and potentialities of graph-based modeling for life cycle analysis, simulation, and optimization is developed.

13.2　LIFE CYCLE APPROACH IN PRODUCT DESIGN AND MANAGEMENT

The choices made throughout the entire span of the product design and development process have repercussions which, beginning at the production phase when the design idea effectively takes shape and the product is physically realized, extend over all phases making up what can be considered the *life cycle* of the product.

The concept of *life cycle*, derived from studies on biological systems, is widely used as a model for the interpretation and analysis of phenomena characterized by processes of change in relation to the most varied fields of application. In industrial product development, it becomes a key concept in the modeling and prediction of product performance, not only during its phase of use, but also upstream (manufacturing) and downstream (retirement and disposal) and is, therefore, an effective instrument for the analysis and support of decision making. Only a systemic view extended to cover the entire life cycle of a product can ensure that the interventions of the designer and production planner, as well as identifying design criticalities, will also develop ideas and corrective expedients in a truly effective way, rather than simply shifting the effects of criticalities from one phase of the life cycle to another. In this way the design and planning processes can evolve toward a solution that best harmonizes the performance and behavior of the product in relation to the different phases it will pass through from its manufacture on.

Even more, this approach is particularly required with regards to environmental issue. Products must be developed in accordance with a design intervention, known as *Life Cycle Design* (Alting 1993, Keoleian and Menerey 1993, Ishii 1995, Molina et al. 1998, Wanyama et al. 2003, Giudice et al. 2006, Ramani et al. 2010), based on the *life cycle approach*, understood as a systematic approach "from the cradle to the grave," the only approach able to provide a complete environmental profile of products, and to guarantee that the design intervention manages to both identify the environmental criticalities of the product and reduce them efficiently.

13.2.1   LIFE CYCLE CONCEPT

The concept of *product life cycle* is used with different meanings in different contexts. It can be used in the management of product development to mean the entire set of phases from need recognition and design development to production, at the most going so far as to include any possible support services for the product, but usually not taking into consideration the phases of retirement and disposal. This limited view of the life cycle has its origins in an approach conditioned by the competencies and direct interests of different actors involved in the life of manufactured goods. This leads to a fragmentation of the life cycle according to the main actors: manufacturer (design, production, and distribution), consumer (use), and third actors (retirement and disposal). It is clear therefore that the managerial concept of life cycle, which usually does not include those phases subsequent to the distribution of the product, meets the interests of the manufacturer.

Given that the environmental performance of a product over its entire life cycle is influenced by interaction between all the actors involved, an effective approach to the environmental problem must be considered in the context of the entire society, understood as a complex system of actors including government, manufacturers, consumers, and recyclers (Sun et al. 2003). Suppliers also should be included, and environmental factors should be incorporate into the supply chain management (Wang and Gupta 2011). From a more complete perspective, therefore, not limited by the point of view of a specific actor, the life cycle of a product must include, as well as development and production phases, the upstream phases (resources supplying) and downstream phases (product use, retirement, and disposal), fully interpreting the life cycle approach perspective.

The considerations made earlier can be summarized in a holistic vision of the product and its life cycle, where the latter is no longer thought of as a series of independent processes, expressed exclusively by their technological aspects, but rather as a complex product-lifecycle system set in its environmental and socio-technological context, spiking of *product-system*, a concept that in its most complete sense, includes the product, understood as integral with its life cycle, within the environmental, social, and technological context in which the life cycle evolves (Giudice et al. 2006).

13.2.2   LIFE CYCLE MODELING AND ENVIRONMENTAL IMPACT

From the specific viewpoint of environmental analysis, the product-system is characterized by flows of resources transformed through the various processes constituting the physical life cycle. The environmental impact of this product-system is the result of life cycle processes that exchange substances, materials, and energy with the ecosphere. With these premises, life cycle modeling for environmental investigation must outline a fundamentally physical model.

Furthermore, the modeling of a system generally tends to reduce its complexity. Such simplifications become necessary in the case of environmental evaluations because of the elevated complexity of the real systems. With a physical–technical based approach, the behavior of the model can be described and simulated using mathematical models of limited complexity, in that they refer to the analysis of a system with static, linear behavior.

Such considerations justify the choice of the physical–technological viewpoint in modeling product life cycle, as has been generally proposed in the literature (Keoleian and Menerey 1993, Vigon et al. 1993, Billatos and Basaly 1997, Hundal 2002, Graedel and Allenby 2009).

13.2.2.1 Modeling by Elementary Activities

With these premises, product life cycle is subdivided into elementary functions (Zust and Caduff 1997, Hundal 2002), also represented by *activity models* (Navin-Chandra 1991, Tipnis 1998, Giudice et al. 2006), which summarize the elementary processes characterizing the main phases of the cycle.

In general terms, activities can be the transformation, handling, generation, use or disposal of material resources, energy, data, or information. The reference activity model in the environmental analysis perspective can be of the type shown in Figure 13.1a, characterized by input flows of physical resources, by output flows, and by a possible input flow of information when there is a margin of choice in how the activity is performed. For the input flows, given that they are physical resources (they can consist of materials and forms of energy), it is possible to distinguish between resources produced by preceding activities and resources coming directly from the ecosphere. For the output flows, consisting of products of the activity, it is possible to distinguish between main products, secondary by-products, and various types of emissions into the ecosphere. Having defined the reference activity model, the product life cycle can be translated into a system model by the following procedure: define the boundaries of the system; identify the elementary processes and functionalities; identify and quantify the connections between elementary activities, and between them and the ecosphere.

The reference activity model of Figure 13.1a can be read in different ways. In the case where the aim is to develop a life cycle model which supports the analysis of the material resources in play, a more simplified representation of activity model is possible, such as that represented in Figure 13.1b. It takes only the flows of material resources into account, considering as input the resources fueling the activity, and as output the product of the activity and any possible discards and waste. Regarding the input resources, however, it is necessary to make a distinction between: primary or virgin resources, coming directly from the ecosphere; secondary or recycled resources. The latter can in turn be divided into preconsumption secondary resources, i.e., originating from discards and waste generated by the activity itself and postconsumption secondary resources, i.e., originating from recycling the product after use and retirement.

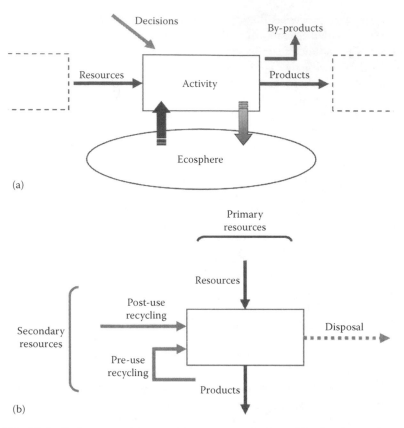

FIGURE 13.1 Reference activity model: (a) generalized model; (b) flows of material resources. (From Giudice, F. et al., *Product Design for the Environment: A Life Cycle Approach*, CRC Press/Taylor & Francis Group, Boca Raton, FL, 2006.)

13.2.2.2   Reference Model for Product Life Cycle

The previous considerations concerning the concept of product life cycle, the appropriateness and the modality of considering the physical life cycle in environmental analysis, and the basic principles of modeling for elementary activities are interpreted by the reference life cycle model described later (Giudice et al. 2006). According to this model, all the processes of transformation of resources involved in the product's entire physical life cycle can be grouped according to the following main phases:

- Preproduction, where materials and semifinished pieces are prepared to be supplied
- Production, involving the transformation of materials, production of components, product assembly, and finishing
- Use, including any possible servicing operations
- Retirement, corresponding to the end of the product's useful life and consisting of various options from product reuse to disposal as waste

Each of these phases interacts with the ecosphere, since it is fuelled by input flows of resources, and produces not only by-products or intermediate products that fuel the successive phases, but also emissions and waste.

Developing each main phase according to the primary activities it encompasses, it is possible to obtain a vision of a product's entire physical life cycle and of the resource flows that characterize it, such as that of Figure 13.2, where the flows of material resources are shown according to the activity model of Figure 13.1b. The first phase of preproduction consists of the production of materials and semifinished pieces required for the subsequent production of components. It includes, therefore, the production phases of all the materials which will go to make up the final product. Once the product is manufactured, distributed, and used (including all service interventions, from inspection and maintenance to repair and component replacement), it arrives at the final phase of retirement and disposal.

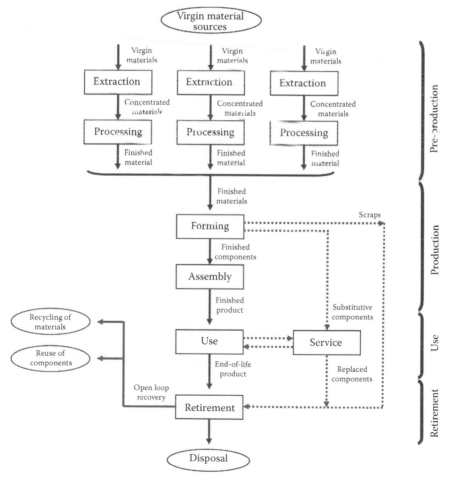

FIGURE 13.2 Physical life cycle of product and flows of material resources: open loop model.

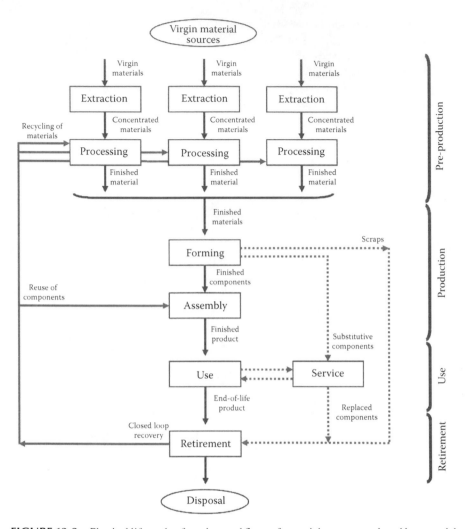

FIGURE 13.3   Physical life cycle of product and flows of material resources: closed loop model.

Figure 13.2 offers a complete picture of the alternatives to disposing of the product (and scraps flow also) as waste at the end of its life, when the recovery flows are directed outside the cycle. Alternatively, Figure 13.3 describes how the recovery flows can be distributed within the same life cycle that generated them, providing the postconsumption secondary resources for various activities. The two models differ in the typology of recovery flows:

1. External recovery (*open loop*)—At the end of the product's life, some of its parts are directed to the production processes of other materials or products external to the cycle under examination. This can result in recovering part of the resources and reducing the volumes disposed of as waste, and in saving virgin materials in other production cycles.

2. Internal recovery (*closed loop*)—The resources recovered reenter the life cycle of the same product which generated the flows, replacing the input of virgin resources. This can occur by directly reusing some components at product end of life, or by recycling materials. From the viewpoint of the environmental consequences, these recovery processes add to the environmental benefits of open loop processes a decrease in the consumption of virgin resources, replaced by postconsumption resources.

13.2.3 Environmental Strategies in the Life Cycle Approach

A design intervention intended to take account of product's behavior, in environmental terms, during its life cycle must, in general, have the aim of optimizing the distribution of the flows of resources and emissions, establishing the conditions which favor

- Reducing the volumes of materials used and extending their working life
- Closing the cycles of resource flows through recovery processes
- Minimizing the emissions and resources consumption in production, use, and disposal

Focusing on the physical dimension of the product, and therefore on the flows of material resources in the life cycle, the environmentally beneficial conditions described earlier can be achieved through the application of two main strategy types (Giudice et al. 2006):

1. *Useful life extension strategies*, directed at extending the product useful life and so conferring increased value on the materials used and on all the other resources employed in its manufacture. They consist in product maintenance, repair, upgrading, and adaptation.
2. *End-of-life strategies*, directed at recovering material at the end of the product useful life, closing the cycle of materials and recovering, at least in part, the other resources used in its manufacture. They consist in reusing systems and components, recycling materials in the primary production cycle or in external cycles.

Figures 13.2 and 13.3 describe the life cycle when both types of strategies are applied. Although these strategies must already be taken into consideration during the design phase, in order to facilitate them, clearly they do not come into effect until after the product has been manufactured. A third important typology of environmental strategy, the *resource reduction strategies*, becomes operational before the production phase. Again associated with the product's material dimension, these strategies are directed at reducing the resources used in its manufacture. Thus, in general terms, they are referable to a wide spectrum of expedients that regard not only product design but also supply chain management and production process planning.

13.2.4   Life Cycle Methods and Techniques: LCM, LCA, and LCC

Managing the complexity of a product design and development process, which takes account of the entire life cycle, and of the consequent wide range of performances required constitutes the principal obstacle to widespread adoption of the life cycle approach in design and production practice. One response to this problem is given by the increasing use of a new approach to the management of the production activity that, taking into consideration the entire life cycle of the product, has the objective of optimizing the interaction between product design and the activities correlated with the various phases of the cycle. Known as *Life Cycle Management (LCM)*, this new approach prescribes the realization of a life cycle oriented management structure whose mission is the planning and operation of strategic decisions, explicitly taking into consideration the costs and other fundamental company metrics, together with quality, safety, environmental factors, and other technological aspects (Fiksel 2009).

These objectives are pursued using various instruments, ranging from data management systems to tools for the analysis of various aspects of the life cycle. Among the latter, two techniques in particular have undergone considerable development and now are supported by well-delineated methodological structures. The first, known as *Life Cycle Costing (LCC)* (Fabrycky and Blanchard 1991, Asiedu and Gu 1998), is oriented toward economic analysis and is thus a valuable aid to the management of the interaction between the life cycle and the economic level of the external environment in which the cycle evolves. This tool has been fully developed for some time, such that it becomes a key resource in product development practice (Dhillon 2009). Instead, the second technique, known as *Life Cycle Assessment (LCA)* (Curran 1996, Guinèe 2002), has been conceived as a response to increasing sensitivity toward environmental problems and makes it possible to analyze and correct the interactions of the life cycle with the domain of the external environment, associated with the consumption of resources and with the impact on the ecosphere. It consists of an objective procedure to assess the consumption of resources and the generation of wastes and emissions of various types associated with a generic industrial activity, or rather, with the entire life cycle of the product of this activity. Through the quantitative determination of the exchange flows between the product life cycle and the ecosphere, involved in all the processes of transformation concerned, from the extraction of raw materials until their return to the ecosphere in the form of waste, LCA allows the quantification of all of the impact factors in play, differentiating them according to the diverse categories of environmental impact, and provides an estimation of the product's overall impact in each phase of its life cycle.

13.2.5   Life Cycle Simulation

Traditional product design is based on a process of reiteration revising the design solutions, guided by a continuous evaluation of the results. The alternative to this evolutionary approach based on the feedback of information used to improve the design intervention, consists of already predicting the consequences of design choices on the product life cycle during the solution synthesis phase itself. This is equivalent to simulating the life cycle of the product, varying the design choices and taking account of the interactions with the supporting systems (from production to disposal systems) and environment in which the life cycle unfolds. The crucial question is that of how to evaluate the consequences that design choices, and supporting system arrangement also, will have on the product entire life cycle, already in the earliest phases of the design and planning processes.

As observed before, modeling the product-system in order to evaluate its environmental impact is based on breaking down the system into activities requiring flows of resources in input and output. The models of flows of material resources shown in Figures 13.2 and 13.3 interpret this approach, evidencing the potential of service and different recovery interventions within the life cycle. To evaluate the environmental behavior of this system of elementary processes and resource flows, and therefore of the product-system it represents, it is necessary to describe the circulation of resources including the service and recovery flows. This can be realized through a simulation of the life cycle, understood as a simulation of the behavior of the product, its parts and constituent materials, expressed by the distribution and circulation of the flows of material resources, in relation with life cycle supporting systems and its environment.

Relating the product's behavior to its constructional characteristics (materials, geometric properties and shapes, system architecture), life cycle simulation becomes an effective instrument for life cycle design, because it already allows, in the design phase, a prediction of the product's performance as a function of the main design choices. Several significant studies have already confirmed the potential of life cycle simulation approaches in life cycle design (Kimura et al. 1998, Sakai et al 2001, Takata et al. 2003, Giudice et al. 2004).

13.3 GRAPH-BASED MODELING OF SYSTEMS

Graph theory and its developments have been acknowledged as powerful means in the modeling and analysis of different types of systems, with regard to various fields of science (Bondy and Murty 1982, Chen 1997, Gross and Yellen 2004).

A *graph* G{V, E, ψ} consists of a set V = {v_1, v_2, ..., v_{nv}} of *vertices*, a set E = {e_1, e_2, ..., e_{ne}} of *edges*, and an *incidence function* $\psi(e_k)$ = {v_i, v_j} that associates with each edge of G an unordered pair of vertices (not necessarily distinct). This mathematical entity is so named because it can be translated into a graphical representation (each vertex is indicated by a point, and each edge by a line joining the points that represent its end), which is helpful to understand its properties.

13.3.1 DIRECTED GRAPHS

A *directed graph* (or *digraph*) is a graph in which each edge has an assigned orientation (becoming an *arc*). So a digraph D{V, A, ψ} consists of a set V = {v_1, v_2, ..., v_{nv}} of *vertices*, a set A = {a_1, a_2, ..., a_{na}} of *arcs*, and an *incidence function* $\psi(a_k)$ = {v_i, v_j} that associates with each arc of D an unordered pair of vertices (not necessarily distinct). If $\psi(a_k)$ = {v_i, v_j}, the arc a_k is said to *join* v_i to v_j, v_i is the *tail* of a_k, and v_j is its *head*.

A digraph D'{V', A', ψ'} is a *subdigraph* of D{V, A, ψ} if V' $\subseteq$ V, A' $\subseteq$ A, and ψ' is the restriction of ψ to A'. The relation can be written as D' $\subseteq$ D.

A *directed walk* in D is a finite sequence DW = $\{v_0, a_1, v_1, ..., a_{nw}, v_{nw}\}$, whose terms are alternately vertices and arcs such that, for i = 1, 2, ..., nw, the arc a_i has head v_i and tail v_{i-1}. If the arcs of a directed walk DW are distinct, DW is called a *directed trail*. If in addition the vertices of DW are distinct, DW is called a *directed path*. If there is a $v_i \rightarrow v_j$ directed path in D, vertex v_j is said to be *reachable* from vertex v_i. If a directed path is closed, i.e., its origin and terminus are the same, it is called a *directed cycle*.

Digraphs have analytical representations as adjacency and incidence matrices. The *adjacency matrix* $M^A = \left\{m_{ij}^A\right\}_{nv \times nv}$ is a square matrix, whose rows and columns are indexed by the vertices vector $(v_1, v_2, ..., v_i, ..., v_{nv})$, and whose elements take values $m_{ij}^A = 1$ if, and only if, an arc directed from column j to row i exists, otherwise $m_{ij}^A = 0$. Each column denotes the presence (value 1) or absence (value 0) of arcs directed from vertex v_j to vertex v_i (which is an arc with tail v_j and head v_i) in the corresponding digraph. The orientation from columns (initial vertices) to rows (terminal vertices) is the same as that of the digraph arcs. Therefore, the *indegree* $d^-(v_i)$ of a vertex v_i in D, which is the number of arcs with head v_i, and the *outdegree* $d^+(v_j)$ of a vertex v_j in D, which is the number of arcs with tail v_j, can be calculated respectively as follows:

$$d^-(v_i) = \sum_{j=1}^{nv} m_{ij}^A \quad d^+(v_j) = \sum_{i=1}^{nv} m_{ij}^A \quad (13.1)$$

The *incidence matrix* $M^I = \left\{m_{ik}^I\right\}_{nv \times na}$ is a matrix whose rows are indexed by the vertices vector $(v_1, v_2, ..., v_i, ..., v_{nv})$, columns are indexed by the arcs vector $(a_1, a_2, ..., a_k, ..., a_{na})$, and whose elements take values $m_{ik}^I = 1$ if v_i is the tail of a_k, $m_{ik}^I = -1$ if v_i is the head of a_k, otherwise $m_{ik}^I = 0$.

13.3.2 Network Flows

A particular branch of graph theory, that of network flows, lies on the border between diversified fields of investigation, such as applied mathematics, operations research, linear programming, engineering, and management, and established network graphs as useful mathematical objects for representing various physical systems and simulating their behavior and performance (Bazaraa and Jarvis 1977, Ahuja et al. 1993, Chen 2003).

In its mathematical formulation, a network is a digraph that possesses additional structure. So a network $N\{X, Y, Z, A, \psi, \chi\}$ is a digraph $D\{V, A, \psi\}$ with the set of vertices $V = X \cup Y \cup Z$ consisting three distinguished subsets of vertices $X = \{x_1, x_2, ..., x_{nx}\}$, $Y = \{y_1, y_2, ..., y_{ny}\}$, $Z = \{z_1, z_2, ..., z_{nz}\}$, and a non-negative real-valued function χ defined on its arc set A. The vertices in X and Y are, respectively, the *sources* and the *sinks* of N. Vertices in Z, which are neither sources nor sinks, are called *intermediate vertices*. The function χ is the *capacity function* of N, and its value $\chi(a_k)$ in an arc is the *capacity* of the arc a_k.

A flow ϕ in a network N is a real-valued function defined on A such that

$$0 \le \phi(a_k) \le \chi(a_k) \quad \text{for all } a_k \in A \tag{13.2}$$

$$\sum_{j:v_j \in V \text{ and } (v_i,v_j) \in A} \phi(v_i, v_j) = \sum_{j:v_j \in V \text{ and } (v_j,v_i) \in A} \phi(v_j, v_i) \quad \text{for all } v_i \in Z \tag{13.3}$$

The first condition expresses the *capacity constraint* that imposes the flow along an arc cannot exceed the capacity of the arc. The *residual capacity* $\chi_R(a_k) = \chi(a_k) - \phi(a_k)$ in an arc is the difference between the capacity and the flow in the arc. This defines a *residual network* $N_R\{X, Y, Z, A, \psi, \chi_R\}$.

The second condition express the *conservation condition*, that requires for any vertex the resultant flow is zero, except for a source, which produces flow, and a sink, which consumes flow. Specifically, this condition means that for any intermediate vertex v_i the *total outflow* $\psi^+(v_i) - \sum_j \phi(v_i, v_j)$, i.e., the flow emanating from the vertex, with the summation extended to all j-th vertex for which an arc a'_k exists such that $\psi(a'_k) = \{v_i, v_j\}$, is equal to the *total inflow* $\phi^-(v_i) = \sum_j \phi(v_j, v_i)$, i.e., the flow entering the vertex, with the summation extended to all j-th vertex for which an arc a''_k exists such that $\psi(a''_k) = \{v_j, v_i\}$. The conservation condition can be generalized, to be extended to all vertices, including sources and sinks, by associating to each vertex $v_i \in V(V = X \cup Y \cup Z)$ a real number $\beta(v_i)$ quantifying its production/consumption of flow. If $\beta(v_i) > 0$ the vertex is a source (β represents its produced flow or *supply*), if $\beta(v_i) < 0$ the vertex is a sink (β represents its consumed flow or *demand*), if $\beta(v_i) = 0$ the vertex is an intermediate vertex and its resultant flow is zero. The conservation condition becomes

$$\sum_{j:v_j \in V \text{ and } (v_i,v_j) \in A} \phi(v_i, v_j) - \sum_{j:v_j \in V \text{ and } (v_j,v_i) \in A} \phi(v_j, v_i) = \beta(v_i) \quad \text{for all } v_i \in V \tag{13.4}$$

Introducing the vector $\phi = \{\phi(a_1), \phi(a_2), ..., \phi(a_k), ..., \phi(a_{na})\}$ of flows in N, and the vector $\beta = \{\beta(v_1), \beta(v_2), ..., \beta(v_i), ..., \beta(v_{nv})\}$ collecting the values assumed by β for each vertex, the conservation condition can be expressed in matrix form as

$$\left\{m^I_{ik}\right\}_{nv,na} \cdot {}^T\{\phi(a_k)\}_{na} = {}^T\{\beta(v_i)\}_{nv} \tag{13.5}$$

where $M^I = \left\{m^I_{ik}\right\}_{nv,na}$ is the incidence matrix of the network N.

If $N'\{X', Y', Z', A', \psi', \chi'\}$ is a *subnetwork* of $N\{X, Y, Z, A, \psi, \chi\}$, which means $X' \subseteq X$, $Y' \subseteq Y$, $Z' \subseteq Z$, $A' \subseteq A$, ψ' and χ' are the restrictions, respectively, of ψ and χ to A', and $N''\{X'', Y'', Z'', A'', \psi'', \chi''\}$ is the subnetwork of $N\{X, Y, Z, A, \psi, \chi\}$ such that $N' \cap N'' = 0$ and $N' \cup N'' \equiv N$, the conservation condition (13.4) can be applied to inflows and outflows of N' as to N. If $\phi^+(V')$ is the *total outflow* of N' as to N, i.e., the summation of flows in the arcs emanating from vertices of subset $V' = X' \cup Y' \cup Z'$ and entering in vertices of subset $V'' = X'' \cup Y'' \cup Z''$, and $\phi^-(V')$ is the *total inflow* of N' as to N, i.e., the summation of flows in the arcs emanating from vertices of subset $V'' = X'' \cup Y'' \cup Z''$ and entering in vertices of subset $V' = X' \cup Y' \cup Z'$, the *resultant flow out of N'* and *the resultant flow into N'* are, respectively,

$$\phi_R^+(V') = \phi^+(V') - \phi^-(V') \quad \phi_R^-(V') = \phi'(V') - \phi^+(V') \quad (13.6)$$

Applying the conservation condition:

$$\phi_R^+(Z') = \phi^+(Z') - \phi^-(Z') = 0 \quad \phi_R^-(Z') = \phi'(Z') - \phi^+(Z') = 0 \quad (13.7)$$

which means that if the total outflow $\phi^+(V')$ is the summation of flows in the arcs emanating from vertices of subset Z' and entering in vertices of subset Z'', and the total inflow $\phi^-(V')$ is the summation of flows in the arcs emanating from vertices of subset Z'' and entering in vertices of subset Z' (all intermediate vertices), then the resultant flows are zero.

Another consequence of the conservation condition (13.4) is related to flows out of sources and into sinks of the network N. Since the condition (13.4) requires that the resultant flow of any intermediate vertex is zero, the resultant flow out of set X is equal to the resultant flow into set Y:

$$\phi_R^+(X) = \phi^+(X) - \phi^-(X) = \phi^-(Y) - \phi^+(Y) = \phi_R^-(Y) \quad (13.8)$$

This common quantity Val_ϕ is the *value* of the flow function ϕ in the network N.

The capacity constraint (13.2) and the conservation condition in its generalized expression (13.4) constitute the two constraints of the mathematical formulation of the *minimum cost flow problem*, which is known as the fundamental of all network flow problems. If each arc a_k of the network N has an associated *cost value* $c(a_k)$ that quantifies the cost per unit flow on that arc, assuming that the flow cost varies linearly with the amount of flow, the minimum cost flow problem can be formulated as the following optimization problem with arc flows as variables:

Minimize $\sum_{a_k \in A} c(a_k) \cdot \phi(a_k)$ subject to conditions (13.2) and (13.4), with $\sum_{i=1}^{nv} \beta(v_i) = 0$.

13.4   LIFE CYCLE MODELING AND ANALYSIS

The network perspective conceptualizes a system as a net of interactions. These interactions can be interpreted in various ways, according to the type of system to be analyzed and the issue to be investigated. With regard to the proposed application, two types of interactions commonly used in other fields, such as ecological science (Fath and Patten 1998), are particularly significant, because they are suitable for modeling both physical and non-physical interactions: transactions and relations. In this perspective, a directed path $v_i \rightarrow v_j$ in a digraph D can be interpreted as a sequence of vertices and arcs over which material resources flow (transaction type interactions) or relationships (relation type interactions) emerge from vertex v_i to vertex v_j.

Concerning this aspect of multilevel modeling, first of all transaction type interactions will be extensively treated, focusing on product life cycle and its flows of material resources. Not physical interactions will be outlined subsequently, introducing a second level of analysis, focused on technological and organizational systems supporting life cycle phases.

13.4.1 DEFINITION OF PRELIMINARY LIFE CYCLE PARAMETERS

In its physical dimension, the product can be thought of as constituted by the whole of its components. If n is their number, and V_i is the *volume* of the i-th component, the physical dimension of product is expressed by the summation of the material quantities constituting it, i.e., $\sum_{i=1}^{n} V_i$.

With regards to production quantities, to contemplate the scraps in manufacturing processes, the quantities V_i can be put up by means of *scrap coefficients* $\beta_i > 1$.

End-of-life quantities depend on some factors such as reusability and separability of components, recyclability of materials. These factors can be quantified by the following parameters:

- *Reusability* r_i—Its value is 1 if the component can be reused, otherwise it is 0. It depends on the durability of the component. If t_i is its *required working life*, and pd_i its *predictable duration*, its *durability* can be defined as $d_i = \text{int}(pd_i/t_i)$. If $d_i > 2$, then the component is reusable and $r_i = 1$, otherwise $r_i = 0$.
- *Separability* s_i—Its value is 1 if the component can be separated from the product, otherwise it is 0. It depends on the disassemblability of the component.
- *Recyclability* ξ_i—Its value expresses the recyclable fraction of the material constituting the component; therefore, $0 \leq \xi_i \leq 1$. It depends on several factors, primarily the properties of material, the efficiency of reverse logistics, and recycling technologies.

The volume V_i component can be reused if $r_i s_i = 1$, which means it is reusable ($r_i = 1$) and separable from the product ($s_i = 1$). When the component is not reusable ($r_i s_i = 0$), its fraction $\xi_i V_i$ can be recycled if $s_i \xi_i > 0$, which means it is separable ($s_i = 1$) and the material constituting it is recyclable ($\xi_i > 0$).

Service quantities depend on the factors determining the necessities for inspection, maintenance, repair, and replacement interventions. These factors can be quantified by the following parameters:

- *Number of replacements* nr_i—Its value can be expressed as the inverse of durability, i.e., $nr_i = \text{int}(t_i/pd_i)$. If $t_i > pd_i$, at least one failure incurs, so $nr_i \geq 1$.
- *Number of services* ns_i—Its value can be expressed as a function of reliability parameters, e.g., $ns_i = \text{int}(t_i/MTBM_i)$, where $MTBM_i$ is the mean time between maintenance of the component. If $t_i > MTBM_i$, at least one maintenance intervention is necessary, so $ns_i \geq 1$. It excludes replacement intervention.

If $nr_i \geq 1$ then the volume V_i component must be replaced nr_i times during its working life. Similarly, if $ns_i \geq 1$, it must be serviced ns_i times. It is supposed that a component requiring replacement or service interventions is separable from the product, which means it can be disassembled. Furthermore, it seems reasonable to assume that replaced components cannot be reused. They can be recycled if $\xi_i > 0$ ($s_i = 1$, as supposed).

13.4.2 STATEMENT OF GRAPH-BASED MODELING

Applying graph theory, the product life cycle in the form synthesized in Figure 13.2 (open loop life cycle) can be modeled by means of the network $N\{X, Y, Z, A, \psi, \chi\}$ in Figure 13.4, defined by

- $X = \{x_1\}$, set of sources (one source of virgin materials from ecosphere)
- $Y = \{y_1, y_2, y_3\}$, set of sinks (three sinks of destination at end of life: disposal, materials recycling, components reuse)
- $Z = \{z_0, z_1, ..., z_5\}$, set of intermediate vertices (each of them corresponding to a life cycle phase, from preproduction to retirement)

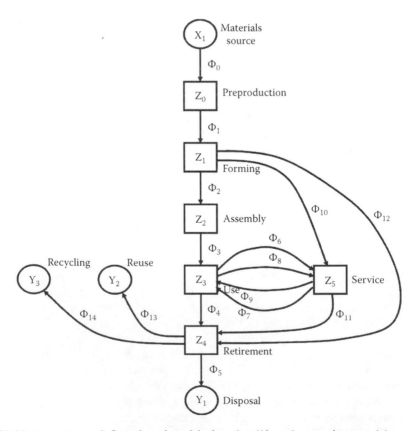

FIGURE 13.4   Network flows-based model of product life cycle: open loop model.

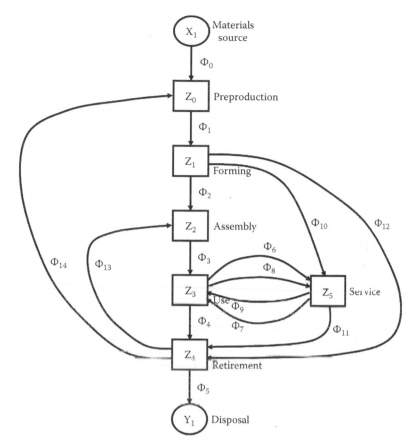

FIGURE 13.5 Network flows-based model of product life cycle: closed loop model.

- $A = \{a_0, a_1, \ldots, a_{14}\}$, set of arcs (each of them characterized by a flow of material resources Φ)
- $\Phi = \{\Phi_0, \Phi_1, \ldots, \Phi_{14}\}$, set of flows of material resources related to product life cycle $\left(\text{the physical dimension of product is expressed by } \Phi_3 = \sum_{i=1}^{n} V_i \right)$

Similarly, the product life cycle in the form synthesized in Figure 13.3 (closed loop life cycle) can be modeled by means of the network N^* in Figure 13.5. The only structural difference between N and N^* lies in sinks set $Y^* = \{y_1\}$, and in the heads of arcs a_{13} and a_{14}, which get into vertices z_2 and z_0, respectively.

13.4.3 ANALYSIS OF FLOWS DISTRIBUTION

Flows definition on network N in Figure 13.4 can be conducted thinking of N as a superposition of three networks: N' (Figure 13.6), which supports the main flow of material resources, related to simple product produced, used, and retired at end of life; N'' (Figure 13.7), which supports the flows related to service operation (maintenance,

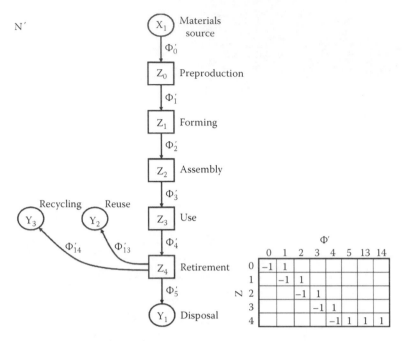

FIGURE 13.6   Subnetwork of main flow in the life cycle.

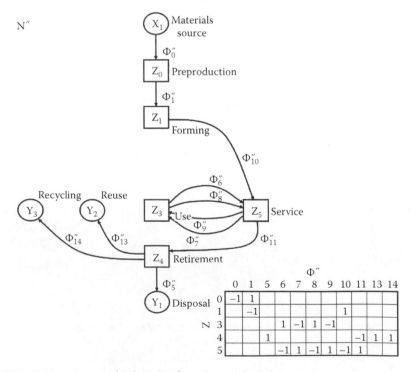

FIGURE 13.7   Subnetwork of service flows in the life cycle.

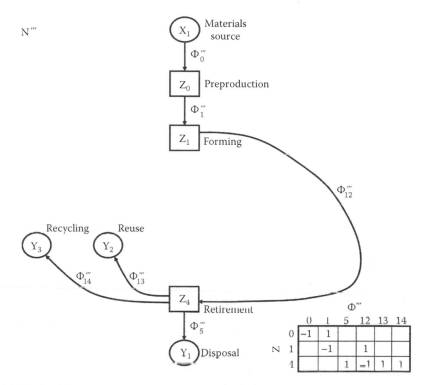

The table within the figure:

Φ'''

N	0	1	5	12	13	14
0	-1	1				
1		-1		1		
1			1	-1	1	1

FIGURE 13.8 Subnetwork of scrap flows in the life cycle.

repair, components replacement); N''' (Figure 13.8), which supports the flows related to all scraps in product life cycle. Thus, flows definition is conducted for each of the three networks separately, applying the conservation condition at the intermediate vertices, and then the results will be superimposed to obtain the overall flows in network N.

As well as practical, this distinction is significant because it allows to preliminarily distinguish between three different flows of material resources that characterize the life cycle (main flow, service, scraps).

With regard to network N' in Figure 13.6, from (13.5) with the incidence matrix shown in the same figure, extended to the intermediate vertices only, the conservation condition provides the following balance equations:

$$\begin{cases} \Phi'_0 - \Phi'_1 = 0 \\ \Phi'_1 - \Phi'_2 = 0 \\ \Phi'_2 - \Phi'_3 = 0 \\ \Phi'_3 - \Phi'_4 = 0 \\ \Phi'_4 - \Phi'_5 - \Phi'_{13} - \Phi'_{14} = 0 \end{cases} \qquad (13.9)$$

For the main flow system,

$$\Phi'_3 = \sum_{i=1}^{n} V_i \qquad (13.10)$$

As end-of-life conditions, the following relations are effective for reuse flow and recycling flow, respectively:

$$\Phi'_{13} = (r_i s_i) \cdot \Phi'_4 \quad \Phi'_{14} = (1 - r_i s_i)(\xi_i s_i) \cdot \Phi'_4 \tag{13.11}$$

From equation system (13.9) and conditions (13.10), (13.11), the following flows expressions are obtained:

$$\left[\begin{array}{l} \Phi'_0 = \Phi'_1 \equiv \Phi'_2 = \Phi'_3 = \Phi'_4 = \sum_{i=1}^{n} V_i \\[2ex] \Phi'_{13} = \sum_{i=1}^{n} (r_i s_i) V_i \\[2ex] \Phi'_{14} = \sum_{i=1}^{n} (1 - r_i s_i)(\xi_i s_i) V_i \\[2ex] \Phi'_5 = \sum_{i=1}^{n} (1 - r_i s_i)(1 - \xi_i s_i) V_i \end{array} \right. \tag{13.12}$$

In the first equation of (13.12), the equivalence $\Phi'_1 \equiv \Phi'_2$ defines the transformation of flow from materials to formed components.

Similarly, with regard to network N'' in Figure 13.7, from (13.5) with the incidence matrix shown in the same figure, extended to the intermediate vertices only, the conservation condition provides the following balance equations:

$$\left\{ \begin{array}{l} \Phi''_0 - \Phi''_1 = 0 \\ \Phi''_1 - \Phi''_{10} = 0 \\ -\Phi''_6 + \Phi''_7 - \Phi''_8 + \Phi''_9 = 0 \\ -\Phi''_5 + \Phi''_{11} - \Phi''_{13} - \Phi''_{14} = 0 \\ \Phi''_6 - \Phi''_7 + \Phi''_8 - \Phi''_9 + \Phi''_{10} - \Phi''_{11} = 0 \end{array} \right. \tag{13.13}$$

For the service flow system,

$$\Phi''_{10} = \sum_{i=1}^{n} n r_i V_i \tag{13.14}$$

As end-of-life condition, the following relations are effective for reuse flow and recycling flow, respectively:

$$\Phi''_{13} = 0 \quad \Phi''_{14} = \xi_i \cdot \Phi''_{11} \tag{13.15}$$

Furthermore, as service conditions, the following relations are effective for maintenance flow and replacement flow, respectively:

$$\Phi_6'' = \Phi_7'' = \sum_{i=1}^n ns_i V_i \quad \Phi_8'' = \Phi_9'' = \sum_{i=1}^n nr_i V_i \tag{13.16}$$

From equation system (13.13) and conditions (13.14) through (13.16), the following flows expressions are obtained:

$$\begin{bmatrix}
\Phi_0'' = \Phi_1'' \equiv \Phi_{10}'' = \sum_{i=1}^n nr_i V_i \\[2ex]
\Phi_6'' = \Phi_7'' = \sum_{i=1}^n ns_i V_i \\[2ex]
\Phi_8'' = \Phi_9'' = \sum_{i=1}^n nr_i V_i \\[2ex]
\Phi_{11}'' = \Phi_{10}'' = \sum_{i=1}^n nr_i V_i \\[2ex]
\Phi_{13}'' = 0 \\[2ex]
\Phi_{14}'' = \sum_{i=1}^n \xi_i nr_i V_i \\[2ex]
\Phi_5'' = \sum_{i=1}^n (1 - \xi_i) nr_i V_i
\end{bmatrix} \tag{13.17}$$

Finally, with regard to network N''' in Figure 13.8, from (13.5) with the incidence matrix shown in the same figure, extended to the intermediate vertices only, the conservation condition provides the following balance equations:

$$\begin{cases}
\Phi_0''' - \Phi_1''' = 0 \\
\Phi_1''' - \Phi_{12}''' = 0 \\
-\Phi_5''' + \Phi_{12}''' - \Phi_{13}''' - \Phi_{14}''' = 0
\end{cases} \tag{13.18}$$

For the scrap flow system,

$$\Phi_{12}''' = \sum_{i=1}^n (\beta_i - 1)(1 + nr_i) V_i \tag{13.19}$$

As end-of-life condition, the following relations are effective for reuse flow and recycling flow, respectively:

$$\Phi_{13}''' = 0 \quad \Phi_{14}''' = \xi_i \cdot \Phi_{12}''' \tag{13.20}$$

From equation system (13.18) and conditions (13.19), (13.20), the following flows expressions are obtained:

$$
\left[
\begin{aligned}
&\Phi_0''' = \Phi_1''' \equiv \Phi_{12}''' = \sum_{i=1}^{n} (\beta_i - 1)(1 + nr_i)V_i \\
&\Phi_{13}''' = 0 \\
&\Phi_{14}''' = \sum_{i=1}^{n} \xi_i(\beta_i - 1)(1 + nr_i)V_i \\
&\Phi_5''' = \sum_{i=1}^{n} (1 - \xi_i)(\beta_i - 1)(1 + nr_i)V_i
\end{aligned}
\right.
\tag{13.21}
$$

The overall flows in network N is obtained by superposition of the results obtained for N', N'', N''', separately; e.g., for the generic k-th flow results $\Phi_k = \Phi_k' + \Phi_k'' + \Phi_k'''$:

$$
\left[
\begin{aligned}
&\Phi_0 = \Phi_1 \equiv \Phi_2' + \Phi_{10}'' + \Phi_{12}''' = \sum_{i=1}^{n} \beta_i(1 + nr_i)V_i \\[2mm]
&\Phi_2 = \Phi_3 = \Phi_4 = \sum_{i=1}^{n} V_i \\[2mm]
&\Phi_5 = \sum_{i=1}^{n} (1 - r_i s_i)(1 - \xi_i s_i)V_i + \sum_{i=1}^{n} (1 - \xi_i)nr_i V_i + \sum_{i=1}^{n} (1 - \xi_i)(\beta_i - 1)(1 + nr_i)V_i \\[2mm]
&\Phi_6 = \Phi_7 = \sum_{i=1}^{n} ns_i V_i \\[2mm]
&\Phi_8 = \Phi_9 = \sum_{i=1}^{n} nr_i V_i \\[2mm]
&\Phi_{10} = \Phi_{11} = \sum_{i=1}^{n} nr_i V_i \\[2mm]
&\Phi_{12} = \sum_{i=1}^{n} (\beta_i - 1)(1 + nr_i)V_i \\[2mm]
&\Phi_{13} = \sum_{i=1}^{n} (r_i s_i)V_i \\[2mm]
&\Phi_{14} = \sum_{i=1}^{n} (1 - r_i s_i)(\xi_i s_i)V_i + \sum_{i=1}^{n} \xi_i nr_i V_i + \sum_{i=1}^{n} \xi_i(\beta_i - 1)(1 + nr_i)V_i
\end{aligned}
\right.
$$

$$
\tag{13.22}
$$

The first equation of (13.22) expresses the equivalence between the flow of materials necessary to manufacture and service the product (including scraps), and the corresponding flow of material resources in the form of product components. The flow of materials $\Phi_0 = \Phi_1$ can be expressed as

$$\Phi_0 = \Phi_1 = \sum_{j=1}^{m} Vm_j \qquad (13.23)$$

where

Vm$_j$ is the volume of the j-th material necessary to manufacture and service the product components

m is the total number of materials

The difference between Φ_0 and Φ_1 consists only in the state of materials (raw and finished state, respectively).

Analysis of flows distribution through network N^* in Figure 13.5 (closed loop life cycle model) can be conducted in the same way as on N. In this case, there is just one sink (disposal y_1), and Φ_{13} and Φ_{14} are inflows for z_2 and z_0, respectively.

13.4.4 IDENTIFICATION OF FLOW PROPERTIES AND SIGNIFICANT STRUCTURE ELEMENTS

According to Equation 13.8, the resultant flow out of set X is equal to the resultant flow into set Y, and the common quantity is the value of the flow Val$_\Phi$ in the network N:

$$Val_\Phi = \Phi^+(X) - \Phi^-(X) = \Phi_0 \equiv \Phi^-(Y) - \Phi^+(Y) = \Phi_5 + \Phi_{13} + \Phi_{14} = \sum_{i=1}^{n} \beta_i(1 + nr_i)V_i \qquad (13.24)$$

This is a significant function because it quantifies the material resources in play for a product life cycle.

Several directed paths and cycles can be identified in network N. These elements are particularly significant with regard to the analysis of service and end-of-life flows of material resources. The following directed paths can be identified:

- $DP1_{REUSE} = \{x_1, a_0, z_0, a_1, z_1, a_2, z_2, a_3, z_3, a_4, z_4, a_{13}, y_2\}$, path of reused components (from which Φ'_{13} comes).
- $DP1_{RECYCL} = \{x_1, a_0, z_0, a_1, z_1, a_2, z_2, a_3, z_3, a_4, z_4, a_{14}, y_3\}$, path of recycled components (from which Φ'_{14} comes).
- $DP1_{DISP} = \{x_1, a_0, z_0, a_1, z_1, a_2, z_2, a_3, z_3, a_4, z_4, a_5, y_1\}$, path of disposed components (from which Φ'_5 comes).
- $DP2 = \{x_1, a_0, z_0, a_1, z_1, a_{10}, z_5, a_9, z_3\}$, path of substitutive components (from which Φ''_9 comes).
- $DP3_{RECYCL} = \{x_1, a_0, z_0, a_1, z_1, a_2, z_2, a_3, z_3, a_8, z_5, a_{11}, z_4, a_{14}, y_3\}$, path of recycled replaced components (from which Φ''_{14} comes).
- $DP3_{DISP} = \{x_1, a_0, z_0, a_1, z_1, a_2, z_2, a_3, z_3, a_8, z_5, a_{11}, z_4, a_5, y_1\}$, path of disposed replaced components (from which Φ''_5 comes).

- $DP4_{RECYCL} = \{x_1, a_0, z_0, a_1, z_1, a_{12}, z_4, a_{14}, y_3\}$, path of recycled scraps (from which Φ'''_{14} comes).
- $DP4_{DISP} = \{x_1, a_0, z_0, a_1, z_1, a_{12}, z_4, a_5, y_1\}$, path of disposed scraps (from which Φ'''_5 comes).

The following directed cycles can be identified:

- $DC1_{REPL} = \{z_3, a_8, z_5, a_9, z_3\}$, cycle of component replacing (from which Φ''_8 and Φ''_9 come).
- $DC1_{SERV} = \{z_3, a_6, z_5, a_7, z_3\}$, cycle of component service (from which Φ''_6 and Φ''_7 come).

If network N evolves in network N^*, which means life cycle evolves from open loop to closed loop model (Figures 13.4 and 13.5), the following directed paths previously identified form directed cycles:

- $DP1_{REUSE} \to DC1_{REUSE} = \{z_2, a_3, z_3, a_4, z_4, a_{13}, z_2\}$, cycle of reused components.
- $DP1_{RECYCL} \to DC1_{RECYCL} = \{z_0, a_1, z_1, a_2, z_2, a_3, z_3, a_4, z_4, a_{14}, z_0\}$, cycle of recycled components.
- $DP3_{RECYCL} \to DC3_{RECYCL} = \{z_0, a_1, z_1, a_2, z_2, a_3, z_3, a_8, z_5, a_{11}, z_4, a_{14}, z_0\}$, cycle of recycled replaced components.
- $DP4_{RECYCL} \to DC4_{RECYCL} = \{z_0, a_1, z_1, a_{12}, z_4, a_{14}, z_0\}$, cycle of recycled scraps.

Subnetworks are other elements that can be identified to analyze specific parts of the life cycle. They are particularly useful to focus on the main phases of the life cycle (Figures 13.2 and 13.4): production, use, and retirement. Conservation condition in the form (13.7) of balance between total outflow and inflow of the subnetworks can be used for the definition of flows distribution.

Production phase is modeled by subnetwork defined by $Z = \{z_0, z_1, z_2\}$, which total outflow is $\Phi^+(Z) = \Phi_3 + \Phi_{10} + \Phi_{12}$, and total inflow is $\Phi^-(Z) = \Phi_0$, under the conservation condition $\Phi^+(Z) = \Phi^-(Z)$. Similarly, use phase is modeled by subnetwork defined by $Z = \{z_3, z_5\}$, which total outflow is $\Phi^+(Z) = \Phi_4 + \Phi_{11}$ and total inflow is $\Phi^-(Z) = \Phi_3 + \Phi_{10}$, and retirement phase is modeled by subnetwork defined by $Z = \{z_4\}$, which total outflow is $\Phi^+(Z) = \Phi_5 + \Phi_{13} + \Phi_{14}$ and total inflow is $\Phi^-(Z) = \Phi_4 + \Phi_{11} + \Phi_{12}$, in both cases under the same conservation condition $\Phi^+(Z) = \Phi^-(Z)$.

13.4.5 ROLE OF NETWORK CAPACITIES

The network flows-based modeling of life cycle can be enriched by introducing the capacity function χ that imposes to each arc of the network the capacity constraints (13.2), particularly the constraint conditions $\Phi(a_k) \leq \chi(a_k)$. The values assumed by χ in each arc a_k can be intended as static or dynamic constraints for the related flows $\Phi(a_k)$. In static terms, χ expresses the capacity of each arc to support quantitatively the flow on the arc. Therefore capacity $\chi(a_k)$ is a function of the factors determinant in order that the flow $\Phi(a_k)$ could actually be realized. These factors reside in and emerge from the vertices v_i and v_j of the arc (when $\psi(a_k) = \{v_i, v_j\}$), which represent activities of the life cycle, and they are an expression of the properties and conditions of the systems supporting and managing these activities. With regards to arcs a_0 and a_1, χ depends on supply chain potential and efficiency. Capacity of arcs a_2 and a_3 depends on manufacturing technology and potential, and production process efficiency. Capacity of arcs a_4, a_{11}, a_{12} depends on potential and efficiency of scrap management and postconsumption logistics. Capacity of arcs a_6, a_7, a_8, a_9, a_{10} depends on service planning and management efficiency. Capacity of arcs a_{13}, a_{14} depends on potential and efficiency of reverse logistics and recycling technologies. Finally, capacity of arc a_5 depends on waste management efficiency.

In this perspective, the capacities χ express the potentials of technological and organization systems supporting the life cycle phases (*life cycle supporting systems*) with regard to actual realization of flows Φ. These potentials are results not only of properties and conditions of the systems supporting and managing life cycle activities, but of their relation type interactions also. The capacity constraints relate flows to capacities, becoming the bridge between product physical life cycle and its technological and organizational supporting systems. These systems in a perspective of environmental conscious life cycle planning and management assume key roles with regard to strategic issues such as remanufacturing, closed-loop supply chains, and reverse logistics (Ilgin and Gupta 2010).

13.4.6 MODELING EXTENSION

The potentials of life cycle supporting systems are not exclusively depending on technological level implemented and organizational choices. They depend on other factors also, external from the life cycle, and related to socio-economic phenomena, that express themselves in market demand, legislative restrictions, environmental awareness, and other factors determinant for the actual definition of flow of material resources in the life cycle, related or not at the environmental strategies implemented. This consideration extends the multilevel interpretation of the interactions that can be modeled by means of the graph-based life cycle model: material interactions at product physical dimension level (transaction type interactions, expressed by flows of material resources), organizational interactions at life cycle supporting systems level (relation type interactions, expressed by network capacities), and phenomenal interactions at socio-economic systems level (relation type interactions, influencing network capacities).

This extended life cycle modeling meets the necessities of facing up to the environmental issue related to production of industrial products resorting to the concept of product-system previously introduces (Section 13.2.1), that is a system characterized by complex dynamics, since the various actors involved interact through the application of reciprocal pressures, dependent on political, economic, and cultural factors (Young et al. 1997).

Concerning this aspect, the importance of conceiving flows as dynamic entities arises. As a dynamic entity, flow on an arc can be likened to the rate at which material resources are transported, or more in general, change their state. Similarly, the capacity of the arc can be thought as the maximum rate at which material resources can flow. In this perspective, the network flows-based modeling becomes a useful instrument to manage the time-dependent phenomena affecting life cycle development and temporal progression.

13.5 LIFE CYCLE SIMULATION AND OPTIMIZATION

The network flows-based model presented has wide potentialities. Since the flows of material resources depend on product physical properties (volumes, durability, separability, reusability, recyclability, all referable to materials, geometric properties, and shapes of component, product architecture, and junction systems), and capacities depend on supporting systems functional properties, flows are related to choices at product design and development level (including environmental strategies planning), and capacities are related to decision making and planning at systems organization and management level. Therefore, interventions on flows are product life cycle design interventions; interventions on capacities are LCM interventions. The proposed life cycle modeling approach can support both types of intervention, allowing to interrelate them also.

In consequence of the versatility of the model, and its analysis potentialities, this modeling approach can be used for several types of investigation, diversified in the following:

- Analysis domain—It can be diversified from single paths or cycles, to main flows (such as the three flows identified for the analysis of open loop life cycle model in Figure 13.4), from specific life cycle phases (by means of subnetworks) to the whole life cycle.
- Intervention level domain—It can be diversified from interventions on flows of material resources, which means at product life cycle design level, to interventions on technological and organization systems supporting life cycle phases (supply chain, manufacturing, service, postconsumption and reverse logistics, reuse and recycling, waste), which means at LCM level.
- Intervention typology domain—It can be diversified from life cycle simulation, oriented toward the matching and harmonization between product life cycle and systems supporting the life cycle, to life cycle optimization, oriented toward product design optimization and optimal support systems planning (configuration and arrangement).

13.5.1 USE OF THE NETWORK FLOWS-BASED MODELING FOR LIFE CYCLE SIMULATION

Defined the set of starting conditions, that is the product, the environmental strategies to be implemented in its life cycle, and the life cycle supporting systems, product life cycle can be simulated by means of the proposed modeling approach. The main objective consists in the matching and harmonization between product life cycle and systems supporting the life cycle.

The simulation process can be summarized in the following steps:

- Definition of flows distribution over the life cycle on the basis of product physical properties and environmental strategies characteristics.
- Definition of capacities distribution over the life cycle on the basis of supporting systems characteristics and environmental strategies requirements.

- Analysis of capacity constraints by means of comparison between flows and capacities.
- Changes in starting conditions of the simulation by means of interventions on flows (product design level interventions) and/or interventions on capacities (supporting systems planning).

The step of capacity constraints analysis has a key role, because within this step the potential interventions can be outlined. Capacity constraints that are not verified identify the critical elements of the network, i.e., the critical states in the interaction between product life cycle and its supporting systems. These critical states are expressed by lack of capacities of supporting systems compared with the expected flows of material resources. To remedy them it is possible to intervene at level of supporting systems planning and arrangement, to increase capacities, or at level of product design (including environmental strategies planning at life cycle design stage), to reduce exceeding flows of material resources in the life cycle.

On the contrary, if capacities exceed the expected flows, supporting systems potentialities can be further exploited. In these cases the network-based model allows to resort to the concept of residual capacity in an arc (the difference between the capacity and the flow), and the definition of the residual network (the network characterized by the residual capacities), both introduced in Section 13.3.2. To exploit the residual capacities is possible to intervene at level of product life cycle design, increasing product properties or strengthening environmental strategies, or to analyze the opportunities of using the residual network to support, partly or fully, the activities of life cycles of other products (e.g., this is the case of a life cycle that takes up open loops from another life cycle). This last option implies an intervention at the further level of systemic planning and management of interactions between different life cycles, typical of the industrial ecology perspective (Ayres and Ayres 1996, Graedel and Allenby 2009).

13.5.2 Use of the Network Flows-Based Modeling for Life Cycle Optimization

The life cycle simulation process previously described can be integrated with an optimization approach. The objective of the optimization can be a cost function, or an environmental impact function. The first case interprets a LCC perspective, the second one a LCA perspective (Section 13.2.4). Both cases can be supported by the network flows-based modeling and the minimum cost flow investigation (Section 13.3.2).

With these premises, the optimization problem can be formulated as follows: if each arc of the network N has an associated cost value c (or environmental impact value ei) that quantifies the cost (environmental impact) per unit flow on that arc, assuming that the flow cost (environmental impact) varies linearly with the amount of flow, minimize the objective function $\sum c_k \cdot \Phi_k \left(\sum ei_k \cdot \Phi_k \right)$, subject to capacity constraints (13.2) and conservation condition (13.4).

Usually, in this optimization problem flows are the variables. In the perspective proposed here, the variables of the problem can be extended to the capacities.

Again, as before, two different levels of interventions emerge: product design level, operating with network flows as variables; supporting systems planning level, operating with network capacities as variables.

13.6 CONCLUSIONS

The analysis and management of distribution of flows of material resources through the life cycle of a product constitute a powerful approach to the environmental issues in product design, production planning, and LCM. In an environmentally conscious perspective, the use of resources must be planned taking account of the environmental efficiency of the distribution of all the resource flows in the entire life cycle of the product. Furthermore, all the technological and organizational system supporting the various phases of the life cycle, from supply chain to reverse logistics, must be arranged and configured to exploit at best the potentialities of the strategies most effective to improve environmental efficiency of the life cycle, from resources reduction, to useful life extension and recovery at end-of-life strategies.

System modeling based on graph theory and network flows applications reveals itself as a versatile and powerful instrument that can be used for several types of investigation on material flows distribution, diversified in the analysis domain (from single flow cycles to main flows, from specific life cycle phases to the whole life cycle), in the intervention level domain (from interventions on flows of material resources, at product life cycle design level, to interventions on technological and organization systems supporting life cycle phases, at LCM level), in the intervention typology domain (from life cycle simulation, oriented toward the matching and harmonization between product life cycle and systems supporting the life cycle, to life cycle optimization, oriented toward product design optimization and optimal support systems planning).

Furthermore, the graph-based life cycle model lends itself to be extended according a multilevel interpretation of the interactions that can be modeled: material interactions at product physical dimension level; organizational interactions at life cycle supporting systems level; phenomenal interactions at socio-economic systems level. Concerning this last aspect, the importance of conceiving life cycle flows as dynamic entities arises, outlining an interesting potentiality of the network flows-based modeling, to be further investigated, as an instrument to manage the time-dependent phenomena affecting life cycle of industrial products.

REFERENCES

Ahuja, R. K., T. L. Magnanti, and J. B. Orlin. 1993. *Network Flows. Theory, Algorithms, and Applications*. Upper Saddle River, NJ: Prentice Hall.

Alting, L. 1993. Life-cycle design of products: A new opportunity for manufacturing enterprises. In *Concurrent Engineering: Automation, Tools and Techniques*, Ed. A. Kusiak, pp. 1–17. New York: John Wiley & Sons.

Asiedu, Y. and P. Gu. 1998. Product life cycle cost analysis: State of the art review. *International Journal of Production Research* 36: 883–908.

Ayres, R. U. and L. W. Ayres. 1996. *Industrial Ecology: Towards Closing the Materials Cycle*. Cheltenham, U.K.: Edward Elgar.

Bazaraa, M. S. and J. J. Jarvis. 1977. *Linear Programming and Network Flows*. New York: John Wiley & Sons.

Billatos, S. B. and N. A. Basaly. 1997. *Green Technology and Design for the Environment*. Washington, DC: Taylor & Francis Group.

Bondy, J. A. and U. S. R. Murty. 1982. *Graph Theory with Applications*. New York: North Holland.

Chen, W.-K. 1997. *Graph Theory and Its Engineering Applications*. Singapore: World Scientific Publishing.

Chen, W.-K. 2003. *Net Theory and Its Applications. Flows in Networks*. London, U.K.: Imperial College Press.

Curran, M. A. 1996. *Environmental Life-Cycle Assessment*. New York: John Wiley & Sons.

Dhillon, B. S. 2009. *Life Cycle Costing for Engineers*. Boca Raton, FL: CRC Press.

Fabrycky, W. J. and B. S. Blanchard. 1991. *Life Cycle Cost and Economic Analysis*. Englewood Cliffs, NJ: Prentice Hall.

Fath, B. D. and B. C. Patten. 1998. Network synergism: Emergence of positive relations in ecological systems. *Ecological Modelling* 107: 127–143.

Fiksel, J. 2009. *Design for the Environment: A Guide to Sustainable Product Development*, 2nd edn. New York: McGraw Hill.

Giudice, F., G. La Rosa, and A. Risitano. 2004. Simulation of product life cycle: Methodological basis and analysis models. In *Proceedings of Design 2004—Eighth International Design Conference*, 2004, Dubrovnik, Croatia, pp. 1527–1538.

Giudice, F., G. La Rosa, and A. Risitano. 2006. *Product Design for the Environment: A Life Cycle Approach*. Boca Raton, FL: CRC/Taylor & Francis Group.

Graedel, T. E. and B. R. Allenby. 2009. *Industrial Ecology and Sustainable Engineering*. Englewood Cliffs, NJ: Prentice Hall.

Gross, J. L. and J. Yellen. 2004. *Handbook of Graph Theory*. Boca Raton, FL: CRC Press.

Guinée, J. B. 2002. *Handbook on Life Cycle Assessment: Operational Guide to the ISO Standards*. Dordrecht, the Netherlands: Kluwer Academic Publisher.

Hundal, M. S. 2002. Introduction to design for the environment and life cycle engineering. In *Mechanical Life Cycle Handbook*, Ed. M. S. Hundal, pp. 1–26. New York: Marcel Dekker.

Ilgin, M. A. and S. M. Gupta. 2010. Environmentally conscious manufacturing and product recovery: A review of the state of the art. *Journal of Environmental Management* 91: 563–591.

Ishii, K. 1995. Life-cycle engineering design. *ASME Journal of Mechanical Design* 117: 42–47.

Keoleian, G. A. and D. Menerey. 1993. *Life Cycle Design Guidance Manual*. Cincinnati, OH: U.S. Environmental Protection Agency.

Kimura, F., T. Hata, and H. Suzuki. 1998. Product quality evaluation based on behavior simulation of used product. *Annals of the CIRP* 47: 119–122.

Molina, A., J. M. Sánchez, and A. Kusiak. 1998. *Handbook of Life Cycle Engineering: Concepts, Models and Technologies*. Dordrecht, the Netherlands: Kluwer Academic Publisher.

Navin-Chandra, D. 1991. Design for environmentability. In *Proceedings of ASME Conference on Design Theory and Methodology*, 1991, Miami, FL, DE-31, pp. 119–125.

Ramani, K. et al. 2010. Integrated sustainable life cycle design: A review. *Journal of Mechanical Design* 132: 1–15.

Sakai, N., T. Tomiyama, and Y. Umeda. 2001. Life cycle simulation with prediction of incoming flow of discarded products. In *Proceedings of Eighth CIRP International Seminar in Life Cycle Engineering*, 2001, Varna, Bulgaria, pp. 65–73.

Sun, J. et al. 2003. Design for environment: Methodologies, tools, and implementation. *Journal of Integrated Design and Process Science* 7: 59–75.

Takata, S. et al. 2003. Framework for systematic evaluation of life cycle strategy by means of life cycle simulation. In *Proceedings of EcoDesign 2003: Third International Symposium on Environmentally Conscious Design and Inverse Manufacturing*, 2003, Tokyo, Japan, pp. 198–205.

Tipnis, V. A. 1998. Evolving issues in product life cycle design: Design for sustainability. In *Handbook of Life Cycle Engineering: Concepts, Models and Technologies*, Eds. A. Molina, J. M. Sánchez, and A. Kusiak, pp. 413–459. Dordrecht, the Netherlands: Kluwer Academic Publisher.

Vigon, B. W. et al. 1993. *Life-Cycle Assessment: Inventory Guidelines and Principles*. Cincinnati, OH: U.S. Environmental Protection Agency.

Wang, H.-F. and S. M. Gupta. 2011. *Green Supply Chain Management: Product Life Cycle Approach*. New York: McGraw Hill.

Wanyama, W. et al. 2003. Life-cycle engineering: Issues, tools and research. *International Journal of Computer Integrated Manufacturing* 16: 307–316.

Young, P., G. Byrne, and M. Cotterell. 1997. Manufacturing and the environment. *International Journal of Advanced Manufacturing Technology* 13: 488–493.

Zust, R. and G. Caduff. 1997. Life-cycle modeling as an instrument for life-cycle engineering. *Annals of the CIRP* 46: 351–354.

14 Delivery and Pickup Problems with Time Windows
Strategy and Modeling

Ying-Yen Chen and Hsiao-Fan Wang

CONTENTS

14.1 INTRODUCTION

As global resources are rapidly decreasing, implementing reverse logistics has shown that it can significantly reduce the cost of returned merchandise, improve customer satisfaction, and increase the profit of enterprises. Recently, many enterprises have incorporated reverse logistics with conventional forward-only logistics to form a closed-loop supply chain. A state-of-the-art survey of reverse- and close-loop supply chains can be found in Ilgin and Gupta (2010). A comprehensive overview of a green supply chain can be found in Wang and Gupta (2011). Within a close-loop supply chain, the logistics between the distribution/collection center and the customers is the most complicated part because it is related to a bi-directional logistics regarding delivery and pickup activities. In the literature, such problems have been referred to as delivery and pickup problem (DPP).

The DPP has been widely applied. For example, it is frequently encountered in the distribution system of grocery store chains. Each grocery store may have a demand for both delivery (cf. fresh food or soft drinks) and pickup (cf. outdated items or empty bottles). The foundry industry is another example in Dethloff (2001). Collection of used sand and delivery of purified reusable sand at the same customer location are carried out.

In order to achieve low carbon emission and high resource productivity, enterprises need to incorporate the reverse logistics into the regular forward logistics to perform deliveries and pickups when the demands of the customers have to be satisfied in the most economic way. One way to reduce both the carbon emissions toward the environment and the operational cost of an enterprise is to lower the total traveling distance and the number of vehicles when the demands of the customers are satisfied. This win–win situation benefits the enterprises, the governments, and human beings; consequently, more efficient and effective bi-directional logistics are desired. Based on this requirement, four models derived from three logistic strategies are reviewed in this chapter. Through these reviews, practitioners can have a deeper understanding of the logistic strategies and the derived schemes.

This chapter is organized as follows: Section 14.2 reviews the literature related to the issues in interest. Sections 14.3 through 14.6 review and discuss four main categories of DPPs. Section 14.7 presents an illustrative example to verify those derived models. Section 14.8 provides the comparisons among these schemes. Finally, conclusions are drawn in Section 14.9.

14.2 RELATED WORK

The DPP was developed from the vehicle routing problem (VRP). The VRP originally focused on how to dispatch a group of vehicles to serve a group of customers with a given demand when the minimum operational cost is desired. This section will first review a variety of VRPs that are related to the DPP as shown in Figure 14.1. In the capacitated vehicle routing problem (CVRP), the vehicles are constrained by limited capacities (Prins 2004). Within the framework of the CVRP, two major categories regarding routing activities are extended: the single activity CVRP, which considers only delivery or pickup, and the multiple activities CVRP, which consider both delivery and pickup.

One common extension of the single activity CVRP is the multi-depot capacitated vehicle routing problem (MDVRP), which allows multiple depots (Cordeau et al. 1997). Another common extension is the vehicle routing problem with time windows (VRPTW) in which customers can request that their service be performed within some specific time windows (Cordeau et al. 2001). A further extension is the vehicle routing problem with multiple time windows (VRPMTW), allowing each customer to have multiple time windows (Wang and Chiu 2009).

With regard to the multi-activity CVRP, it also includes two subclasses: the pickup and delivery problems (PDPs) and the DPPs. In the PDP, each customer's request consists of picking up an amount of goods at one location and delivering

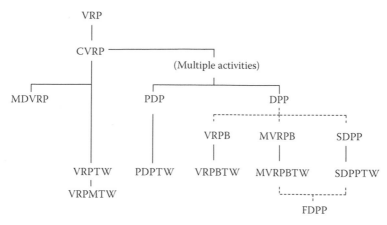

FIGURE 14.1   A variety of VRPs that are related to the DPP.

it to another location (Renaud et al. 2002). The PDP then has been extended to the pickup and delivery problem with time windows (PDPTW) to allow the customers being served within some specific time windows (Ropke and Pisinger 2006a).

While the PDP is often referred to as a mail express system, the DPP can be regarded as a bi-directional logistic problem (Dethloff 2001). In the general DPP, two types of customers are served from a single depot by a fleet of vehicles. The first type of customers is known as "linehaul" customers, who require deliveries of goods to their specific locations. The second type is known as "backhaul" customers, who require pickups from their specific locations. Parragh et al. (2008) provided a survey on both the PDP and the DPP.

There are three main strategies for the DPP: (1) delivery-first, pickup-second; (2) mixed deliveries and pickups; and (3) simultaneous deliveries and pickups.

Delivery-first, pickup-second strategy: vehicles can only pick up goods after they have finished delivering their entire load. One reason for this is that it may be difficult to rearrange delivery and pickup goods on the vehicles. Such an assumption makes the implementation issue easier because accepting pickups before finishing all deliveries results in a fluctuating load. This may cause the vehicle to be overloaded during its trip (even if the total delivery and the total pickup loads are not above the vehicle capacity), resulting in an infeasible vehicle tour.

Mixed deliveries and pickups strategy: linehauls and backhauls can occur in any sequence on a vehicle route. This strategy releases the constraints that pickups are only accepted after finishing all deliveries. When there are no difficulties in rearranging the load on the vehicle, this strategy is more attractive to backhaul customers and enterprises. The satisfaction of backhaul customers can be higher since they can be served earlier. Moreover, the enterprises can save the transportation cost since the sequence of deliveries and pickups can be arranged in a more economical way.

Simultaneous pickups and deliveries: simultaneously performing delivery and pickup services with a single stop for each customer. In some applications, customers can have both a delivery and a pickup demand. They may not accept to be serviced separately for the delivery and pickup they require because handling effort is caused by both activities. In this situation, simultaneous pickups and deliveries are the only choice.

Referring to these service strategies, the DPP is divided into three categories: the vehicle routing problem with backhauls (VRPB), the mixed vehicle routing problem with backhaul (MVRPB), and the simultaneous delivery and pickup problem (SDPP). In some literature, the SDPP was called the vehicle routing problem with simultaneous delivery and pickup (VRPSDP).

In some practical applications that the pickup services are performed only after the last delivery is made, the delivery-first, pickup-second strategy is applied and the VRPB was modeled (Ropke and Pisinger 2006b). In theory, this restriction reduces the complexity of the problem, and, in practice, it avoids the problems that may arise because of rearranging goods on the vehicle and supports the fact that linehaul customers have priority over backhaul customers. However, it can easily be realized that ignoring this restriction may reduce the total traveling cost. Therefore, the mixed deliveries and pickups strategy is applied and the MVRPB was derived from the VRPB by relaxing the precedence restriction. Please refer to Wade and Salhi (2002), Nagy and Salhi (2005), Crispim and Brandao (2005), and Tütüncüa et al. (2009).

For the environmentally motivated distribution/redistribution systems where most customers have both a pickup and a delivery demand, the simultaneous deliveries and pickups strategy is more attractive. To reduce the service effort and the interference to customers, performing simultaneous delivery and pickup services with a single stop for each customer is favored from both the service suppliers' and customers' viewpoints. Based on the simultaneous deliveries and pickups strategy, the SDPP was proposed and has been studied extensively. Please refer to Min (1989), Dethloff (2001), Nagy and Salhi (2005), Crispim and Brandao (2005), Chen and Wu (2006), Dell'Amico et al. (2006), Montané and Galvao (2006), Berbeglia et al. (2007), Bianchessi and Righini (2007), Hoff et al. (2009), and Ai and Kachitvichyanukul (2009).

By allowing customers to appoint time windows, the VRPB, the MVRPB, and the SDPP have been extended to the vehicle routing problem with backhauls and time windows (VRPBTW), the mixed vehicle routing problem with backhauls and time windows (MVRPBTW), and the simultaneous delivery and pickup problem with time windows (SDPPTW). The SDPPTW has the advantage that simultaneously performing delivery and pickup reduces the accessing time to the customers; on the contrary, the MVRPBTW has the advantage of flexible delivery and pickup. These three problems will be reviewed in Sections 14.3 through 14.5. To take both advantages, a newer model that retains the flexibility of mixing delivery and pickup operations while carrying out time saving from simultaneously performing delivery and pickup is reviewed in Section 14.6. This new scheme is named the flexible delivery and pickup problem with time windows (FDPPTW).

14.3 VEHICLE ROUTING PROBLEM WITH BACKHAULS AND TIME WINDOWS

The objective of the VRPBTW is to find a set of routes with the following features that minimizes both the number of vehicle and the total traveling distance:

- All customers demand a delivery service (linehaul) or a pickup service (backhaul).
- Each customer can request his or her delivery or pickup demand to be served within a specific time window.
- Each demand (whether pickup demand or delivery demand) cannot be split and must be served by only one vehicle.
- Each vehicle is restricted by capacity constraints.
- Each service consumes a service time.
- Both the distribution center and the collection center have their time horizon.
- A vehicle is allowed to arrive at a customer before the relevant time window, but cannot serve the customer until the time window opens.
- In any route, all linehaul customers must precede all backhaul customers.

The infrastructure of the VRPBTW network can be seen in Figure 14.2. The black and the white squares indicate the distribution center and the collection center, respectively. The white circles and black triangles indicate linehaul and backhaul customers, correspondingly. The solid arrows indicate the movements.

To formulate such problems, let us first assume that there are n_D linehaul customers and n_p backhaul customers served by m available vehicles; therefore, if a customer requires both services of delivery and pickup, the customer will be labeled twice in the network. Furthermore, if 0 denotes a distribution center (DC), then we shall denote k as a collection center (CC) with $k = n_D + n_P + 1$.

Before the model is defined, other notations are given as follows:

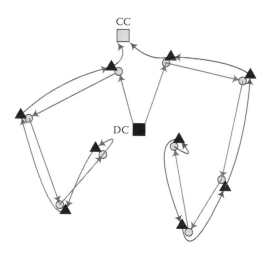

FIGURE 14.2 Infrastructure of the VRPBTW network.

Notations

Sets

J set of all customers, $J = \{j \mid j = 1,\dots,n_D + n_P\}$

J_0 set of all customers plus DC, $J_0 = \{0\} \cup J$

J_k set of all pickup customers plus CC, $J_k = J \cup \{k\}$

J_{0k} set of all nodes, $J_{0k} = \{0\} \cup J \cup \{k\}$

J_D set of all delivery customers, $J_D = \{j \mid j = 1, \dots, n_D\}$

J_P set of all pickup customers, $J_P = \{j \mid j = n_D + 1, \dots, n_D + n_P\}$

J_{0D} set of all delivery customers plus DC, $J_{0D} = \{0\} \cup J_D$

J_{Pk} set of all pickup customers plus CC, $J_{Pk} = J_P \cup \{k\}$

V set of all vehicles, $V = \{v \mid v = v_1, \dots, v_m\}$

Coefficients

a_0 earliest departure time of any vehicle from DC, $a_0 \in R^+$

a_j earliest service time of customer $j \in J, a_j \in R^+$

b_j latest service time of customer $j \in J, b_j \in R^+$

b_k latest arrival time that a vehicle must return CC, $b_k \in R^+$

c_{ij} distance between nodes $i \in J_0, j \in J_k; i \neq j, c_{ij} \in R^+$

d_j delivery demand of customer $j \in J_D, d_j \in Z^+$

g_v dispatching cost of vehicle $v, g_v \in R^+$

M an arbitrary large constant

p_i pickup demand of customer $i \in J_P, p_i \in Z^+$

q_v capacity of vehicle $v, q_v \in R^+$

s_j service time of customer $j \in J, s_j \in R^+$

t_{ij} traveling time between nodes $i \in J_0, j \in J_k; i \neq j, t_{ij} \in R^+$

α a parameter indicating the trade-off between dispatching cost and traveling cost, $\alpha \in [0,1]$

Decision variables

L_{0v} load of vehicle $v \in V$ when leaving DC, $L_{0v} \in Z^+$

L_{kv} load of vehicle $v \in V$ when arriving CC, $L_{kv} \in Z^+$

T_{0v} departure time of vehicle $v \in V$ at DC, $T_{0v} \in R^+$

T_j time to begin service at customer $j \in J, T_j \in R^+$

T_{kv} arrival time of vehicle $v \in V$ at CC, $T_{kv} \in R^+$

x_{ijv} = 1 if vehicle v travels directly from node $i \in J_0$ to node $j \in J_k$; = 0 otherwise.

The VRPBTW is then formulated into a mixed binary integer programming model denoted by Model VRPBTW as follows.

14.3.1 Model VRPBTW

$$\text{Minimize } z = \alpha \sum_{v \in V} \sum_{j \in J} g_v x_{0jv} + (1-\alpha) \sum_{(i,j) \in (J_0,J_k) \setminus (J_P,J_D)} \sum_{v \in V} c_{ij} x_{ijv} \qquad (14.1)$$

(Minimize total vehicle dispatching cost and total traveling cost) subject to

$$\sum_{i \in J_0} \sum_{v \in V} x_{ijv} = 1 \quad \forall j \in J \tag{14.2}$$

(Service all customer nodes exactly once)

$$\sum_{i \in J_0} x_{ihv} = \sum_{i \in J_k} x_{hjv} \quad \forall h \in J, \forall v \in V \tag{14.3}$$

(Arrive at and leave each customer with the same vehicle)

$$\sum_{j \in J} x_{0jv} = \sum_{i \in J} x_{ikv} \quad \forall h \in J, \forall v \in V \tag{14.4}$$

(Vehicles that depart from DC should finally return CC)

$$L_{0v} = \sum_{i \in J_{0D}} \sum_{j \in J_D} d_j x_{ijv} \quad \forall v \in V \tag{14.5}$$

(Initial vehicle loads)

$$L_{kv} = \sum_{i \in J_P} \sum_{j \in J_{Pk}} p_i x_{ijv} \quad \forall v \in V \tag{14.6}$$

(Final vehicle loads)

$$L_{0v} \leq q_v \quad \forall v \in V \tag{14.7}$$

$$L_{kv} \leq q_v \quad \forall v \in V \tag{14.8}$$

(Vehicle capacity constraints)

$$T_j \geq T_{0v} + t_{0j} - M(1 - x_{0jv}) \quad \forall j \in J, \forall v \in V \tag{14.9}$$

$$T_j \geq T_i + s_i + t_{ij} - M\left(1 - \sum_{v \in V} x_{ijv}\right) \quad \forall i \in J, \forall j \in J \tag{14.10}$$

$$T_{kv} \geq T_i + s_i + t_{ik} - M(1 - x_{ikv}) \quad \forall i \in J, \forall v \in V \tag{14.11}$$

$$a_0 \leq T_{0v} \quad \forall v \in V \tag{14.12}$$

$$a_j \leq T_j \leq b_j \quad \forall j \in J \tag{14.13}$$

$$T_{kv} \leq b_k \quad \forall v \in V \tag{14.14}$$

(Ensure feasibility of the time schedule)

$$x_{ijv} \in \{0, 1\} \quad \forall (i, j) \in (J_0, J_k) \setminus ((J_P, J_D) \cup (0, k)), \forall v \in V \tag{14.15}$$

For the VRPBTW, Gelinas et al. (1995) constructed 45 test problems of size 25, 50, and 100, based on the data provided by Solomon (1987). Thangiah et al. (1996) introduced a data set containing 24 large problems of size 250 and 500. Gelinas et al. (1995) proposed an exact algorithm, a branch-and-bound approach based on column generation. Thangiah et al. (1996) proposed a route construction heuristic with some local search heuristics improving the initial solutions. Duhamel et al. (1997) proposed a tabu search heuristic. Reimann et al. (2002) proposed an ant system algorithm based on insertion procedure. Zhong and Cole (2005) proposed a cluster-first, route-second heuristic with adapted sweep algorithm, a guided local search heuristic, and a section planning technique.

The VRPBTW scheme is very useful for some practical situations that line-haul customers have precedence over backhaul customers. However, a solution that allows mixed routes can produce significantly lower route costs than a solution to the VRPBTW. In Section 14.4, we will discuss the MVRPBTW scheme that is without the customer precedence.

14.4 MIXED VEHICLE ROUTING PROBLEM WITH BACKHAULS AND TIME WINDOWS

The description of the MVRPBTW is the same as the description of the VRPBTW except the last feature statement. In the MVRPBTW, there is no customer precedence. The infrastructure of the MVRPBTW network can be seen in Figure 14.3. One can see that backhaul customers are allowed to be served before linehaul customers.

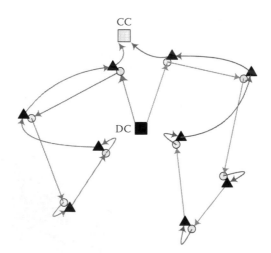

FIGURE 14.3 Infrastructure of the MVRPBTW network.

The difference between the VRPBTW scheme and the MVRPBTW scheme is whether they are with or without customer precedence. The formulation of the MVRPBTW is just slightly different from the formulation of the VRPBTW. The objective function is modified as follows:

$$\text{Minimize } z = \alpha \sum_{v \in V} \sum_{j \in J} g_v x_{ojv} + (1-\alpha) \sum_{(i,j) \in (J_0, J_k) \setminus (0,k)} \sum_{v \in V} c_{ij} x_{ijv} \quad (14.16)$$

(Minimize total vehicle dispatching cost and total traveling cost.)

Since we can mix linehaul and backhaul customers freely within a route, it is a little bit complicated to examine load feasibility. The combined loads associated with delivery and pickup of each vehicle must not exceed the given capacity. In order to cope with this restriction, we need to introduce a new type of auxiliary variables as follows.

L_j Remaining load of a vehicle after having served customer $j \in J$, $L_j \in Z^+$

With the addition of new variables, the constraints of load (constraints (14.5) through (14.8)) then could be modified as follows:

$$L_{0v} = \sum_{i \in J_0} \sum_{j \in J} d_j x_{ijv} \quad \forall v \in V \quad (14.17)$$

(Initial vehicle loads)

$$L_j \geq L_{0v} - d_j + p_j - M(1 - x_{0jv}) \quad \forall j \in J, \forall v \in V \quad (14.18)$$

(Vehicle loads after first customer)

$$L_j \geq L_i - d_j + p_j - M\left(1 - \sum_{v \in V} x_{ijv}\right) \quad \forall i \in J, \forall j \in J \quad (14.19)$$

(Vehicle loads "en route")

$$L_{0v} \leq q_v \quad \forall v \in V \quad (14.20)$$

$$L_j \leq q_v + M(1 - x_{ijv}) \quad \forall j \in J, \forall v \in V \quad (14.21)$$

(Vehicle capacity constraints)

The binary constraints of decision variables are modified as follows:

$$x_{ijv} \in \{0, 1\} \quad \forall (i, j) \in (J_0, J_k) \setminus (0, k)), \forall v \in V \quad (14.22)$$

Consequently, objective function (14.16), constraints (14.2) through (14.4), (14.17) through (14.21), (14.9) through (14.14), and (14.22) constitute the formulation of the MVRPBTW. This formulation is called Model MVRPBTW.

14.4.1 MODEL MVRPBTW

$$\text{Minimize } z = \alpha \sum_{v \in V} \sum_{j \in J} g_v x_{0jv} + (1-\alpha) \sum_{(i,j) \in (J_0, J_k) \backslash (0,k)} \sum_{v \in V} c_{ij} x_{ijv} \qquad (14.16)$$

(Minimize total vehicle dispatching cost and total traveling cost)

subject to

$$\sum_{i \in J_0} \sum_{v \in V} x_{ijv} = 1 \quad \forall j \in J \qquad (14.2)$$

(Service all customer nodes exactly once)

$$\sum_{i \in J_0} x_{ihv} = \sum_{i \in J_k} x_{hjv} \quad \forall h \in J, \forall v \in V \qquad (14.3)$$

(Arrive at and leave each customer with the same vehicle)

$$\sum_{j \in J} x_{0jv} = \sum_{i \in J} x_{ikv} \quad \forall h \in J, \forall v \in V \qquad (14.4)$$

(Vehicles that depart from DC should finally return CC)

$$L_{0v} = \sum_{i \in J_0} \sum_{j \in J} d_j x_{ijv} \quad \forall v \in V \qquad (14.17)$$

(Initial vehicle loads)

$$L_j \geq L_{0v} - d_j + p_j - M(1 - x_{0jv}) \quad \forall j \in J, \forall v \in V \qquad (14.18)$$

(Vehicle loads after first customer)

$$L_j \geq L_i - d_j + p_j - M\left(1 - \sum_{v \in V} x_{ijv}\right) \quad \forall i \in J, \forall j \in J \qquad (14.19)$$

(Vehicle loads "en route")

$$L_{0v} \leq q_v \quad \forall v \in V \qquad (14.20)$$

$$L_j \leq q_v + M(1 - x_{ijv}) \quad \forall j \in J, \forall v \in V \qquad (14.21)$$

(Vehicle capacity constraints)

$$T_j \geq T_{0v} + t_{0j} - M(1 - x_{0jv}) \quad \forall j \in J, \forall v \in V \qquad (14.9)$$

$$T_j \geq T_i + s_i + t_{ij} - M \left(1 - \sum_{v \in V} x_{ijv} \right) \quad \forall i \in J, \forall j \in J \tag{14.10}$$

$$T_{kv} \geq T_i + s_i + t_{ik} - M(1 - x_{ikv}) \quad \forall i \in J, \forall v \in V \tag{14.11}$$

$$a_0 \leq T_{0v} \quad \forall v \in V \tag{14.12}$$

$$a_j \leq T_j \leq b_j \quad \forall j \in J \tag{14.13}$$

$$T_{kv} \leq b_k \quad \forall v \in V \tag{14.14}$$

(Ensure feasibility of the time schedule)

$$x_{ijv} \in \{0, 1\} \quad \forall (i, j) \in (J_0, J_k) \backslash (0, k)), \forall v \in V \tag{14.22}$$

For the MVRPBTW, Kontoravdis and Bard (1995) constructed 27 new problems from Solomon's VRPTW problems and proposed a greedy randomized adaptive search procedure to solve them. Zhong and Cole's method was also applied to the MVRPBTW and tested among Kontoravdis and Bard's test problems. Cheung and Hang (2003) coped with an extended problem of the MVRPBTW, which was applied to a real application "land transportation of air-cargo freight forwarders."

In some cases, to reduce the service effort and interference to the customers, simultaneously performing delivery and pickup services with a single stop for each customer is favored from both the service suppliers' and customers' viewpoints. In Section 14.5, we will discuss the SDPPTW scheme where delivery and pickup services are performed simultaneously.

14.5 SIMULTANEOUS DELIVERY AND PICKUP PROBLEM WITH TIME WINDOWS

The description of the SDPPTW is the same as the description of the VRPBTW except the first, the third, and the last feature statements. These three statements should be modified as follows:

- All customers demand a delivery service (linehaul), or a pickup service (backhaul), or both.
- Each demand (whether pickup demand or delivery demand) cannot be split and each customer must be served by only one vehicle.
- The delivery service and the pickup service for the same customer should be performed simultaneously.

The infrastructure of the SDPPTW network can be seen in Figure 14.4. One can see that some arrows are dotted and not solid. Since the driver will not need to reaccess

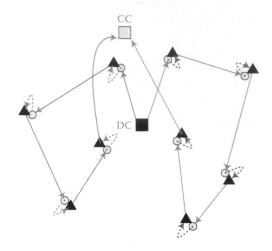

FIGURE 14.4 Infrastructure of the SDPPTW network.

a customer if he or she picks up stuff right after delivering goods, we use a dot arrow to describe that a pickup service for a customer is performed right after his or her delivery service.

If an enterprise wants to change their logistic scheme from the MVRPBTW scheme to the SDPPTW scheme, there are some adjustments that need to be made. First, a linehaul customer and a backhaul customer in the same location should be aggregated into a customer. Also, the service times and the time windows of two different services need to be integrated into single service time and single time window. Pure linehaul customers and pure backhaul customers should add a zero pickup or delivery demand. After the aforementioned aggregations and augmentations, the formulation of the SDPPTW could be arrived at by just adjusting some assumptions and the definitions of some notations of the MVRPBTW and then totally copying Model MVRPBTW. Let us see how it is done. Let us assume that there are n customers who demand both delivery and pickup service. Let k $(k = n + 1)$ denote the collection center. The customer set J should be modified as follows:

J set of all customers, $J = \{j \mid j = 1, \ldots, n\}$

In this formulation, customer sets J_0, J_k, and J_{0k} are not changed and remain, but customer sets J_D, J_P, J_{0D}, and J_{Pk} are not used. The demand coefficients should be modified as follows:

d_j delivery demand of customer $j \in J$, $d_j \in \mathbf{Z}^+$

p_i pickup demand of customer $i \in J$, $p_i \in \mathbf{Z}^+$

With the preceding adjustments, objective function (14.16), constraints (14.2) through (14.4), (14.17) through (14.21), (14.9) through (14.14), and (14.22) also constitute the formulation of the SDPPTW. This formulation is called Model SDPPTW. Note that once a linehaul customer and a backhaul customer in the same location are aggregated into a customer j, the service time of this customer (s_j) will be less than the summation of the service times of the original linehaul and backhaul customers. For the SDPPTW, Angelelli and Mansini (2002) constructed

29 test problems with size 20, based on the data provided by Solomon (1987). They implemented a branch-and-price approach based on a set covering formulation for the master problem. A relaxation of the elementary shortest-path problem with time windows and capacity constraints is used as pricing problem. Wang and Chen (2012a) constructed 56 test problems with size 100 by modifying Solomon's benchmark and proposed a coevolutionary algorithm to solve them.

The SDPPTW has the advantage that simultaneously performing delivery and pickup reduces the accessing time to the customers; on the contrary, the MVRPBTW has the advantage of freely mixing deliveries and pickups. To take both advantages, a newer model that retains the flexibility of mixing pickup and delivery operations while carrying out the time saving from simultaneously performing delivery and pickup, is reviewed in the next section. This kind of problem is named the flexible delivery and pickup problem with time windows (FDPPTW).

14.6 FLEXIBLE DELIVERY AND PICKUP PROBLEM WITH TIME WINDOWS

The description of the FDPPTW is the same as the description of the MVRPBTW except the first feature statement. The first feature statement should be modified as the following two statements:

1. All customers demand a delivery service (linehaul), or a pickup service (backhaul), or both.
2. If the delivery service and the pickup service for the same customer are performed simultaneously, the accessing time should not be double counted in service times.

The infrastructure of the FDPPTW network can be seen in Figure 14.5. One can see that some arrows are dotted but solid. Since a driver will not need to reaccess to a customer if he/she picks up stuff right after delivering goods, we use a dot arrow to

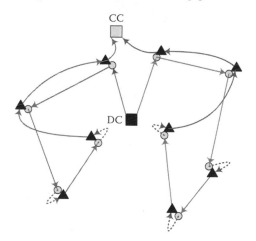

FIGURE 14.5   Infrastructure of the FDPPTW network.

describe that a pickup service for a customer is performed right after his or her delivery service. In Figure 14.5, there are five customers that are served delivery and pickup simultaneously. The other four customers are served delivery first but pickup later.

For the formulation of the FDPPTW, pure linehaul customers and pure backhaul customers should add a zero pickup or delivery demand. Then assume that there are n customers, and each customer demands both a delivery and a pickup service. When modeling, $2n$ customers are generated with n new customer i, $i = 1, \ldots, n$, demanding only delivery service, and n new customer $n + i$, $i = 1,\ldots, n$, demanding only pickup service. Let k $(k = 2n + 1)$ denote the collection center. The customer sets J and J_D should be modified as follows:

J set of all customers, $J = \{j \mid j = 1, \ldots, 2n\}$

J_D set of all delivery customers, $J_D = \{j \mid j = 1, \ldots, n\}$

If the delivery and the pickup service for a customer are performed simultaneously, the accessing time should not be double counted in service times. New variables should be introduced as follows:

r_j reduced accessing time if customer $n + j$ is serviced next to customer $j, j \in J_D$. Consider constructing the model for the SDPP by revising from Model MVPBTW. With the addition of new variables, the constraints (14.11) could be modified as follows:

$$T_i + s_i + t_{ij} - M\left(1 - \sum_{v \in V} x_{ijv}\right) \leq T_j \quad \forall i \in J, \forall j \in J - \{n + i\} \tag{14.23}$$

$$T_i + s_i - r_i + t_{i,n+i} - M\left(1 - \sum_{v \in V} x_{i,n+i,v}\right) \leq T_{n+i} \quad \forall i \in J \tag{14.24}$$

With these adjustments, objective function (14.16), constraints (14.2) through (14.4), (14.17) through (14.21), (14.9), (14.10), (14.23), (14.24), (14.12) through (14.14), and (14.22) constitute the formulation of the FDPPTW. This formulation is called Model FDPPTW.

14.6.1 Model FDPPTW

$$\text{Minimize } z = \alpha \sum_{v \in V} \sum_{j \in J} g_v x_{ojv} + (1 - \alpha) \sum_{(i,j) \in (J_0, J_k) \setminus (0,k)} \sum_{v \in V} c_{ij} x_{ijv} \tag{14.16}$$

(Minimize total vehicle dispatching cost and total traveling cost.)

subject to

$$\sum_{i \in J_0} \sum_{v \in V} x_{ijv} = 1 \quad \forall j \in J \tag{14.2}$$

(Service all customer nodes exactly once)

$$\sum_{i \in J_0} x_{ihv} = \sum_{i \in J_k} x_{hjv} \quad \forall h \in J, \forall v \in V \tag{14.3}$$

(Arrive at and leave each customer with the same vehicle)

$$\sum_{j \in J} x_{0jv} = \sum_{i \in J} x_{ikv} \quad \forall h \in J, \forall v \in V \tag{14.4}$$

(Vehicles which depart from DC should finally return CC)

$$L_{0v} = \sum_{i \in J_0} \sum_{j \in J} d_j x_{ijv} \quad \forall v \in V \tag{14.17}$$

(Initial vehicle loads)

$$L_j \geq L_{0v} - d_j + p_j - M(1 - x_{0jv}) \quad \forall j \in J, \forall v \in V \tag{14.18}$$

(Vehicle loads after first customer)

$$L_j \geq L_i - d_j + p_j - M\left(1 - \sum_{v \in V} x_{ijv}\right) \quad \forall i \in J, \forall j \in J \tag{14.19}$$

(Vehicle loads "en route")

$$L_{0v} \leq q_v \quad \forall v \in V \tag{14.20}$$

$$L_j \leq q_v + M(1 - x_{ijv}) \quad \forall j \in J, \forall v \in V \tag{14.21}$$

(Vehicle capacity constraints)

$$T_j \geq T_{0v} + t_{0j} - M(1 - x_{0jv}) \quad \forall j \in J, \forall v \in V \tag{14.9}$$

$$T_j \geq T_i + s_i + t_{ij} - M\left(1 - \sum_{v \in V} x_{ijv}\right) \quad \forall i \in J, \forall j \in J \tag{14.10}$$

$$T_i + s_i + t_{ij} - M\left(1 - \sum_{v \in V} x_{ijv}\right) \leq T_j \quad \forall i \in J, \forall j \in J - \{n+i\} \tag{14.23}$$

$$T_i + s_i - r_i + t_{i,n+i} - M\left(1 - \sum_{v \in V} x_{i,n+i,v}\right) \leq T_{n+i} \quad \forall i \in J \tag{14.24}$$

$$a_0 \leq T_{0v} \quad \forall v \in V \tag{14.12}$$

$$a_j \leq T_j \leq b_j \quad \forall j \in J \tag{14.13}$$

$$T_{kv} \leq b_k \quad \forall v \in V \tag{14.14}$$

(Ensure feasibility of the time schedule)

$$x_{ijv} \in \{0, 1\} \quad \forall (i, j) \in (J_0, J_k)\backslash(0, k)), \forall v \in V \tag{14.22}$$

The FDPPTW applies the mixed deliveries and pickups strategy, yet retains the advantage of the simultaneous deliveries and pickups strategy, saving the accessing time when simultaneously performing a delivery and a pickup. For the FDPPTW, Wang and Chen (2012b) also constructed 56 test problems with size 100 by modifying Solomon's (1987) benchmark and also proposed a coevolutionary algorithm to solve them. In the next section, an illustrative example is created to examine these four schemes.

14.7 ILLUSTRATIVE EXAMPLE

Assume that there are four customers with delivery and pickup demands (30, 40), (10, 20), (50, 20), and (10, 20). The locations of these customers (nodes 1–4) can be drawn as Figure 14.6. The distribution center and the collection center are located in nodes 0 and 9. In Figure 14.6, the numbers lying on the arcs indicate the distances between end nodes.

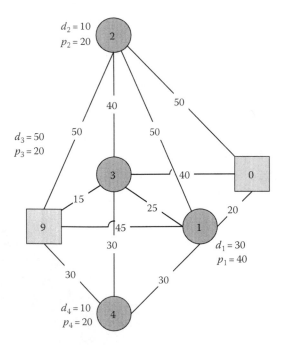

FIGURE 14.6   Network of the illustrative example.

In order to depict the network in a simple and clear way, we do not augment the network to eight customers. Let us just use the original network for illustration. The information of the original network is listed in Table 14.1. For each customer, both the delivery and pickup service times include an accessing time. Once the delivery and pickup services for a customer are performed simultaneously, the associated accessing time can be reduced once. Service times and accessing times are in hours. Time windows are recorded in 24 h format. For example, the pickup time window of customer 1 is from 12:00 to 17:00. The traveling times between nodes are also in hours and listed in Table 14.2. Assume that there are three vehicles available and each vehicle has a dispatching cost of 2000. The capacity limit of each vehicle is 100.

First, let us consider the VRPBTW. The optimal VRPBTW solution of this example can be also arrived at by Cplex and is illustrated in Figure 14.7. All linehaul customers precede all backhaul customers. In this scheme, one vehicle cannot finish all services and therefore an additional vehicle is dispatched. The objective value of the optimal VRPBTW solution is 4255.

Second, let us consider the MVRPBTW. The optimal MVRPBTW solution of this example can be got by Cplex and is illustrated in Figure 14.8. One can see three backhaul customers 2, 3, and 1 precede linehaul customers 4 in this solution. With this flexibility, one vehicle can finish all services in the VRPBTW scheme. The

TABLE 14.1

Information of the Original Network

Node	0	1	2	3	4	9
Delivery demand		30	10	50	10	
Pickup demand		40	20	20	20	
Delivery service time		0.3	0.1	0.5	0.1	
Pickup service time		0.4	0.2	0.2	0.2	
Accessing time		0.2	0.05	0.1	0.05	
Delivery time window		7	8	8	12	7
		10	12	12	17	17
Pickup time window	7	12	8	8	12	
	17	17	15	15	17	

TABLE 14.2

Traveling Times between Nodes (h)

	0	1	2	3	4	9
0	0	0.6	1.5	1.2	1.5	M
1	0.6	0	1.5	0.75	0.9	1.35
2	1.5	1.5	0	1.2	2.1	1.5
3	1.2	0.75	1.2	0	0.9	0.45
4	1.5	0.9	2.1	0.9	0	0.9
9	M	1.35	1.5	0.45	0.9	0

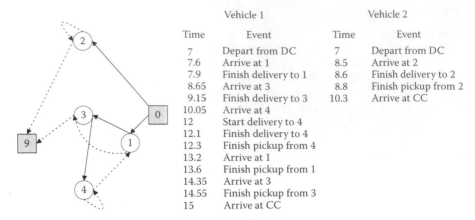

Vehicle 1		Vehicle 2	
Time	Event	Time	Event
7	Depart from DC	7	Depart from DC
7.6	Arrive at 1	8.5	Arrive at 2
7.9	Finish delivery to 1	8.6	Finish delivery to 2
8.65	Arrive at 3	8.8	Finish pickup from 2
9.15	Finish delivery to 3	10.3	Arrive at CC
10.05	Arrive at 4		
12	Start delivery to 4		
12.1	Finish delivery to 4		
12.3	Finish pickup from 4		
13.2	Arrive at 1		
13.6	Finish pickup from 1		
14.35	Arrive at 3		
14.55	Finish pickup from 3		
15	Arrive at CC		

FIGURE 14.7 Optimal VRPBTW solution of the illustrative example.

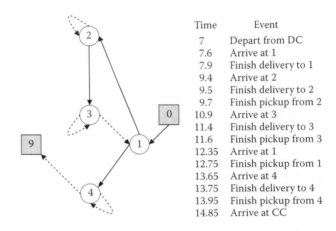

Time	Event
7	Depart from DC
7.6	Arrive at 1
7.9	Finish delivery to 1
9.4	Arrive at 2
9.5	Finish delivery to 2
9.7	Finish pickup from 2
10.9	Arrive at 3
11.4	Finish delivery to 3
11.6	Finish pickup from 3
12.35	Arrive at 1
12.75	Finish pickup from 1
13.65	Arrive at 4
13.75	Finish delivery to 4
13.95	Finish pickup from 4
14.85	Arrive at CC

FIGURE 14.8 Optimal MVRPBTW solution of the illustrative example.

objective value of the optimal MVRPBTW solution is 2195, which is much less than the one of the optimal VRPBTW solution.

Third, let us consider the SDPPTW. The information in Table 14.1 is not applicable to the SDPPTW scheme. Since the delivery and the pickup services should be performed simultaneously, there should be only one service time and only one time window for each customer. However, there is no intersection of the delivery and the pickup time window of customer 1. Assume that the pickup time window of customer 1 is abandoned. Then, the aggregated service times and time windows are given in Table 14.3.

The optimal SDPPTW solution of the revised example can be achieved by Cplex and is illustrated in Figure 14.9. The delivery and the pickup services are all performed simultaneously in this scheme. The objective value of the SDPTW solution is 2215.

Last, let us consider the FDPPTW. The information in Table 14.1 is applicable to the FDPPTW. The optimal FDPPTW solution of this example can be achieved by Cplex and is illustrated in Figure 14.10. One can see that the delivery and pickup services for customers 2, 3, and 4 are performed simultaneously; therefore, the accessing

TABLE 14.3
Aggregated Service Times and Time
Windows of the SDPPTW Network

Node	0	1	2	3	4	9
Service time		0.5	0.25	0.6	0.25	
Time window	7	7	8	8	12	
	17	10	12	12	17	

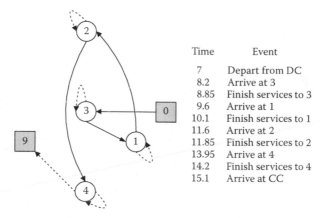

Time	Event
7	Depart from DC
8.2	Arrive at 3
8.85	Finish services to 3
9.6	Arrive at 1
10.1	Finish services to 1
11.6	Arrive at 2
11.85	Finish services to 2
13.95	Arrive at 4
14.2	Finish services to 4
15.1	Arrive at CC

FIGURE 14.9 Optimal SDPPTW solution of the illustrative example.

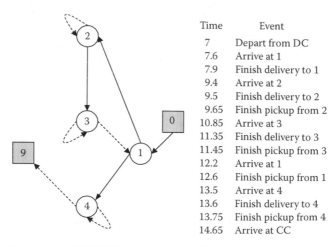

Time	Event
7	Depart from DC
7.6	Arrive at 1
7.9	Finish delivery to 1
9.4	Arrive at 2
9.5	Finish delivery to 2
9.65	Finish pickup from 2
10.85	Arrive at 3
11.35	Finish delivery to 3
11.45	Finish pickup from 3
12.2	Arrive at 1
12.6	Finish pickup from 1
13.5	Arrive at 4
13.6	Finish delivery to 4
13.75	Finish pickup from 4
14.65	Arrive at CC

FIGURE 14.10 Optimal FDPPTW solution of the illustrative example.

times to customers 2, 3, and 4 should be reduced once. Take customer 2, for example: the driver arrives at customer 2 at 9.4 o'clock and spends 0.1 h to deliver goods. However, the pickup service does not take the driver 0.2 h but just $0.2 - 0.05 = 0.15$ h. The accessing time 0.05 h is reduced since we do not have to access the customer twice. The objective value of the FDPPTW optimal solution is also 2195. Comparing this result to MVRPBTW's result, the objective values are the same, but the service duration of FDPPTW's solution is 0.2 h less than the one of MVPRBTW's solution. If the number of customers becomes larger, the FDPPTW scheme may use fewer vehicles and cause lower transportation cost than the MVRPBTW scheme may.

14.8 DISCUSSIONS

In some practical cases, if the linehaul customers have precedence over the backhaul customers then the VRPBTW scheme is favored. In some practical cases, if there are no difficulties in rearranging the load on the vehicle and freely mixing deliveries and pickups is the most concerned then the MVRPBTW scheme is favored. In some practical cases, if performing simultaneous delivery and pickup services with a single stop for each customer to reduce the service effort and the interference to customers is most concerned then the SDPPTW scheme is favored. These schemes have their advantages but also disadvantages.

The VRPBTW scheme avoids the problems that may rise because of rearranging goods on the vehicle and supports the fact that linehaul customers have priority over backhaul customers in some cases. However, there is no absolute precedence for linehaul customers in most cases and this model becomes inconvenient. The SDPPTW scheme reduces the service effort and the interference to customers and hence saves the accessing times. However, mixing pickup services and delivery services freely is not allowed. In the MVRPBTW scheme, freely mixing deliveries and pickups provides conveniences for customers and enterprises. However, Model MVRPBTW proposed in the literature does not consider the advantage of simultaneous deliveries and pickups: if the delivery and pickup services of a customer are performed simultaneously, the accessing time will not be double counted.

This defect is adjusted in Model FDPPTW. Furthermore, corresponding to delivery and pickup services, each customer can request two time windows (delivery time window and pickup time window) in the FDPPTW. It is very common in reality that customer requests that his or her incoming goods be delivered in the morning and his or her outgoing goods be picked up in the evening. This brings out the design of two time windows.

In summary, if there is no specific consideration when choosing a logistic model to a delivery and pickup environment, the result of Section 14.7 suggests that Model FDPPTW is the most flexible, efficient, and economical one.

14.9 CONCLUSIONS

As the reverse logistics and the closed-loop supply chain networks have been adopted by enterprises, the DPPs with time windows have drawn much attention and have been studied extensively in recent years. At the very beginning of developing a delivery and pickup logistic system, choosing a suitable scheme is essential and sustainably has huge effects over time. This chapter reviews three strategies and four derived schemes and provides models for those schemes.

If linehaul customers have precedence over backhaul customers, the delivery-first, pickup-second strategy is favored and the VRPBTW scheme is suitable. When there are no difficulties in rearranging the load on the vehicle, the mixed deliveries and pickups strategy is favored. The MVRPBTW scheme and the FDPPTW scheme both apply this strategy but the FDPPTW scheme is more reasonable than the MVRPBTW scheme. If customers are not willing to accept to be serviced separately for the delivery and pickup they require, the simultaneous deliveries and pickups strategy is favored and the SDPPTW scheme is suitable. If there is no specific consideration, the FDPPTW scheme is the most flexible, efficient, and economical one.

ACKNOWLEDGMENT

The authors acknowledge the financial support from the National Science Council with Project no. NSC97-2221-E007-092.

REFERENCES

Ai, T.J. and V. Kachitvichyanukul. 2009. A particle swarm optimization for the vehicle routing problem with simultaneous pickup and delivery. *Computers and Operations Research* 36: 1693–1702.

Angelelli, E. and R. Mansini. 2002. The vehicle routing problem with time windows and simultaneous pickup and delivery. In *Quantitative Approaches to Distribution Logistics and Supply Chain Management*, Lecture Notes in Economics and Mathematical Systems, Eds. A. Klose, M.G. Speranza, and L.N. Van Wassenhove, pp. 249–267. New York: Springer.

Berbeglia, G., J.F. Cordeau, I. Gribkovskaia, and G. Laporte. 2007. Static pickup and delivery problems: A classification scheme and survey. *TOP* 15: 1–31.

Bianchessi, N. and G. Righini. 2007. Heuristic algorithms for the vehicle routing problem with simultaneous pick-up and delivery. *Computers and Operations Research* 34: 578–594.

Chen, J.-F. and T.-H. Wu. 2006. Vehicle routing problem with simultaneous deliveries and pickups. *Journal of the Operational Research Society* 57: 579–587.

Cheung, R.K. and D.D. Hang. 2003. Multi-attribute label matching algorithms for vehicle routing problems with time windows and backhauls. *IIE Transactions* 35: 191–205.

Cordeau, J.F., G. Desaulniers, J. Desrosiers, M.M. Solomon, and F. Soumis. 2001. VRP with time windows. In *The Vehicle Routing Problem*, Eds. P. Toth and D. Vigo, pp. 157–193. Philadelphia, PA: SIAM.

Cordeau, J.F., M. Gendreau, and G. Laporte. 1997. A tabu search heuristic for periodic and multi-depot vehicle routing problems. *Networks* 30: 105–19.

Crispim, J. and J. Brandao. 2005. Metaheuristics applied to mixed and simultaneous extensions of vehicle routing problems with backhauls. *Journal of the Operational Research Society* 56: 1296–1302.

Dell'Amico, M., G. Righini, and M. Salani. 2006. A branch-and-price approach to the vehicle routing problem with simultaneous distribution and collection. *Transportation Science* 40: 235–247.

Dethloff, J. 2001. Vehicle routing and reverse logistics: The vehicle routing problem with simultaneous delivery and pick-up. *OR Spectrum* 23: 79–96.

Duhamel, C., J.-Y. Potvin, and J.-M. Rousseau. 1997. A tabu search heuristic for the vehicle routing problem with backhauls and time windows. *Transportation Science* 31: 49–59.

Gelinas, S., M. Desrochers, J. Desrosiers, and M.M. Solomon. 1995. A new branching strategy for time constrained routing problems with application to backhauling. *Annals of Operations Research* 61: 91–109.

Hoff, A., I. Gribkovskaia, G. Laporte, and A. Løkketangen. 2009. Lasso solution strategies for the vehicle routing problem with pickups and deliveries. *European Journal of Operational Research* 192: 755–766.

Ilgin, M.A. and S.M. Gupta. 2010. Environmentally conscious manufacturing and product recovery (ECMPRO): A review of the state of the art. *Journal of Environmental Management* 91: 563–591.

Kontoravdis, G. and J.F. Bard. 1995. A GRASP for the vehicle routing problem with time windows. *ORSA Journal on Computing* 7: 10–23.

Min, H. 1989. The multiple vehicle routing problem with simultaneous delivery and pick-up points. *Transportation Research A* 23(5): 377–386.

Montané, F.A.T. and R.D. Galvao. 2006. A tabu search algorithm for the vehicle routing problem with simultaneous pick-up and delivery service. *Computers and Operations Research* 33(3): 595–619.

Nagy, G. and S. Salhi. 2005. Heuristic algorithms for single and multiple depot vehicle routing problems with pickups and deliveries. *European Journal of Operational Research* 162(1): 126–141.

Parragh, S.N., K.F. Doerner, and R.F. Hartl. 2008. A survey on pickup and delivery problems Part II: Transportation between pickup and delivery locations. *Journal of Betriebswirtschaft* 58: 81–117.

Prins, C. 2004. A simple and effective evolutionary algorithm for the vehicle routing problem. *Computers and Operations Research* 31: 1985–2002.

Reimann, M., K. Doerner, and R.F. Hartl. 2002. Insertion based ants for vehicle routing problems with backhauls and time windows. *Lecture Notes in Computer Science* 2463: 135–148.

Renaud, J., F.F. Boctor, and G. Laporte. 2002. Perturbation heuristics for the pickup and delivery traveling salesman problem. *Computers and Operations Research* 29: 1129–1141.

Ropke, S. and D. Pisinger. 2006a. An adaptive large neighborhood search heuristic for the pickup and delivery problem with time windows. *Transportation Science* 40: 455–472.

Ropke, S. and D. Pisinger. 2006b. A unified heuristic for a large class of vehicle routing problems with backhauls. *European Journal of Operational Research* 171: 750–775.

Solomon, M.M. 1987. Algorithms for the vehicle routing and scheduling problems with time window constraints. *Operations Research* 35(2): 254–265.

Thangiah, S.R., J.-Y. Potvin, and T. Sun. 1996. Heuristic approaches to vehicle routing with backhauls and time windows. *Computers and Operations Research* 23: 1043–1057.

Tütüncüa, G.Y., C.A.C. Carreto, and B.M. Baker. 2009. A visual interactive approach to classical and mixed vehicle routing problems with backhauls. *Omega* 37: 138–154.

Wade, A.C. and S. Salhi. 2002. An investigation into a new class of vehicle routing problem with backhauls. *Omega* 30: 479–487.

Wang, H.-F. and Y.-Y. Chen. 2012a. A genetic algorithm for the simultaneous delivery and pickup problems with time window. *Computer and Industrial Engineering* 62: 84–95.

Wang, H.-F. and Y.-Y. Chen. In press. A coevolutionary algorithm for the flexible delivery and pickup problem with time windows. *International Journal of Production Economics*.

Wang, H.-F. and Y.-C. Chiu. 2009. A vehicle routing and scheduling model for a distribution center. In *Web-Based Green Products Life Cycle Management Systems: Reverse Supply Chain Utilization*, Ed. H.-F. Wang, Chapter XIV. New York: Information Science Reference.

Wang, H.-F. and S.M. Gupta. 2011. *Green Supply Chain Management: Product Life Cycle Approach*. New York: McGraw Hill.

Zhong, Y. and M.H. Cole. 2005. A vehicle routing problem with backhauls and time windows: A guided local search solution. *Transportation Research Part E* 41: 131–144.

15 Materials Flow Analysis as a Tool for Understanding Long-Term Developments

A.J.D. Lambert, J.L. Schippers,
W.H.P.M. van Hooff, H.W. Lintsen,
and F.C.A. Veraart

CONTENTS

15.1 INTRODUCTION

Materials flow analysis (MFA) (Brunan and Rechberger, 2004) is the study of physical flows through the technosystem (also frequently called "the economy"). This method is also referred to as industrial metabolism (Ayres and Simonis, 1994). In contrast to many economic studies, industrial metabolism is not primarily focused on value, but rather on physical quantities such as mass (expressed in kilotons, or 10^6 kg) and energy (expressed in GWh, which equals 3.6×10^6 MJ). The technosystem refers to the part of reality that is—more or less—controlled by humans and that is transformed to serve human-defined purposes. A concise discussion of MFA can be found in Gupta and Lambert (2008).

As the emphasis of this study is on quantity rather than on value, precious or rare materials that are used in small quantities, such as silver, gold, diamonds, and elements such as indium and tellurium, are completely neglected. The same is true for precious biological products, such as spices and silk, and for some luxury goods and artwork. In contrast, bulk goods are dominant in this study. The production of services, although essential from an economic point of view, is also not taken into account.

Industrial metabolism is analogous to metabolism in biological systems, where living organisms transform feed in order to obtain energy and building materials, and release waste (CO_2, H_2O, feces, dead organisms, etc.) that are reused by other organisms. Similarly, one can also define metabolism in abiotic systems, in which transformation and transportation processes are also continuously occurring. Mimicking the wasteless biotic system by the anthropogenic system (technosystem) via closing the loops is usually referred to as industrial ecology (Frosch and Gallopoulos, 1989) and more recently popularized as cradle to cradle (McDonough and Braungart, 2002). Still earlier studies on physical flows came in the wake of the 1973 oil crisis. Kneese et al. (1974) introduced the concept of total materials requirement (TMR) and coarsely determined the mass of the materials that were affected by the U.S. population. Study on the connection between materials and energy was proposed by IFIAS (1973) and elaborated for metals by Chapman and Roberts (1983).

Studies that have been performed so far usually result in figures that refer to physical flows, in systems such as the world, a definite country, a region, or an industrial branch. These studies are useful for comparison of, e.g., artificial flows with natural flows, for comparison of a set of countries with each other, and for comparison of the technosystem in different years.

This has been performed, e.g., by Adraanse et al. (1997), where the material flows of four industrialized countries (United States, Germany, the Netherlands, and Japan) were compared with each other and a period of time 1975–1993 was selected. A comparative study on waste production was made by Matthews et al. (2000). Here the same set of countries, with Austria added, was considered over the period 1975–1996.

Studies that refer to processes over periods of multiple years have been done by using instruments such as materials input–output analysis. This kind of research is focused on the exchange of materials and energy between the various branches of industry and is usually characterized by a high level of aggregation. Topics such as the role of investment versus final consumption can be studied with such an instrument, and also, e.g., the production of waste in relationship with the structure of industry in a specific country.

If recently published MFA studies are considered, we can classify them according to three types of approach: scenario analysis, dynamic stock and flow modeling, and historical trend analysis. The time span considered (with the future included sometimes), the region or country, and the material considered are also characteristics of the research. A listing of some papers is given as follows:

Reference	Method	Time Span	Region	Materials
Steubling et al. (2010)	Scenario analysis	1994–2020	Chile	Wasted computers
Gerst (2009)	Scenario analysis	2000–2100	Global	Copper
Weiqiang et al. (2010)	Dynamic stock and flow	2001–2007	China	Aluminum
Hatayama et al. (2010)	Dynamic stock and flow	2005–2050	Global	Steel
Hu et al. (2010)	Dynamic stock and flow	1900–2100	China	Steel in buildings
Cochran and Townsend (2010)	Dynamic stock and flow	1900–2050	United States	Construction and demolition waste
Daigo et al. (2009)	Dynamic stock and flow	1950–2005	Japan	Copper
Daigo et al. (2010)	Dynamic stock and flow	1980–2005	Japan	Cr and Ni in stainless steel
Bader et al. (2011)	Dynamic stock and flow	1850–2050	Switzerland	Copper
Xueyi et al. (2010)	Historical trends	1998–2006	China	Zinc
Krausmann et al. (2009)	Historical trends	1900–2005	Global	All materials
Wood et al. (2009)	Historical trends	1975–2005	Australia	All materials
Dittrich and Bringezu (2010)	Historical trends	1962–2005	Global	All materials
Kovanda and Hak (2011)	Historical trends	1855–2005	Czechoslovakia	All materials
Tachibana et al. (2008)	Historical trends	1980–1996	Part of Japan	All materials
This work (2012)	Historical trends	1850–2005	Netherlands	All materials

In the present chapter, the method of industrial metabolism is applied to a single country. Starting points are historical data about mass flows. Relationships are established with the use of technical data on processes. The study is extended over a period of 155 years. It is a comparative study in economy during four selected years, in order to cover the transition from a mainly rural country to the present-day industrialized society. To this purpose, we focus in on the years 1850, 1910, 1970, and 2005.

Aim of the study is to gain insight in the evolution of the sustainability of the society over a long period of time. Sustainability is considered both quantitatively—by the magnitude of material flows, efficiency of processes, waste production and so on—, and qualitatively—by studying and interpreting relevant documents from the years of reference that refer to sustainability.

15.2 SYSTEM OF REFERENCE

The Netherlands is a relatively small country with an area of about $40,000\,km^2$ and rather densely populated with a population that increased from about 3.5×10^6 inhabitants in 1850 to about 16.5×10^6 inhabitants in 2005.

The size of the country, its nearly absence of rocky soil, and its relatively long seashore destined it to become a maritime nation in which trade became dominant.

Industry was mainly involved with processing of merchandises from abroad, and with shipbuilding. Agriculture was also relatively well developed. New developments in industrial production, however, started relatively late. The industrial revolution based on coal and iron emerged in England in the course of the eighteenth century with a sequence of inventions, followed by Belgium (which was then part of France) about 1800. The Netherlands was lagging behind because coal was relatively expensive, and domestic iron ore was virtually absent. Steam engines were introduced in a slow pace and real industrial plants were restricted in 1850 to a few textile processing plants, a ceramics and glass processing plant, and some foundries and forges.

Industry based on coal, iron, and steam emerged quickly in the years 1850–1910, and in 1910 many factories were operational. Surprisingly, the figures do not show a variety of novel products: Portland cement, fertilizer, beet sugar, and margarine being some of the few exceptions. The multitude of chemicals and complex products that characterizes present times was small in mass in comparison with the more traditional commodities. The leap forward is mainly quantitative in this period. The quantity of both production and international trade dramatically increased.

The year 1910 can be considered as the start of the electrification in the Netherlands. This was accompanied by the emergence of considerable electrotechnical and electronic industries, based on power generation and the production of incandescent lamps, respectively. About the year 1910, domestic coal production started. The use of mineral oil still was in its infancy state, although oil was reclaimed in present-day Indonesia, which was a Dutch colony in those times. Although a primitive refinery already existed in Rotterdam, mineral oil products were mainly restricted to kerosene for lighting purposes and gasoline for textile cleaning. Refinery products such as gasoline mainly found their way to other countries, such as Germany. The number of cars still was negligible, in contrast to, e.g., the United States. In spite of this, the first commercial seagoing motor ship was built in the Netherlands in 1910. It was propelled by a domestically produced Diesel engine.

Starting about 1900, agriculture also changed drastically. Co-operative organizations were erected that enabled individual farmers to buy seeds, fertilizer, and feed, and to fit their products to the food processing industry. By this, farmers were gaining access to the world market.

The period 1910–1970 was characterized by major disruptions. Two World Wars and the Great Depression deeply influenced the economic development. Although not directly involved in World War I (1914–1918), there was a shortage on raw materials in those years, which boosted the tendency toward autarky. This resulted in the erection of branches of industry that have not been earlier in the Netherlands on that scale, such as integrated steel works and cement industries. Domestic coal reclamation also expanded. New resources were discovered, such as rock salt and crude oil, and this stimulated industry as well.

When World War II ended in 1945, a period of reconstruction and rapid economic growth started. This came along with Marshall Aid and European integration. Particularly, the petrochemical industry became important. This also boosted the production and application of large quantities of synthetic materials. In agriculture, rationalization was introduced on a large scale. The port of Rotterdam enabled the import of vast quantities of feed ingredients—often originating from developing countries—that enabled the rapidly increasing husbandry of pigs and poultry. The year 1970 marked multiple changes. Traditional industries disappeared nearly completely, starting with textile and shipbuilding, followed by engine works, artificial fiber, and production of electronic devices. About the same year, the coal mines were closed and a new domestic resource, natural gas, became exploited on a large scale. Global oil crises started putting a challenge on energy conservation and the urgency to find alternatives for mineral oil. Simultaneously, a noticeable environmental awareness emerged, starting with topics such as hazardous emissions, acid precipitation, and phosphate saturation of the soil. Apart from this, 1970 can be considered the beginning of the information era with the introduction of personal computers and, later, cell phones and internet. This caused the introduction and proliferation of a lot of small electronic devices, fed by batteries.

The year 2005 has been selected as most closely to present, with a set of elaborated data already available.

15.3   DATA COLLECTION

The most detailed data that are available over a long period of time are the figures on *import and export*, such as published by the customs authorities (Statistiek, 1850, 1910; Maandstatistiek, 1970; Statline, 2005). Qualitatively, these figures reflect the variety of commodities that are traded. Quantitatively, these figures are crucial for calculating the available amount of products. As many commodities are not domestically produced, the available amount is even known. Disadvantage is the variety of units that is applied in the figures. Although many commodities are expressed in weight, some are expressed in value, and others in number of items, volume, area, etc. This means that conversions must be made based on estimated conversion factors.

Many figures exist on *domestic extraction*. Agricultural production is relatively well documented over long periods of time. These figures reflect a dramatic increase in productivity. Figures on extraction of natural gas, oil, coal, and rock salt, and limestone are also available as well as for most of the surface minerals (clay, sand, and gravel). For peat and bog iron, however, no univocal figures exist. These have to be estimated. Data on energy products, including petrochemical feedstock, are available over the postwar period. Data on hunting, fisheries, and forestry are partly available.

Data on *secondary agricultural production*, including animal husbandry, are available (Land-en Tuinbouwcijfers, 2007; Statline, 2010).

A complete set of data on *industrial production* with regard to physical production is not available. There is, of course, the fundamental problem that many industrial outputs are inputs for other industries, which give rise to ambiguity in integrated plants and which can potentially be double counted. When the basic processes in industry are known, these can be used for calculating purposes. Figures on production of selected industrial products appeared available for 1913, 1968, 1970, 1992, but not for 2005. Data on construction of buildings and infrastructure is only partly available.

Data on *final consumption* is still more incomplete, although estimates can be made on topics such as per capita food consumption. Presently, figures on final consumption in some industrial branches are available.

Time series of data on *waste production* started after 1970. Trade on waste with a market value, such as scrap, is reflected by import and export figures.

Particularly in the year 1850, when a considerable part of the economy was based on self subsistence, official data do not reflect the complete economy.

Because of the interrelationship between extraction, production, consumption, and waste generation, even an incomplete set of figures can be tested on consistency and can be used to construct a reliable picture of the economy in those different years.

15.4 CATEGORIZATION

Categorization forms an essential challenge in industrial metabolism. The import and export tables show 461 different commodities in 1850; 627 in 1910; about 6,300 in 1970 and about 13,000 in 2005. Categorization proceeded along different systems, in 1850 and 1910 there was no categorization at all, but the commodities were listed alphabetically. Although there are different systems for categorization of commodities, we had to follow a slightly different method that was adapted to the purpose of industrial metabolism and that could be applied to all the years under consideration.

We applied the categorization in different steps, each with increasing level of aggregation. By this, a scheme is created in which all types of commodity could be placed. Apart from this, the data on different years are studied according to the same scheme making them comparable.

At the top level of aggregation, the following categories have been used:

1. Minerals and ores, and products of it, including construction materials, fertilizers and inorganic chemicals, glass and ceramics, and some groups of substances of various origins. Metals are excluded
2. Metals and products that mainly contain metals (semifinished products, instruments, machinery, electric and electronic equipment, vehicles)
3. Fossil fuels and products from it (organic chemicals, polymers, etc.)
4. Wood and timber and its products (paper, furniture)
5. Vegetable products, other than wood (including textile)
6. Animal products (including leather products)

It is self-evident that many commodities are a mix of products of each of these categories. Textile, e.g., comprises both from vegetable (linen, cotton), animal (silk, wool), and synthetic origin.

At the second highest level, we made a categorization into 60 principal groups, and at the third highest level, 354 groups are distinguished and still more groups of commodities were at the level below it.

15.5 RESULTS

On the highest level of aggregation, the following results are obtained (Table 15.1):

Import and export figures alone are no valid indicators in industrial ecology, for those can be contaminated by figures that include some transit. When considered by mass flows, we find (Table 15.2):

TABLE 15.1
Yearly Import and Export

Commodity	1850 kton/year	1850 %	1910 kton/year	1910 %	1970 kton/year	1970 %	2005 kton/year	2005 %
				Year				
Import								
Minerals	196	*13*	11,315	*35*	51,150	31	59,845	16
Metals	29	2	1,805	6	7,759	5	20,488	6
Fossil fuels	557	*36*	10,848	*33*	84,453	51	183,793	57
Wood	320	*21*	827	3	6,999	4	15,109	5
Vegetable	413	27	7,298	22	13,476	8	38,765	12
Animal	15	*1*	227	*1*	722	0	4,750	1
Total	1530		32,320		164,559		322,750	
Export								
Minerals	73	*13*	8,571	*40*	19,388	20	27,857	*10*
Metals	10	2	1,342	6	6,440	7	20,488	8
Fossil fuels	37	6	4,566	21	60,818	62	167,938	*63*
Wood	18	3	695	3	1,722	2	8,730	3
Vegetable	290	*51*	5,612	26	7,036	*7*	33,554	*13*
Animal	145	25	540	3	2,460	3	7,616	3
Total	573		21,326		97,864		266,183	

Italics denote main categories, such as import and export.

TABLE 15.2
Evolution of Ratio between Import and Export

	Year 1850	1910	1970	2005
Import/export (kton/kton)	2.67	1.52	1.68	1.21

These figures show that the import and export figures tend to approach each other. This means that the excess import is not reliably represented by import figures that are growing over time much faster than the economy does.

From this, a list can be derived that presents the net import of a group of commodities. Negative figures refer to net export (Table 15.3).

15.6   EXTRACTION

Extraction is a way of production, in which the product is straightly extracted from the nature as a raw material. This includes minerals reclamation, agriculture, fisheries, hunting and gathering, and forestry.

TABLE 15.3
Yearly Net Import

	Year							
	1850		1910		1970		2005	
Commodity	kton/year	%	kton/year	%	kton/year	%	kton/year	%
Minerals	114	12	2,744	25	31,762	48	31,988	57
Metals	19	2	463	4	1,319	2	−891	−2
Fossil fuels	526	55	6,282	57	23,635	35	15,855	28
Wood	305	32	132	1	5,277	8	6,379	11
Vegetable	123	13	1,686	15	6,440	10	5,211	9
Animal	−129	−13	−313	−3	−1,738	−3	−2,866	−5
Total	958		10,994		66,695		55,676	

From the minerals extracted there are some categories that are not included in the statistics. The first one is *combustion air*. Its quantity can be directly calculated by considering the fossil fuel mix, but the reservoir of it is considered huge.

The second one is *embankment sand*. This is sand that is displaced but not processed and neither chemically nor physically modified. Estimates are up to 77,400 kton in the year 2005. Quantities like this would dominate the statistics and therefore they are not included there. Their main importance lays in the destruction or modification of existing landscape forms.

The third one is *water*. Particularly, cooling water is used in huge quantities, about 14,000,000 kton in 2005. This water is not consumed, however, but residual heat is added to it. Surface water and groundwater, used for tap water, irrigation, and process water in industry, also refers to a large quantity, about 1,500,000 kton yearly.

Not the complete agricultural production can be considered extraction. Animal husbandry is secondary production because it consumes feed. Feed consists of both roughage and concentrates, such as compound feed. Roughage is either directly browsed from the meadows by horses and cattle or it is processed as hay or silage. Much of the feed, however, consists of other primary agricultural products, and of byproducts from the processing of agricultural products. In the Netherlands, nearly half of the arable area is grassland and the greater part of the imported vegetable materials is also consigned to feed. How extremely important grass production is, can be calculated for the year 1970, in which about 75,000 kton grasses were produced and consumed by cattle either directly as meadow grass or in processed form.

Another consideration refers to the often high percentage of water in biomass, which makes mass balancing questionable. For instance, the dry matter content of grass is about 20%, that of silage is 44%, and that of hay is 80%. These and similar percentages have to be accounted for in the analysis of the results.

When extraction figures are listed, one must account for unavailable or incomplete figures. Therefore, estimations (educated guesses) must be made, see indicated cells in Table 15.4.

TABLE 15.4
Yearly Extraction

	Year			
	1850	1910	1970	2005
Commodity	kton/year	kton/year	kton/year	kton/year
Clay	1400[b]	3,000	7,600	3,200
Sand[a]	1000[b]	5,000[b]	21,855	20,750
Gravel	44[b]	400[b]	14,380	3,500
Limestone	1[b]	8[b]	2,695	1,583
Shells	80[b]	166[b]	0	0
Stone	0[b]	20[b]	0	0
Rock salt	0[b]	0[b]	2,871	6,443
Bog iron	10[b]	10[b]	1[b]	0
Minerals	**2535**	**8,604**	**49,402**	**35,476**
Peat	1000[b]	2,150[b]	1,000	0
Coal	17	1,500	4,334	0
Mineral oil	0	0	1,919	1,492
Natural gas	0	0	25,651	60,313
Fossil fuels	**1017**	**3,650**	**32,904**	**61,805**
Wood	**20[b]**	**300[b]**	**475[b]**	**475[b]**
Cereals	670	901	1,350	1,706
Legumes	65	70	148	18
Oilseeds	70	14	29	12
Tuberous plants	366	5,871	11,748	13,910
Spices	10	12	3	0
Vegetables	100	200	822	1,516
Veg. fruit	0	0	660	1,500
Fruit	200	300	639	595
Decorative plants	20[b]	40[b]	500[b]	1,300[b]
Fibers	7	53	35	27
Vegetable products	**1508**	**7,461**	**15,934**	**20,584**
Fish	**100[b]**	**350[b]**	**350[b]**	**355**
Total	5189	20,365	99,065	118,695

Values in bold refer to the sum of the mass flow of this aggregated flow.
[a] Embankment sand excluded.
[b] Estimated values.

15.7 AVAILABILITY

The available amount of some commodity Av is defined as the amount of this commodity that is set available for either further processing or final consumption. It equals the following:

$$Av = Import - Export + Domestic\ production/extraction$$

TABLE 15.5
Yearly Available Commodities

Commodity	Year							
	1850		1910		1970		2005	
	kton/year	%	kton/year	%	kton/year	%	kton/year	%
Minerals	2535	43	11,348	36	81,164	49	67,464	39
Metals	19	0	463	1	1,319	2	−891	−1
Fossil fuels	1543	26	9,932	32	56,539	34	77,660	45
Wood	325	5	432	1	5,752	3	6,854	4
Vegetable	1624	27	9,094	29	22,374	13	25,795	15
Animal	−100	−2	37	0	−1,388	−1	−2,511	−1
Total	5946		31,306		165,760		174,372	

This can be done for both raw materials and processed products. If one is focused on the mass flows that are entering a country, one has to deal only with

$$Av = Import - Export + Domestic\ extraction$$

The problems that might occur can be seen with a simple product type as example. Coffee, e.g., is not extracted in the Netherlands. The coffee available thus equals Import − Export. However, there is an industry in the Netherlands that processes unprocessed coffee (peeling, roasting, packing), which results in processed coffee and losses. When "coffee" is disaggregated in "raw coffee" and "processed coffee," we would arrive at different and more detailed figures (Table 15.5).

In this table production, other than extraction, is not included. This reflects itself in the category "metals," which are already included as an ore in the "minerals" section. It also reflects itself in the category "animal products," in which the domestic animal production was not accounted for. Roughage has been excluded. A more refined calculation is required, in which the share of extracted roughage (grass, maize fodder) of the total amount of feed is accounted for. In this case, part of the animal production can be assigned to extraction.

15.8   DISAGGREGATION

The figures in the tables above contain information about mass flows in a country. For a deeper insight, however, the figures must be disaggregated and the reality behind it must be unveiled. We will do it for the metals—specifically iron—in order to illustrate how this proceeds (Table 15.6).

Ores are placed in the minerals group. It is the only material that can be extracted, which took place in the Netherlands in minor quantities as bog iron. In 1890, the last primitive blast furnace finished its operation. No domestic raw iron production took place in the Netherlands in the 1910, apart from some scrap processing. The first modern domestic blast furnace started its operation in 1924. In 1939 it expanded

TABLE 15.6
Yearly Available Iron-Related Commodities

	Year											
	1850			1910			1970			2005		
Commodity	Imp.	Exp.	Prod.	Imp.	Exp.	Prod.	Imp.	Exp.	Prod.	Imp.	Exp.	Prod.
Iron ore	0	0	10	5941	6354	10	5429	98	0	12,301	1,978	0
Raw iron	7	0	2	342	261	0	173	2	3594	658	392	?
Raw steel	7	1	0	484	295	0	289	175	5042	232	85	6,900
Scrap iron	0	1	?	86	150	?	298	920	?	2,111	4,158	1,969
Semi finished	4	0	?	577	340	?	3566	3490	5847	7,415	8,176	?
Final prod	3	3	?	64	56	?	2926	1573	?	7,262	4,497	?

? indicates that figures are not available.

to an integrated steel plant with the start of the operation of a Siemens Martin steel production plant. Rolling mills came also into operation.

It can be seen that, particularly in 1910, the available figures are not very trustworthy. There is more export than import. This is probably due to delivery from stock.

Semifinished products include rolled steel sheets and plates, castings, wire, rods, tinplate, tubes, profiles, construction parts, and fasteners.

Final products include vehicles, machinery, electric and electronic products, instruments, and tools. Production figures can only be roughly estimated.

The domestic production of scrap does not include all scrap. The iron fractions in industrial waste (740), municipal waste (59), discarded cars (420), and construction- and demolition waste (750) are added.

Although not all relevant quantities are known, this table presents a set of relevant figures on iron use in the Netherlands. The set of figures can further be interpreted by adding some process knowledge, which involves (Statistiek, 1920; Brown et al., 1985):

$$1358 \text{ kton ore} + 82 \text{ kton scrap} \rightarrow 831 \text{ kton raw iron}$$

$$260 \text{ kton scrap} \rightarrow 240 \text{ kton steel (electric arc furnace)}$$

$$200 \text{ kton raw iron} + 110 \text{ kton scrap} + 20 \text{ kton ore} \rightarrow 300 \text{ kton steel}$$
$$\text{(open hearth furnace)}$$

Although these figures are only rough estimates, they provide an indication on how the iron and steel flow is organized.

15.9 SUSTAINABILITY

The figures above are not straightly related to environmental issues. The hidden flows are not accounted for. Many minerals are not mined in the Netherlands, and their environmental burden takes place elsewhere. The same is true for the import of large quantities of oilseeds and feed ingredients.

One of the challenges, e.g., is to replace fossil fuel consumption by the use of renewables, for which biomass is considered a potential candidate. Apart from the discussion whether such a transition is sustainable or not, its feasibility must be assessed. Therefore, some figures are generated.

One of the issues thus is the ratio between the abiotic and biotic materials put available. The evolution of this ratio is as follows (Table 15.7):

TABLE 15.7
Ratio of Abiotic and Biotic Mass Flows

	Year			
	1850	1910	1970	2005
Abiotic/biotic (kton/kton)	2.3	2.3	5.7	4.6

TABLE 15.8
Ratio of Fossil Fuel Consumption versus Biomass Consumption

	Year			
	1850	1910	1970	2005
Fossil fuel/biomass (kton/kton)	0.87	1.07	2.27	2.5

Surprisingly, the industrial revolution, which took place between 1850 and 1910, is not reflected in the figures above. The principal reason is that both the industrial and the agricultural sector were growing fast. An increasing amount of biomass was also converted inside the agricultural system into animal products.

The figures on abiotic materials in this table strongly depend on construction activities. These are rather fluctuating over time, dependent on, e.g., short-term economic condition and the realization of great infrastructural projects. When the ratio of the consumption of fossil fuel and biomass is accounted for, with minerals excluded, the following evolution is revealed (Table 15.8):

In figures of this kind, we must realize that a substantial part of the biomass that is considered here consists of water.

A rough estimate for the year 2005 reveals that 77,660 kton fossil fuel was made available. The available biomass was 30,138 kton. From this, 21,000 kton was domestically extracted. Grass was excluded. The extracted amount, however, consists for a major part of water. Indicative figures for water content are for instance tomatoes (95%), milk (88%), grass (84%), fish (80%), potatoes and sugar beet (77%), wood (20%), and cereals (15%). If a certain area is made void of natural forest and converted to a monoculture of fast-growing wood with a maximum yearly yield of 15 ton/ha, which is true for European black pine (*Pinus nigra*), the Netherlands alone would occupy 51,773 km² of land area for fuel alone. However, the reality is still worse, because the calorific value of wood is not more than half that of oil and coal. Accounting for this, an area of at least three times the land area of the Netherlands (which is about 33,000 km²) is required. We must notice that this is a conservative estimate. Because the complete area of the Netherlands has already a destination, all this area must be found in the exterior. This area should even add to the land area in the exterior that is required for the production of the imported feedstuff. If we account for the fast-growing energy consumption of the complete industrialized world, and the fact that much land area in the world is already intensively used, this would result in the complete destruction of the remaining pristine nature and its accompanied biodiversity.

The rough calculation given earlier illustrates the way how even crude figures can be used for a preliminary feasibility study.

15.10   CONCLUSION

We collected a set of historical data for a rather small and specialized country from the lowest level of aggregation and aggregated this from the bottom up to a high aggregation level. We performed this in the same method for 4 years that are rather distant in the course of time, thus combining MFA with history. This enabled us to derive some basic figures on general trends in the anthropogenic material flows over a long period of time, such as these are traceable in the technosystem. We used MFA as an instrument to perform this task. It has been demonstrated that the set of data that we collected enabled the quantification of some trends in history. It also makes it possible to perform studies on lower levels of aggregation and to focus on partial flows in the economy, without losing a view on the complete picture of materials flows in the country over time.

REFERENCES

Adraanse, A., Bringezu, S., Hammond, A., Moriguchi, Y., Rodenburg, E., Rogich, D., and Schütz, H., 1997. *Resource Flows: The Material Basis of Industrial Economics.* Washington, DC: World Resources Institute.

Ayres, R.U. and Simonis, U.E. (Eds.), 1994. *Industrial Metabolism.* Tokyo, Japan: United Nations University Press.

Bader, H.-P., Scheidegger, R., Wittmer, D., and Lichtensteiger, T., 2011. Copper flows in buildings, infrastructure and mobiles: A dynamic model and its application to Switzerland. *Clean Technology and Environmental Policy* 13, 87–101.

Brown, H.L., Hamel, B.B., and Hedman, B.A., 1985. *Energy Analysis of 108 Industrial Processes.* Atlanta, GA: Fairmont Press.

Brunan, P.H. and Rechberger, H., 2004. *Material Flow Analysis: Advanced Methods in Resource and Waste Management.* Boca Raton, FL: CRC Press.

Chapman, P.F. and Roberts, F., 1983. *Metal Resources and Energy.* London, U.K.: Butterworth.

Cochran, K.M. and Townsend, T.G., 2010. Estimating construction and demolition debris generation using a materials flow analysis approach. *Waste Management* 30, 2247–2254.

Daigo, I., Hashimoto, S., Matsuno, Y., and Adachi, Y., 2009. Material stocks and flows accounting for copper and copper-based alloys in Japan. *Resources, Conservation and Recycling* 53, 208–217.

Daigo, I., Matsuno, Y., and Adachi, Y., 2010. Substance flow analysis of chromium and nickel in the material flow of stainless steel in Japan. *Resources, Conservation and Recycling* 54, 851–863.

Dittrich, M. and Bringezu, S., 2010. The physical dimension of international trade, Part 1: Direct global flows between 1962 and 2005. *Ecological Economics* 69, 1838–1847.

Frosch, R.A. and Gallopoulos, N.E., 1989. Strategies for manufacturing. *Scientific American* 261(September), 144–152.

Gerst, M.D., 2009. Linking material flow analysis and resource policy via future scenarios of in-use stock: An example for copper. *Environmental Science and Technology* 43, 6320–6325.

Gupta, S.M. and Lambert, A.J.D. (Eds.), 2008. *Environment Conscious Manufacturing.* Boca Raton, FL: CRC Press.

Hatayama, H., Daigo, I., Matsuno, Y., and Adachi, Y., 2010. Outlook of the world steel cycle based on the stock and flow dynamics. *Environmental Science and Technology* 44, 6457–6463.

Hu, M., Pauliuk, S., Wang, T., Huppes, G., Van der Voet, E., and Müller, D., 2010. Iron and steel in Chinese residential buildings: A dynamic analysis. *Resources, Conservation and Recycling* 54, 591–600.

IFIAS, 1973. Report of the International Federation of Institutes for advanced study, Workshop no. 6 on Energy Analysis. Guldsmedshyttan, Sweden.

Kneese, A.V., Ayres, R.U., and D'Arge R.C., 1974. Economics and the environment: A materials balance approach. In: *The Economics of Pollution*, Wolozin, H. (Ed.). Morristown, NJ: General Learning Press.

Kovanda, J. and Hak, T., 2011. Historical perspectives of material use in Czechoslovakia in 1855–2007. *Ecological Indicators* 11, 1375–1384.

Krausmann, F., Gingrich, S., Eisenmenger, N., Erb, K.-H., Haberl, H., and Fischer-Kowalski, M., 2009. Growth in global materials use, GDP and population during the 20th century. *Ecological Economics* 68, 2696–2705.

Land-en Tuinbouwcijfers (Agricultural and Horticultural data), 2007. Landbouw Economisch Instituut (LEI) and Dutch Central Bureau of Statistics (CBS).

Maandstatistiek van de Buitenlandse Handel, per Goederensoort (Monthly Statistics of Foreign Trade, listed by Commodity), December 1970. The Hague, the Netherlands: CBS (Dutch Central Bureau of Statistics).

Matthews, E. et al., 2000. *The Weight of Nations: Materials Outflows from Industrial Economies*. Washington, DC: World Resources Institute.

McDonough, W. and Braungart, M., 2002. *Cradle to Cradle: Remaking the Way We Make Things*. New York: North Point Press.

Statistiek van den Handel en de Scheepvaart van het Koningrijk der Nederlanden over het jaar (Statistics of the Dutch International Trade), 1850. The Hague, the Netherlands: Department of Finance.

Statistiek van den Handel en de Scheepvaart van het Koningrijk der Nederlanden over het jaar (Statistics of the Dutch International Trade), 1910. The Hague, the Netherlands: Department of Finance.

Statistiek van de Voortbrenging en het Verbruik der Nederlandsche Nijverheid in 1913 en 1916 (Statistics of the Dutch Industrial Production in 1913 and 1916), 1920. The Hague, the Netherlands: Dutch Central Bureau of Statistics (CBS).

Statline, Foreign Trade, 2005. http://statline.cbs.nl/statweb/CBS (Dutch Central Bureau of Statistics).

Statline, Foreign Trade, 2010. http://statline.cbs.nl/statweb/CBS (Dutch Central Bureau of Statistics). Historical time series on Agriculture, 1851–2008.

Steubling, B., Böni, H., Schluep, M., Silva, U., and Ludwig, C., 2010. Assessing computer waste generation in Chile using material flow analysis. *Waste Management* 30, 473–482.

Tachibana, J., Hirota, K., Goto, N., and Fujie, K., 2008. A method for regional-scale material flow and decoupling analysis: A demonstration case study of Aichi prefecture, Japan. *Resources, Conservation and Recycling* 52, 1382–1390.

Weiqiang, C., Lei, S., and Yi, Q., 2010. Substance flow analysis of aluminum in mainland China for 2001, 2004 and 2007: Exploring its initial sources, eventual sinks and the pathways linking them. *Resources, Conservation and Recycling* 54, 557–570.

Wood, R., Lenzen, M., and Foran, B., 2009. A material history of Australia. *Journal of Industrial Ecology* 13, 847–862.

Xueyi, G., Juya, Z., Yu, S., and Qinghua, T., 2010. Substance flow analysis of zinc in China. *Resources, Conservation and Recycling* 54, 171–177.

Author Index

Subject Index

TABLE 7.1

Summary of Results

		Q (units)	p_s ($/unit)	r_c ($/unit)	p_w ($/unit)	d (units/day)	x (units/day)	Π_r ($/day)	Π_m ($/day)	Π_s ($/day)
Retailer's optimal policy	Given p_w	428.456	30.032	1.486	20.000	29.904	19.283	318.360	342.675	661.035
	Variable p_w	388.392	31.680	1.498	23.286	24.960	19.299	228.215	358.616	586.830
Manufacturer's optimal policy	Given p_w	6083.126	29.992	1.399	20.000	30.227	17.994	275.698	377.596	653.293
	Variable p_w	12145.911	31.893	1.257	24.138	24.138	15.660	113.753	399.039	512.792
Centralized optimal policy		1562.425	23.251	0.519	—	50.246	5.455	400.271[a]	430.842[a]	831.112

[a] Allocated on the basis of proportional shares of total supply chain profit under retailer's optimal policy.

From this table it is clear that if the retailer has sufficient policy implementation power in the supply chain, it attempts to keep the replenishment lot size comparatively small (i.e., 428.456 units), in view of its relatively low fixed ordering cost. Furthermore, through its retail pricing (p_s = \$30.032/unit), in conjunction with a customer return reimbursement price of \$1.486/unit, it prefers to achieve daily market demand and customer return rates of 29.904 and 19.283 units, respectively, that attempt to balance the gains from sales and returns against the ordering and inventory carrying (for both new and used items) costs. The maximum attainable daily profit for the retailer is, thus, \$318.360, resulting in a profit of \$342.675/day for the manufacturer. Note that as every unit of the returned product represents a net gain of \$1.314 (i.e., the difference between the amount, r_m, compensated by the manufacturer and the customer reimbursement, r_c) for the retailer, it attempts to achieve a relatively high used item return rate about 64.483%.

If, on the other hand, the manufacturer is in a position to exert a greater level of negotiating power in the supply chain, its individual optimal policy would dictate a significantly larger replenishment batch of 6083.126 units, due to the relatively high fixed setup and transportation costs. In spite of a more than sixfold increase in the lot size, however, the selling price and return reimbursement, set by the retailer in response, are both only slightly lower than their values under its own optimal policy, i.e., \$29.992 and \$1.399 per unit, respectively. It is interesting to note that, consequently, the retail demand rate increases slightly to 30.227 units/day and the average product returns decline slightly to 17.994 units/day. The returns, however, now decline slightly to 59.530% of sales. Not unexpectedly, implementing the manufacturer's optimal replenishment policy reduces the retailer's profit to \$275.698/day, whereas the manufacturer's profit increases to \$377.596/day. Nevertheless, in terms of total supply chain profitability, the difference between adopting any one party's optimal policy over the other's amount to only about 1.171%.

Table 7.1 shows that if the retailer and the manufacturer decide to cooperate through the sharing of necessary information and adopt a jointly optimal policy that maximizes the total supply chain profit, instead of optimizing either party's position, both parties stand to gain considerably from such an approach. As mentioned earlier, the centralized model attempts to avoid double marginalization, i.e., the manufacturer does not explicitly charge the retailer a wholesale price, nor does it explicitly offer the latter a reimbursement for collecting the returns (implying that $p_w = r_m = 0$). Without these cost factors, the centrally controlled approach results in a maximum supply chain profit of \$831.112/day, representing more than 25.728% improvement in total system profitability, compared to the retailer's optimal policy or over 27.219% improvement *vis-à-vis* the manufacturer's optimal policy. As expected, the jointly optimal replenishment quantity now is 1562.425 units, which is less than the manufacturer's optimal batch size, but larger than the retailer's optimal order quantity. More interestingly, the retail price is reduced to \$23.251/unit and the return reimbursement is decreased to \$0.519/unit, respectively, resulting in a considerably larger demand rate of 50.246 units/day, as well as a smaller average product return rate of 5.455 units/day (i.e., about 10.857% of items sold are returned by customers). The implication of our centralized model is that under a jointly optimal policy, relatively fewer products sold are remanu-factured items. Under the given set of problem parameters, it appears desirable to increase the overall market demand through a lower retail price. Also, there is a lesser emphasis on collecting customer returns for remanufacturing. The central-ized model reduces the incentive for customer returns, which maximizes the total supply chain profitability.

The absence of a wholesale price and an explicit incentive for the retailer to col-lect returned items raises some interesting questions concerning a fair and equitable sharing of the total gain resulting from the centralized cooperative policy shown in Table 7.1. Although this can be achieved in several possible ways, we propose a profit sharing plan under a scenario where the retailer is the more powerful member of the supply chain and can dictate the implementation of its own optimal policy. The task of the manufacturer is then to offer sufficient incentive to the retailer in order for the latter to adopt the results of this procedure. Note that under its own individual optimal policy, the retailer's share is 48.161% of the total profit for both the parties. Therefore, it would be reasonable if the retailer is allocated the same percentage of the total supply chain profit of $831.112/day yielded by the centralized model. In other words, the retailer's share of the total profit is $400.271/day and that of the manufacturer is $430.842/day. With this profit sharing arrangement, each party's daily profit is more than 8% larger than that achieved under the retailer's optimal policy. Thus, it is economically attractive for both parties to adopt the jointly optimal policy yielded by our centralized model. If the manufacturer is more powerful of the two parties, the terms of a corresponding profit sharing arrangement, can also be derived easily along similar lines.

Finally, Table 7.1 also shows the results for the decentralized models where under monopolistic competition, the manufacturer can set its wholesale price, which is now treated as a decision variable. Compared with the results for a given wholesale price, the retailer's individual optimal policy dictates increasing both the selling price from $30.032/unit to $31.680/unit and the customer return reimbursement from $1.485/unit to $1.498/unit. Consequently, the order quantity is reduced from 428.46 units to 388.392 units. These changes indicate that the retailer would expend less effort to increase market demand and would tend to compensate by attempting to increase its revenue from returns. This appears to be a rational response to a higher wholesale price. Also, as expected, the manufacturer's share of the total supply chain profit now increases from 51.839% to 61.111%, while, the total profits for the supply chain declines to $586.830/day. These effects, not unexpectedly, tend to be magnified when the supplier is in a position to dictate the adoption of its own optimal policy by the retailer. Now the total supply chain profit shrinks further to 512.792$/day, although the manufacturer's relative share of this, as well as its own daily profit go up substan-tially, albeit at the expense of the retailer.

7.6 SUMMARY AND CONCLUSIONS

In this study, we have developed mathematical models under deterministic condi-tion, for simultaneously determining the production/delivery lot size, the retail price, and the customer return reimbursement level for a single recoverable product in a two-echelon supply chain consisting of a single retailer and a single lean manufacturer. Items returned by customers at the retail level are refurbished and totally reintegrated into the manufacturer's existing production system for remanufacturing and are sold eventually as new products. As in many lean manufacturing (a JIT) environments, we assume a lot-for-lot operating mode for production, procurement, and distribution, as an effective mechanism for supply chain coordination.

Decentralized models are developed and solved for determining profit maximizing optimal policies from the perspectives of both members of the supply chain. A centralized, jointly optimal procedure for maximizing total supply chain profitability is also presented. A numerical example illustrates that the centralized approach is substantively superior to individual optimization, due to the elimination of double marginalization. The example also outlines a fair and equitable proportional profit sharing scheme, which is economically desirable from the standpoint of either member of the supply chain, for the purpose of implementing the proposed centrally controlled model.

Of necessity, the simplifying assumptions made here (e.g., deterministic parameters and the lot-for-lot modality), are the major limitations of this study. Embellishments by future researchers, such as relaxation of the lot-for-lot assumption, incorporation of uncertainty, more realistic and complex demand and product return functions, multiple products, manufacturers, etc., will, undoubtedly, lead to more refined remanufacturing and related models. Furthermore, future efforts in this area should consider the development of integrated decision models under stochastic conditions, which are likely to be more realistic from an implementation standpoint. Nevertheless, the results obtained in this study are likely to be of some value to practitioners as broad guidelines for integrated pricing, recoverable product collection, production planning and inventory control decisions, as well as for designing more streamlined, well-coordinated supply chains toward gaining competitive advantage. We also hope that our efforts will prove to be useful for researchers in shedding light on some of the intricate and interrelated aspects of product remanufacturing toward developing more effective decision-making models for supply chain and reverse logistics management.

APPENDIX

Note: All the proofs outlined are based on the following concept:

f is concave on D if and only if $D^2 f(x)$ is a negative semidefinite matrix for all $x \in D$. An $n \times n$ symmetric matrix A is negative semidefinite if and only if $(-1)^k |A_k| \geq 0$ for all $k \in \{1, \ldots, n\}$ where A_k is the upper left k-by-k corner of A.

Proof of Proposition 7.1:

If Q^*, r_i^* and p_s^* obtained from (7.3 through 7.5) are local optimum, the sufficient condition is that the objective function (7.2) should be jointly concave in these three variables. The Hessian matrix for (7.2) is